Studia Chaburensia
(StCh)

Edited by Hartmut Kühne

Volume 4

2014
Harrassowitz Verlag · Wiesbaden

Settlement Dynamics and Human-Landscape Interaction in the Dry Steppes of Syria

Edited by Daniele Morandi Bonacossi

2014
Harrassowitz Verlag · Wiesbaden

Cover illustration: Desert-kite in the Palmyra steppe (photo Alberto Savioli, Italo-Syrian Mission in the Western Palmyrena)

The series will consider contributions in the following fields:
- History, with an emphasis on regional, local, and micro-historical approaches.
- Archaeology, with an emphasis on studies on material-cultural phenomena from excavation contexts and on functional analysis.
- Environmental studies, with an emphasis on the reconstruction of rural and urban landscapes and their development in relation to the natural conditions.
- Settlement history, with an emphasis on the development of settlement patterns and systems.
- Social studies, with an emphasis on rural communities, their organization and relationship to the central government; every day life and social systems.
- Publication of dissertations and other theses dealing with topics concerning these subjects.
- Publication of international and interdisciplinary conferences on topics concerning these subjects.

Manuscripts are to be submitted to the editor as word-documents, with figures as single jpg-documents with a resolution of at least 800 dpi. Languages: English, French, German (for other languages please contact the editor).

Address of the editor: Hartmut Kühne, Institut für Vorderasiatische Archäologie,
Hüttenweg 7, 14195 Berlin, Germany.

Bibliografische Information der Deutschen Nationalbibliothek
Die Deutsche Nationalbibliothek verzeichnet diese Publikation in der Deutschen Nationalbibliografie; detaillierte bibliografische Daten sind im Internet über http://dnb.dnb.de abrufbar.

Bibliographic information published by the Deutsche Nationalbibliothek
The Deutsche Nationalbibliothek lists this publication in the Deutsche Nationalbibliografie; detailed bibliographic data are available in the internet at http://dnb.dnb.de.

For further information about our publishing program consult our website http://www.harrassowitz-verlag.de

Printed on permanent/durable paper.
Typesetting: Katia Gavagnin
Printing and binding: Memminger MedienCentrum AG
Printed in Germany

ISSN 1869-845X
ISBN 978-3-447-10166-0

Contents

Preface VII

Bernard GEYER, Marie-Laure CHAMBRADE and Éric COQUEUGNIOT
Water Quest and Subsistence Strategies during the Neolithic in the Arid Margins of Northern Syria 1

Daniele MORANDI BONACOSSI
Desert-kites in an Aridifying Environment. Specialised Hunter Communities in the Palmyra Steppe during the Middle and Late Holocene 33

Marco IAMONI
Late Neolithic Funerary Evidence from Palmyra: the Rujem al-Majdur necropolis and the "desert-kite and tumulus cultural horizons" in Central Syria 49

Sumio FUJII
Chronology of the Bishri Pastoral Prehistory and Protohistory: A Cross-check against the Jafr Chronology in Southern Jordan 63

Minna LÖNNQVIST
The Emergence of Pastoral Transhumance at Jebel Bishri in Central Syria 93

Yoshihiro NISHIAKI
Steppe exploitation by Bronze Age communities in the Middle Euphrates Valley, Syria 111

Atsunori HASEGAWA and Katsuhiko OHNUMA
New Evidence from Tell Ghanem Al-Ali: A Third Millennium Site in the Middle Euphrates, Syria 125

Shogo KUME and Ahmad SULTAN
Burials, nomads, and cities: A perspective to changing nomad-sedentary relations on the Syrian Middle Euphrates during the third and second millennium BC 137

Stefan L. SMITH, Tony J. WILKINSON and Dan LAWRENCE
Agro-pastoral Landscapes in the Zone of Uncertainty: The Middle Euphrates and the North Syrian Steppe during the 4th and 3rd millennia BC 151

Olivier BARGE, Corinne CASTEL and Jacques ÉLIE BROCHIER
Human impact on the landscape around Al-Rawda (Syria) during the Early Bronze IV: evidence for exploitation, occupation and appropriation of the land 173

Bernd MÜLLER-NEUHOF
Desert irrigation agriculture: Evidences for Early Bronze Age Rainwater-Harvesting Irrigation Agriculture at Jawa (NE-Jordan) 187

Maria Giovanna BIGA
The Syrian steppes and the kingdom of Ibal in the third millennium B.C.: new data from the Ebla texts 199

Nele Ziegler and Hervé Reculeau
The Sutean Nomads in the Mari Period 209
Frederick Mario Fales
Hamlets and Farmsteads in the Balīḫ River Valley:
The Middle Assyrian and the Neo-Assyrian Evidence 227
Olivier Rouault and Maria Grazia Masetti-Rouault
From One Valley to Another: Bir el-Haddad, a Neo-Assyrian Trading Post? 243
Dorothée Sack, Martin Gussone and Dietmar Kurapkat
A Vivid City in the 'Syrian Desert': The case of Resafa-Sergiupolis/Rusafat Hisham 257
Elisabetta Randazzo
La Steppe Syrienne à la période omeyyade 275

Preface

If in the past archaeological endeavour has mainly concentrated on urban sites located in the great river valleys of Syria or the agro-ecological zone of stable and long-term settlement in the Jezirah steppe with the highest rainfall and the plains of northern Syria, during the last few decades the marginal semi-arid and dry steppes of the southern Jezirah, northern Syria, the Palmyrena and the Hawran have been brought to the centre of field research. This crucial expansion of the horizons of archaeological research is certainly the result of a more aware and mature affirmation of the archaeology of pastoralism and mobility as a strategic sector in the archaeology of the ancient Near East, which has been made possible by the progressive and increasingly rigorous definition of the objectives and research methodologies specific to this new branch of archaeological investigation.

Nevertheless, at the origin of the recent "archaeological conquest" of Syria's most arid steppe there are also two other extremely important critical factors: on the one hand, an awareness of the limitations, lacunae and biases inherent in the representation of semi-nomadic pastoralism handed down from cuneiform texts and, secondly, our inability to explore the crucial formative periods of the pastoralist phenomenon (the Neolithic and Chalcolithic especially, but also to a lesser extent the Early Bronze Age) through written sources.

Notwithstanding the serious problems posed by the conservation, identification, and dating of the archaeological record left by semi-nomadic communities which the archaeology of pastoralism must face up to, archaeological fieldwork offers important new perspectives and makes it possible to fill gaps left by textual sources, offering especially a direct knowledge of the pastoral world of ancient Syria, its material culture and evolution through time. In this sense, of course, the methods of field research (especially survey techniques) and absolute dating (even with luminescence techniques) still need fine-tuning; but they must also make substantial progress that only the practice of archaeological research in the field can provide.

However, only the relentless cross-disciplinary dialectic between archaeologists of pre-, proto-historic and historic periods and philologists of the ancient Near East can define in their countless facets the occupation of the vast territories of the Syrian semi-arid and dry steppe, the use of their resources, and the interaction of these marginal regions with the world of villages and urban centres in the sustainable agriculture zone.

The seventeen articles collected in this volume are rooted in this world and are the result of the workshop "Settlement dynamics and human-landscape interaction in the steppes and deserts of Syria", held on May 3-4, 2012 at the University of Warsaw during the 8th International Congress on the Archaeology of the Ancient Near East. The chapters in this book offer diachronic and multifaceted perspectives on the state of research on pastoral nomads and tribe-state interaction in the Syrian steppes and present to the interested reader the results of the most up-to-date research methodologies and the most important field projects running in the region before the tragic outbreak of the Syrian crisis.

I am grateful for the commitment and enthusiasm shown by all participants in the workshop. Rafał Koliński deserves a special mention for the support he granted to the organization of the meeting and for his warm hospitality in Warsaw. Hartmut Kühne generously offered to publish the proceedings of the meeting in the series Studia Chaburensia and followed with dedication the progress of this volume. A special debt of gratitude is also owed to Katia Gavagnin, who did the typesetting and solved the numerous technical problems with great competence.

Daniele Morandi Bonacossi

Udine
March 2014

BERNARD GEYER, MARIE-LAURE CHAMBRADE and ÉRIC COQUEUGNIOT
UMR 5133 – Archéorient, Université Lyon 2 – CNRS

Water Quest and Subsistence Strategies during the Neolithic in the Arid Margins of Northern Syria[1]

The region of the arid margins in northern Syria[2] (Fig. 1) suffers, like the Syrian steppe as a whole, from a considerable drawback, the arid climate (Besançon and Geyer 2006). We shall focus more particularly on a sector of the Ǧabal 'Iṯriyā (including the 'Iṯriyā's uplands and the Ǧabal Hrībǧa), situated southeast of Aleppo (Fig. 2), where Neolithic sites appear in large numbers and seem to have found a specific way of acquiring water; nevertheless, the sector is not exempt from those climatic restraints with regard to any attempt at settling, exploitation or improvement. The amount of average annual precipitation underlines their severity: in the extreme west the region enjoys an annual rainfall of 300 to 350 mm, but in the east the amount reaches hardly 150 mm and, due to the altitude, a bit more than 200 mm in the Ǧabal'Iṯriyā (Traboulsi 2004) which from that viewpoint seems to be somewhat favoured, although it is also subject to a long hot season (5 to 6 months, sometimes more) without rain, while the evapotranspiration may by far surmount the total quantity of precipitation and thus cause a serious water deficit.

But the scarcity of rainfall to which must be added its great variability (Traboulsi 2004) are not the only elements influencing the modalities of exploitation: a role of similar importance falls to the soils' degree of fertility and to the quality of the resources, especially of water. For the whole of the region, a transitional area between the bio-climates of the Mediterranean zone and the hyper-arid steppe, those factors turn out to be much diversified.

With a view to the possibilities of occupation, environmental conditions play a decisive role – be it in our time or in the past (Geyer 2011). Thus the limits of rain-fed agriculture, which are usually put at the isohyet of 200 or 250 mm of average annual precipitation, turn out to be theoretical here, because even at less than 200 mm rain-fed agriculture remains possible in specific contexts such as the bottom of *wadis*, small shut-off depressions, *faydas*.[3] Over time, their particular potential has been largely exploited. But any occupation of a territory, if only temporary, presupposes a solution for the water supply – for humans as well as for animals. This becomes

1 Many thanks to Brigitte Finkbeiner who translated this article into English.

2 The region of the survey corresponds to the map of Salamya, scale 1:200 000. The study has been carried out as part of a research programme entitled "Les Marges arides de la Syrie du Nord", directed by B. Geyer and R. Jaubert and promoted by the laboratory ARCHÉORIENT (UMR 5133), Maison de l'Orient et de la Méditerranée, Université Lumière-Lyon 2 – CNRS, by the Institut Universitaire d'Études du Développement (Geneva) and by the Directorate General of Antiquities and Museums of Syria (Damascus). It has been carried out in collaboration with ICARDA (Aleppo), the UMR 5189 – HISOMA (Lyon), the IFEAD (IFPO, Damascus) and the IGARUN (Nantes); it has been financed by the French Ministry of Foreign and European Affairs, the CNRS and the Coopération Suisse. Beside the authors, there have participated: N. al-Awad, M. al-Dbiyat, J. Argant, O. Aurenche, O. Barge, J. Besançon (†), Y. Calvet, B. Chalich, F. Debaine, C. Duvette, P.-L. Gatier, R. Jaubert, H. Morel, J.-C. Moncel, S. Muhesen, H.-G. Naton, M.-O. Rousset, S. Sanz, J. Vaudour.

3 *Faydas* (a local term) are wide opened valley floors profiting from deep and fertile soil adduced by surface runoff, situated either at the confluence of wadis, on structural benches or around endoreic depressions.

Studia Chaburensia 4 (2014), pp. 1–31.

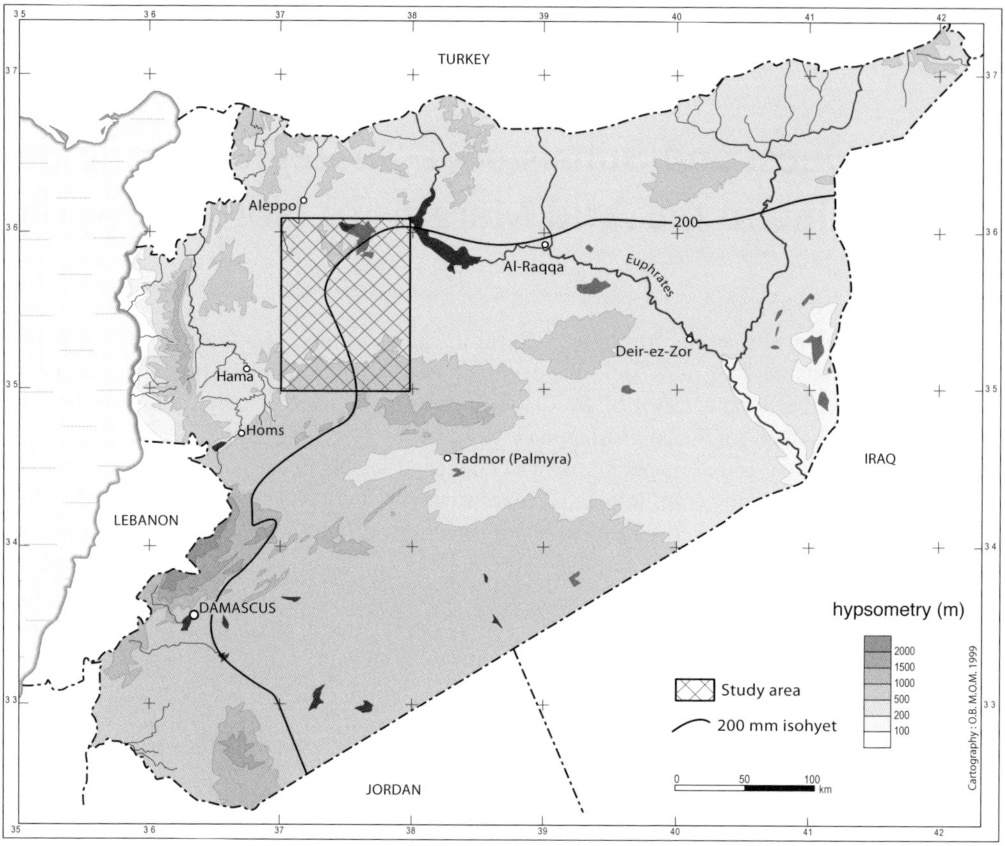

Fig. 1: Map of localization.

even more important when considering older periods that knew at best primitive and rudimentary techniques of water acquisition. In order to understand the occupation one must also come to a grasp of the reasons in favour of an installation, one must pin-point the criteria for choosing the places of a settlement, circumscribe the environments most suitable for the practice of this or that activity providing the means of subsistence, show the access to other resources, besides water.

Interplay of Relief and Aridity

The region is part of the Syrian Badiya, land of arid margins at the inner curve of the Fertile Crescent (Fig. 1).[4]

4 The Fertile Crescent corresponds to a bio-climatic region shaped like a crescent extending from Palestine to the foot of the Zagros mountain via the Jezirah. In the Levant it constitutes the transition between the more or less mountainous Mediterranean zone proper and the true Arabian desert. Originally featuring hardly any trees (pistachio tree, almond tree, etc.), it was covered by herbaceous formations, especially including diverse grasses (wild cereals among them) and wild pulses, and favoured the invention of agriculture in the Neolithic. The domestication of plants (barley, wheat, peas, etc.) was supplemented by that of certain gregarious mammals (Bovidae, Ovicaprinae) that would naturally be grazing there. The potentials for improvement varied and continue to vary, mainly with regard to the amount of rainfall and to the fertility of the soils.

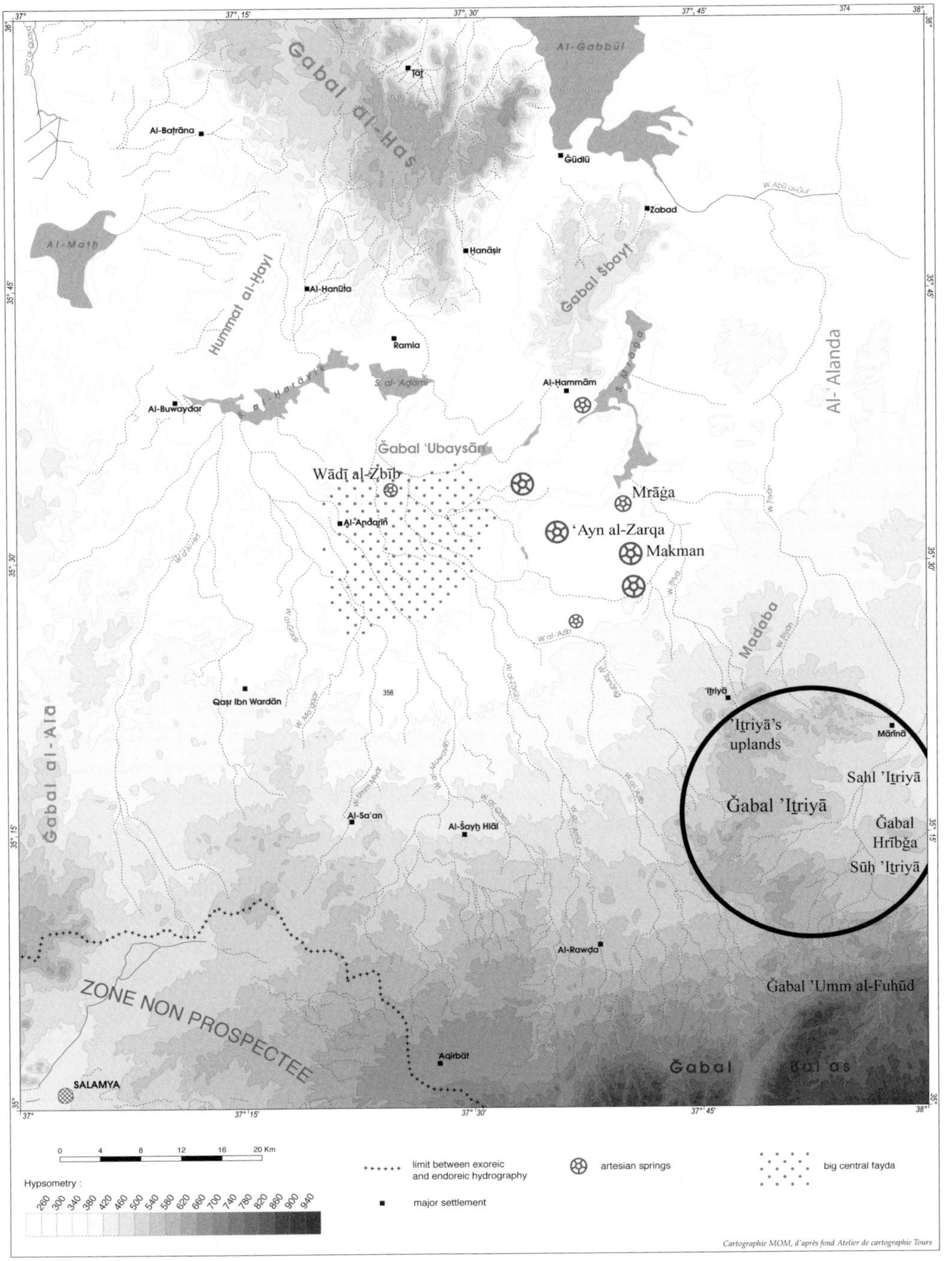

Fig. 2: Topography and hydrography. Enclosed in the circle, the Ǧabal ʾIṯriyā.

Its topography (Fig. 2)[5] is frequently encountered in the dry zones, with a command relief to the south – the northern Palmyra mountains (Ǧabal al-Balʿas, Ǧabal 'Umm al-Fuhūd) and the Ǧabal 'Iṯriyā to which we shall come back –, followed towards the north by a complex piedmont system essentially made up of a succession of glacis, cut by wadis. The erosion of those glacis has brought about level areas (wide valley floors, *faydas*) that take up more and more space in the landscape downhill until a vast plain (the great central *fayda*), which, in its place, is followed by enclosed depressions (*sebkhas*) where the surface runoff comes out.

In the west the picture is completed by a basalt mesa (the Ǧabal al-ʿAlā), in the east by a limestone plateau (Al-ʿAlanda). It is in that region with an apparently simple relief, which turns, however, out to be much more complex in detail, that we have conducted a geo-archaeological study which focused on the characteristics of the archaeological settlements but also on the diverse aspects of their natural context, in order to better understand the environmental impact on the development of land occupation and on the potential for land use.

M. Traboulsi's work concerning the characteristics of the present-day climate in the region (Traboulsi 2010) has brought to the fore facts that had been underestimated so far. It has, for example, turned out that the massifs – including Ǧabal 'Iṯriyā – enjoyed somewhat more precipitation than had been assumed, but that, on the other hand, the plains and low reliefs in the centre are drier than had been expected. Underneath the often thin and eroded soils, the geological substrata and the surface formations are, in most cases, hardly able to retain water. As a consequence the surface runoff is limited in space and, especially, in time. Thus, immediate access to water is only possible at a few places (sources) or for a very short time (floods in *wadis* that are usually dry), or even at places where the water is unusable because salty (salt lakes or *sebkhas*). Those opportunities of access to water are distributed very irregularly leaving vast sectors that are not apt for human installation due to the scarcity of direct resources.

In the absence of sufficient resources, any installation, be it permanent or temporary must be able to come by natural subterranean reserves (wells) or to store the precipitation water. The storing may be carried out on the slopes ("artificial ponds",[6] derivations, cisterns, etc.) or along the thalwegs (weirs, *birkehs*, etc.).

The survey that we have conducted from 1995 till 2010 (fifteen campaigns, directed as much to the search of human installations as to an understanding of their environment) in the complete arid margins region of northern Syria has resulted in the study and description of more than 1000 archaeological sites on a surface of almost 7000 sq. kilometres. The large number made it possible not only to determine more exactly, from the Epipalaeolithic to the present,[7] the various forms of land occupation, the types of settling, the economic choices, but also to arrive at a reasonably close picture of what land occupation in the arid steppe may have meant in different epochs.

In view of the climatic constraints we tried above all to demonstrate the various ways of acquiring water: we have inventoried as systematically as possible the different water resources that are available in the region as well as the different systems of attaining water that have been put into effect, and that since the Neolithic, because – as we will see – the earliest

5 For more detail concerning the physical aspect, see Besançon and Geyer 2006.

6 For a description of the "artificial ponds", see below the section on Tulūl al-Ḥafyān.

7 For studies concerning the present, see especially: Jaubert and Geyer 2006; Chatty and Jaubert 2002; Jaubert *et al.* 1999. For the ancient periods, see Gatier *et al.* 2010; Geyer *et al.* 2004-2005; Geyer 2001. For prehistoric occupation, the determination of the periods is based on the study of flint artefacts gathered in the course of the survey (Coqueugniot 2011).

installations seem definitely to go back to that distant epoch. And it appears to be in the Ǧabal 'Iṯriyā that the first, certainly not permanent, settlement succeeded – founded on techniques of attaining and storing water.

The Case of the Ǧabal 'Iṯriyā

In the southeast of the region, the Ǧabal 'Iṯriyā, situated before the Ǧabal 'Umm al-Fuhūd (Fig. 2), forms semi-mountainous uplands less than 600 m high, mainly consisting of lime stone, partially protected from erosion by its Eocene basement and especially by the hard limestone rocks dating to the Oligocene. Doubtlessly, their massiveness is the cause of a dense net of faults converging towards the famous eponymous site of 'Iṯriyā dating to Roman/ Byzantine times. In part taken away by erosion, the open-lying but relatively thin limestones of the Oligocene are only preserved on the 'Iṯriyā's uplands and in the middle course of Wādī Ḥsayya (Ǧabal Hrībǧa). That environment of resistant limestone explains the remarkable development of a double karstification characterizing this low-mountain region. There is in the first place a deep-going karstification that takes up a good part of the atmospheric waters and thereby lies at the bottom of the marked edaphic aridity of the sector, for which the precipitations – even if enhanced by the altitude – cannot compensate. But there exists also a shallow karstification bringing forth very special shapes, namely the *kamenitsas*: natural basins formed by dissolution, extending over several metres and generally situated on massive limestone rocks. Often they hold back the surface runoff and provide small temporary ponds which are frequently rearranged by man. Those *kamenitsas* also punctuate the moderately steep, stripped slopes on the flanks of the Ǧabal al-Balʿas or the Ǧabal 'Umm al-Fuhūd.

Although a little bit mitigated by the altitude (± 550 m), the whole sector is hit by the aridity to a degree that allows hardly anything but livestock husbandry, as *birkehs* and cisterns add just partially to what the rare wells punctuating the *wadis* can supply. Thus, in the heart of the massif a dozen Byzantine installations are located over a length of about 6 km, each widely surrounded by walls checkering the slopes and tentatively interpreted as farms for the breeding of livestock depending upon the settlement of 'Iṯriyā (Geyer 2000). Beyond this central sector, there are only few sites, either assembled on the borders of the massif or on the margins of the few *faydas* (Sahl 'Iṯriyā, Sūḥ 'Iṯriyā). The important number of PPNB sites (about twenty in this massif, only) is therefore the more remarkable.

The slow maturing of the first installations

The first settlements[8] that we could localize and that might be linked to a possible hydraulic installation go back to the Kebaran (Fig. 3). With 14 identified sites, this period shows a relatively dense occupation for the region. Most of the sites are situated immediately beside or at a short distance from artesian springs – 'Ayn al-Zarqa, Makman, Mrāġa, Wādī al-Zbīb, etc. (Fig. 3) – that is, at places that were certainly among the most attractive ones in the steppe. They constitute in fact some of the few natural watering-places, possibly even permanent at the time and therefore attracting game – a fact that must have made them the more interest-

8 We are aware that despite the care that we have taken with this survey some sites will have escaped us. As we have systematically looked for and gathered lithic artefacts (provided by 254 sites), we take the tendencies observed during this survey to be pertinent.

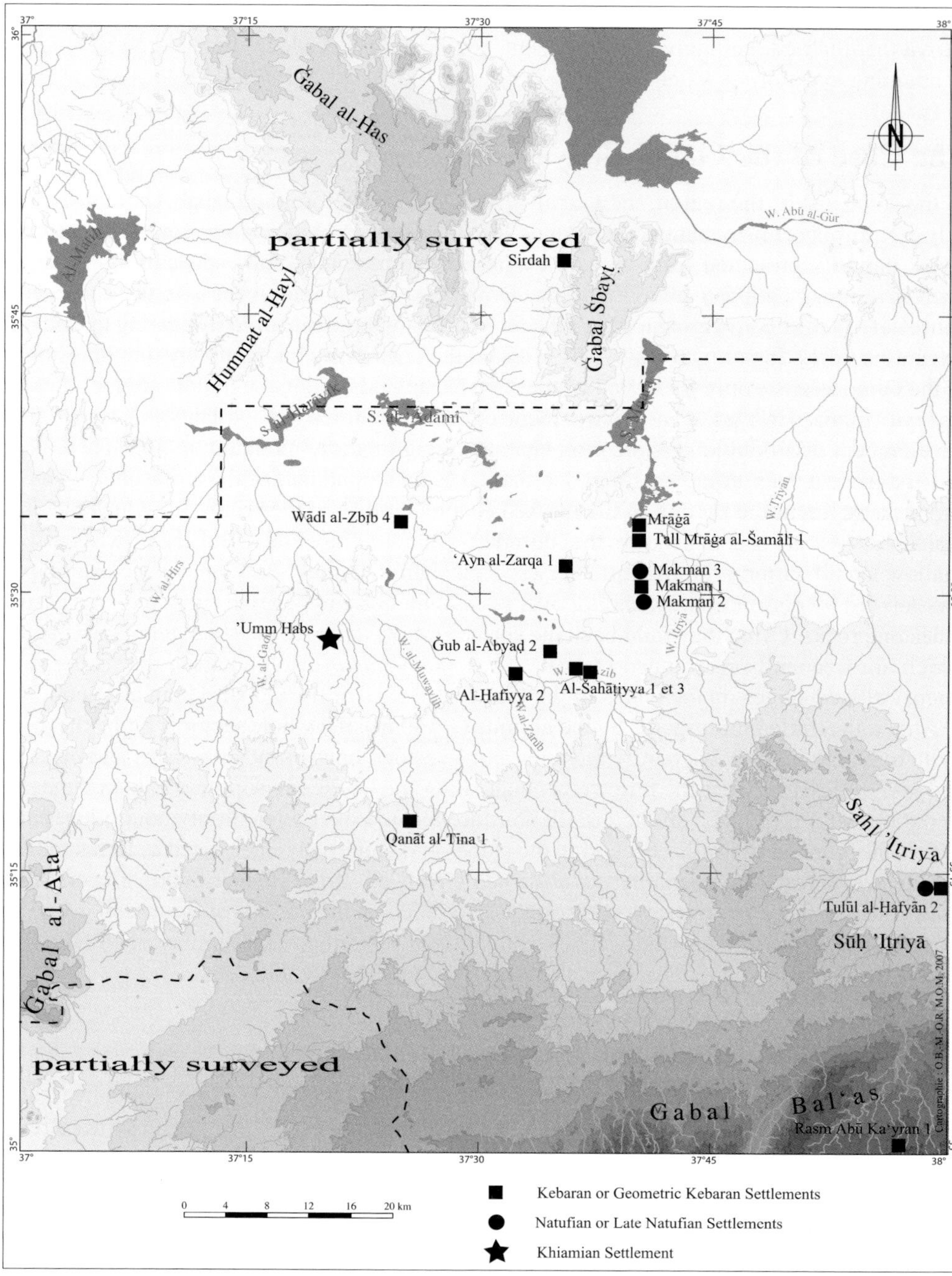

Fig. 3: The pre-Neolithic steppe (© mission Marges arides).

ing. Nevertheless, two sites that have yielded artefacts attributed to the Geometric Kebaran, namely Tulūl al- Ḥafyān 2 and Rasm Abū Kaʿyran 1, are located in an especially arid mountain context, on massive limestone, making the water supply problematic. Close to the former site "artificial ponds"[9] over *kamenitsas* have been found, allowing the supposition that they were used since the Kebaran.[10]

The case is different for Rasm Abū Kaʿyran 1, where we did not observe any pond but where the presence of *kamenitsas* has been noticed. In fact, their presence on both sites leads us to the hypothesis that, in the Geometric Kebaran, they were used for the temporary acquirement of water. Those natural basins may then have served as models for the arranging of "artificial ponds" (Geyer and Coqueugniot 2013).

In the Natufian, represented by three sites, only (Fig. 3), the region seems to have been less attractive. Two sites on spring basins (Makman 2 and 3) have surely profited from the same advantages as in the preceding period – permanent water supply and easy hunting. The third site, Tulūl al-Ḥafyān 2, already settled in the Geometric Kebaran – possibly due to the *kamenitsas* and perhaps to the "artificial ponds" – may, for the Natufian, serve as an example of the connection between *kamenitsa* and "artificial pond". As the climatic conditions had become more restraining than in the Kebaran and also in the final PPNB rendering the quest for water more difficult, the "ponds" destined as reserves – certainly always limited, but vital – might then have been brought into effect.

For the Khiamian only one site could be identified, 'Umm Ḥabs (Fig. 3), an observation that seems to confirm the increasing abandonment of the region. The site is situated next to the vast grassy surfaces of the great central *fayda*, which surely attracted game. That single site also turned out to be situated in close vicinity of an "artificial pond".

The improvement of the climate marking the whole region since the PPNA will grow in the early and middle PPNB but appears to have had no influence on the occupation of the region where not a single site has been determined for those epochs.[11]

Even if the chronological link between, on the one hand, *kamenitsas* and "artificial ponds" and, on the other hand, sites cannot always be formally established, we must underline the fact that – with the exception of two sites occupied in the Geometric Kebaran, namely Qanāt al-Tīna 1 and Sirdaḥ, for which we have not been able to determine the manner of water acquirement in use at the time – from the Kebaran to the Khiamian those settlements that were not situated next to artesian springs lay instead next to *kamenitsas* or to "ponds". The number of cases is certainly not big enough to draw definite conclusions, and only a more profound confrontation with the attributes of the terrain will allow to prove or disprove this preliminary statement, but it is worth noting that this situation will become frequent in the late/final PPNB, accompanied by numerous installations that are, above all, adapted to different contexts.

9 Diverse types of "artificial ponds" adapted to their environmental contexts and forming as many small temporary water reserves have been found on numerous slopes of the Ǧabal 'Iṯriyā, but also – if in smaller amounts – on the slopes of Ǧabals 'Umm al-Fuhūd and al-Balʿas.

10 Just as the well or the cistern, the "artificial pond" is an "ageless" technique for the acquirement of water because easily realisable by small communities. Over time those techniques played a role in transforming the potentials of the natural milieu. The repulsive character of the zone affected by the installation is reduced. In consequence the zone is opened up for occupation, which may become permanent and thus correspond – with regard to occupation and development of the land – to a transformation of the original environmental conditions. There is no need to transmit a special knowledge: even after several centuries of abandonment the evidence imposes itself and the new inhabitants profit "naturally" from a resource of, after all, artificial origin (Geyer 2009).

11 The steppe may have lost its few inhabitants by and by, but that does not apply to the Bal'as mountains, situated immediately south of our region, where F. Abbès and his team verified numerous sites going back to the Khiamian and to the beginning PPN (Abbès 2008 and pers. comm. at the colloquium in Toulouse "Transitions en Méditerranée ou comment des chasseurs devinrent agriculteurs", 14-15 april, 2011).

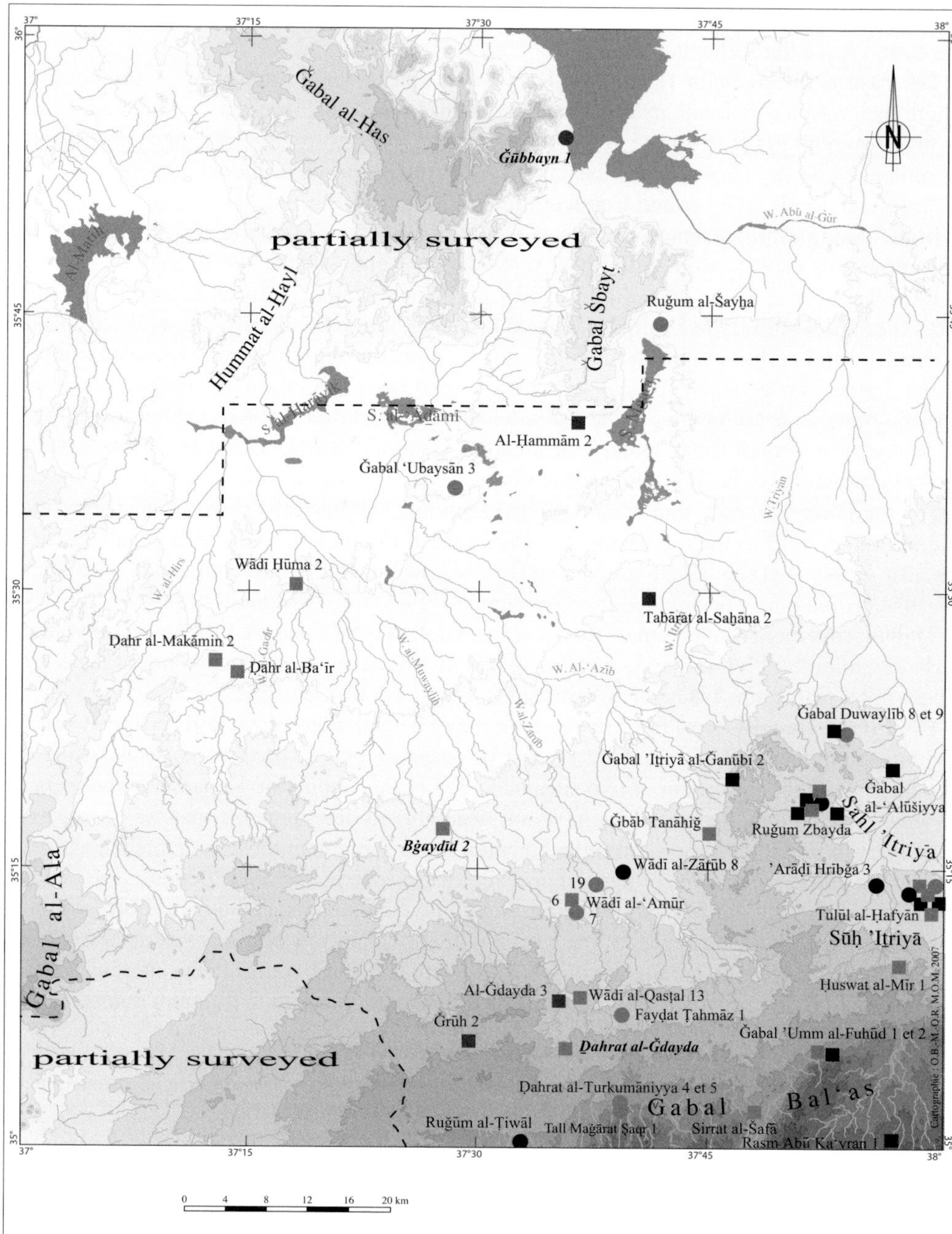

Fig. 4: The steppe in the late and final PPNB.

Legend: circle = site attributed to the PPNB, without further restriction; square = site attributed to the late/final PPNB; dark grey = settlement at a spring; light grey = settlement in a *wadi* or on a slope; black = settlement at an "artificial pond" and/or *kamenitsa*. The place-names in italics mark sites considered as permanent, the place-names in straight characters mark sites considered as temporary.

Installations mastered in the Late/Final PPNB

In the late/final PPNB the situation changes radically (Fig. 4):[12] practically the whole of the region under study gets occupied, with the noteworthy exceptions of the plateau of al-ʿAlanda in the east, where the climate as well as the soil are too dry, and of the surfaces of the basalt mesas, where the problems of water access were likely insoluble at the time (there the access to water would mean to dig wells through high-reaching basalt outcrops).

46 sites have been attributed to the PPNB in the large sense, 16 of which could only be attributed to an undifferentiated PPNB, while 30 belong to the late/final PPNB.[13]

One of the theories able to offer an explanation of this phenomenon is a significant growth of the population causing a move to new territories (Bocquet-Appel and Bar-Yosef 2008). It seems at any rate evident that a need was felt to turn to new areas characterized by a prominent relief: the land near the mountains and the mountains themselves had an attraction for Neolithic man as never before and hardly ever after.

Among the 46 localized sites less than a dozen do not adhere to that "rule", even if – as we have seen – the aridity on the highlands of the southeast is rather severe. The foundation of settlements – certainly just temporary ones in the overwhelming majority of instances – in sectors that due to their edaphic conditions, lack in water resources, meant to install a minimum of water storage measures in order to be able to live there at all. In at least a third of the cases we may affirm that such measures consist on the one hand in *kamenitsas*, on the other hand in "artificial ponds" making it possible for man to settle – if only temporarily – in those dry zones: it appears that either measure has been reused frequently, some have been maintained to this day.[14] We have marked *kamenitsas* and "artificial ponds" only when they were situated in the immediate neighbourhood of sites. It is obvious that those phenomena – natural or artificial – are very frequent in the mountainous zone of the high piedmont and that a number of sites must have profited from the possibilities of water resources available in this way, even if we have not made the formal connection between their presence and the sites.

12 Lacking preserved remains other than lithic artefacts and for reasons to do with the volume of the available sample (the main criterion for an attribution to the PPNB in the large sense lies in the debitage technology and in the typology of the fittings of arrowheads), the distinction between late PPNB and final PPNB could only be established for two sites (late PPNB for Bġaydīd 2, one of the few sites that may be considered as permanently occupied; final PPNB for Al-Ḥammām 2, in the immediate vicinity of one of the most important artesian springs of the region). Therefore all sites have been attributed to the late/final PPNB without further differentiation.

13 17 sites have yielded lithic industry which is perhaps, but not with certainty, attributable to the PPNB. They have not been accounted for as such in the text and are not represented in the corresponding figure. Although not safely datable, they strengthen by their localization the picture imaginable with regard to the diffusion of the PPNB sites in the region: a distinct concentration in the higher sectors towards the south (13) and just a few in the plains and on the lower glacis (4). It should here be underlined that the sites lying in the west of the region are more likely to have escaped our attention, as they have seen a number of reoccupations and as this sector has been more intensely developed.

14 Ceramics (evaluated by Y. Calvet and M.-O. Rousset) have been found – if in very small amount – in the sector of the "artificial ponds" at Tulūl al-Ḥafyān: they belong to the Early Bronze Age, to the Roman, Byzantine, Ayyubid and modern periods. That corresponds to the maximum periods of occupation of the arid steppe by nomads and attests to the frequency if not the permanence of occupation at those few water points.

Diverse installations

Detailed studies of the lithic material gathered in the immediate vicinity of the hydraulic installations and attributable to the PPNB permit to advance the hypothesis that they were originally meant as much to supply humans with water as to attract game. But the mere collection of artefacts from the surface of the sites, necessarily accidental and incomplete, suffices only very seldom to arrive at a characterization of a site's functions. In spite of those limitations it is possible to formulate some hypotheses based on the facts that we have gathered. Beyond knowledge about the natural environment and its potential and beyond the chrono-cultural attribution of the sites, their paleoethnological nature remains obviously hypothetical: simple hunting camps, places for specialized activities (workshops for the knapping of flint…), base camps (temporary or more permanent). In the absence, in most cases, of architectural remains (remains of "houses"…) for those periods, any hypotheses about the kind of occupation must be deducted from the lithic artefacts (generally flint), as they were the only traces to be observed in the course of the surveys. The hunting camps, for example, show little variety in the tools (and the numbers are small due to the short-term occupation) with a domination of arrowheads. In general, the dwelling-sites ("base camps") yield material in much greater abundance and diversity including a large number of domestic tools. Those base camps may at any rate yield – or not – arrowheads to do with the preparation and maintenance of hunting weapons (besides fittings that may have been brought back sticking in the carcasses of hunted animals). The workshop sites provide cores and a huge number of by-products resulting from preparation and knapping of the cores, but few of the intended products (blades…); retouched tools (and weapons) are often lacking, because the transformation of rough supports was carried out elsewhere.

The first sites considered to have been permanently settled go back to that period: Ǧūbbayn 1 in the north, Bġaydīd 2 in the central south, Ḏahrat al-Ǧdayda in the south[15] (Fig. 4). In the low lands (low piedmont and central plains), only five sites have been localized: Bġaydīd 2, which has already been mentioned; Ḍahr al-Baʿīr situated on a high plain where hunting activities dominate; Ḍahr al-Makāmin and Wādī Ḥūma 2 also situated on a high plain; finally Ǧabal ʿUbaysān 3, at the foot of the eponymous massif. Only two sites are linked to artesian springs: Al-Ḥammām 2, according to the material mainly a hunting camp but also yielding domestic tools, and Tabārat al-Saḫāna 2. Ǧūbbayn 1, already mentioned, lies on the banks of "Lake" Ǧabbūl in the immediate vicinity of springs fed by the Ǧabal al-Ḥaṣ. Finally Ruğum al-Šayḫa, thanks to the availability of flint nodules from the lower Eocene a large knapping workshop is located a little bit north of the *sebkha* Mrāġā.

But it is mainly in the south-eastern mountainous part of the region under study that the sites turned up in the greatest numbers. They also reflect the diversity of the occupations, but among the collected material the artefacts relating to hunting activities are often quite prominent. In that very sector have been documented those PPNB sites that are situated close to *kamenitsas*, to "artificial ponds" and to their various aspects.[16]

Rasm Abū Ka'yran 1, the southernmost of those sites, lies at the border of the Ǧabal al-Bal'as. Beside the *kamenitsas* observed in the sector, there exist, in the bed of a secondary *wadi*, "artificial ponds" in the shape of semicircular micro-weirs. In prehistory the cretaceous

15 The assumption of a probable permanent occupation is based, on the one hand, on the very large as well as diversified lithic assemblages and, on the other hand, on the presence of fragments of coloured coatings (red and/or beige) and of gypsum floors indicating the presence of built structures, gypsum crockery, equipment for grinding.

16 Cf. below, Diversified hydraulic installations.

limestone rocks and their rose-coloured flint, of a typical Ǧabal Bal'as quality, have been much exploited: Sirrat al-Šafā, situated on the borders of a polje, is a PPNB workshop for debitage that bears witness to this usage. Further north, in the foothills of the Ǧabal al-Bal'as, Huswat al-Mīr 1 and Ǧabal'Umm al-Fuhūd 1, feature installations pointing to diversified domestic activities, with hunting occupying a prominent place in the subsistence economy. "Artificial ponds" on *kamenitsas* appear at Ǧabal 'Umm al-Fuhūd 1, while a spring is situated near Huswat al-Mīr 1, where no "pond" has been observed. The situation is the same at Ǧbāb Tanāhiǧ, at the foot of the Ǧabal 'Iṯriyā, which encompasses the sites Ǧabal 'Iṯriyā al-Ganubi 2, Ǧabal al-'Alūšiyya and Ǧabal Duwaylīb 8. The first of those is a site of undetermined character dominating the western low lands, where *kamenitsas* and two "artificial ponds" have been noticed. The material from Ǧabal al-'Alūšiyya seems to point to a hunting camp, a hypothesis supported by its location on an outlier dominating the vast embayment of Sahl 'Iṯriyā. Ǧabal Duwaylīb 8, a rock shelter with stone circles, is situated on a slope northeast of the Ǧabal 'Iṯriyā and may constitute an encampment with diversified domestic activities, at any rate characterized by the importance of hunting. Below the shelter there lies an "artificial pond". But the strongest concentration of sites dating to the late/final PPNB is met with in the sectors of Ruǧum Zbayda (1, 2, 3, 4 and 6), an eastern chain of the Ǧabal 'Iṯriyā, and of Tulūl al-Ḥafyān (2, 4, 5, 9 and 12), in the Ǧabal Hrībǧa, in addition associated with a great number of *kamenitsas*, and a range of hydraulic structures like "artificial ponds" in various shapes. Most of those sites seem to be domestic encampments, in some cases probably with hunting as an important activity.

Diversified hydraulic installations

Whatever the function of the documented sites, daily access to water was imperative and necessarily one of the essential elements in settlement choices. We have seen that in this region water seepage proceeds very rapidly, there are only some sectors where the limestone rocks at the surface are massive enough to assure runoff and, at the same time, the persistence of temporary water ponds on the *kamenitsas*, models for the "artificial ponds", which, on their part may appear, according to the context, as "ponds" on slopes, as "ponds" in the thalwegs and on water derivations on slopes (cf. below). The most characteristic instances are to be found on the open-lying limestone rocks dating to the Oligocene: we shall consider two sectors, the first is situated in the Ǧabal Hrībǧa, at Tulūl al-Ḥafyān, where seven PPNB sites have been discovered, the second in the 'Iṯriyā's uplands, at Ruǧum Zbayda, where six PPNB sites have been localized.

Tulūl al-Ḥafyān

On the last foothills of the northern piedmont of the Palmyra mountains, in the Ǧabal Hrībǧa, Tulūl al-Ḥafyān rises as a stripped rocky ridge on the margins of a fayda suitable as a pasture, in an environment of limestone, very arid (Fig. 5), hit as much by low and erratic rainfall as by very hard edaphic conditions. All in all a dozen sites have been discovered in the sector, dated by the lithic artefacts to the middle Paleolithic (2, the dating of 1 is not certain), the Geometric Kebaran (1), the Natufian (1), the PPNB (7 have yielded PPNB artefacts, 5 of which go back to the late/final PPNB) and, by the ceramics, to the Early Bronze Age (4, of which 2 have yielded only little pottery from that period), the Roman (1, uncertain), Byzantine (2, with little pottery) and Ayyubid periods (2, 1 of which with little pottery).

Fig. 5: Tulūl al-Ḥafyān: in the foreground the limestone hillock, at the border of the *fayda* to be seen in the background. In the centre, an "artificial pond": for a close-up of the pond, cf. fig. 7 (© mission Marges arides, negative B. Geyer).

As far as the quantities of material and associated structures are concerned the major PPNB sites are Tulūl al-Ḥafyān 5 and 9.

Tulūl al-Ḥafyān 9 is the most important site: several hundreds of artefacts were collected there, among them more than fifty tools. The site, lying on a crest to the north of the other sites of Tulūl al-Ḥafyān, shows on the surface traces of circular structures and is exclusively late/final PPNB. According to the examined material the site seems to have been inhabited, temporarily or seasonally, and – regarding the structures and the great amount of material – in all likelihood regularly reoccupied.

Tulūl al-Ḥafyān 5 is located in a small valley 500 m south of Tulūl al-Ḥafyān 9, on the margins of the *fayda*. The valley is characterized by the presence of numerous installations (Fig. 6): circular and square structures, tombs, a kite, terraces or enclosures and, above all, "artificial ponds" and other hydraulic structures (cf. below). The artefacts (more than 300) and pottery collected on the site date to the late/final PPNB and the Early Bronze Age. Its industry is characterized by a large quantity of lithic artefacts suggestive of hunting activities as well as of flint knapping. The existence of multiple and diverse structures, the nearness of Tulūl al-Ḥafyān 9 as well as the extension over which material was assembled: by all those indications Tulūl al-Ḥafyān 5 may have served, alternatively and/or simultaneously, as a temporary encampment site, specialized or not, but equally a place for activities related to the residential site Tulūl al-Ḥafyān 9.[17] The large proportion of arrowheads

17 Based on ethnographic examples, D.J. Seymour (Seymour 2009: 266) proposes that for mobile groups, other than for sedentary ones, the areas for activities did not lie in the midst of or very close to the habitat but were dispersed in the environment of the site, at places adapted for the different types of activities.

among the lithic assemblage underlines the importance of hunting. But evaluation of just the lithic material brings to light only part of the complexity or the diversity of the ways of subsistence, above all, it does not permit the identification of pastoral practices, a hypothesis that we are going to consider in detail (cf. below, Micro-regional potentials: territories and strategies of subsistence).

In that sense, the *fayda* bordering the small valley of Tulūl al-Ḥafyān 5 represents a vast pastureland, which is today put to regular use by nomadic herdsmen. But even if the *fayda* is able to provide green fodder in the rainy season, the surface flows, rare and short-lived as they are, do not suffice to ensure a regular water supply, not even for a short span.

In order to be able to profit from the resources seasonally available in the *fayda* and to practice, simultaneously or not, hunting and animal-breeding, it was necessary to store a minimum of water reserves thus making possible at least a temporary stay. To that end the people have either improved the small natural reservoirs – the *kamenitsas* – or they created new ones – the "artificial ponds", making as many small temporary water reservoirs.[18]

Structures of this type have been found at 4 of the 7 PPNB sites, especially at Tulūl al-Ḥafyān 5 (Fig. 6).

Most of these "ponds", established on stripped slopes, are enclosed downhill by a small curved wall made of big limestone blocks, which are partially let into the ground (Fig. 7). They may attain up to twenty metres in "diameter", with the retaining basin extending uphill (Fig. 8). In one of the thalwegs arriving in the *fayda*, equivalent installations have been arranged in succession: two "ponds",[19] possibly extending to 40 m long and 20 m wide, retain the short-lived flows that are concentrated in the *wadi* (Fig. 9), the excess waters of the first pond feed the second, which is situated downhill. Finally, on a lower part of the slope, a small wall of unhewn limestone blocks observed over ca. twenty metres (it is partially dug in), interrupts the surface flow caused by precipitations and leads it towards a small retaining "pond" (Fig. 10): the whole is situated on a water derivation on the slope.

18 Similar installations, though more sophisticated, have been discovered at Ibn el-Ghazzi (Betts and Helms 1989) and at Dhuweila (Betts 1988), in east Jordan, where bio-climatic conditions are very close to those of the region that concerns us here. The complexity of the hydraulic installations discovered at Ibn el-Ghazzi left the authors unable to be more precise about their dating: they propose as "best choices" the "*Pre-Pottery Neolithic period or the Roman/Byzantine centuries before the Umayyads*" (Betts and Helms 1989: 10).
Newer research by S. Fujii (Fujii 2009) in the Wadi Abu Tulayha (Jordan) has demonstrated, in an agro-pastoral "outpost" situated in the northwestern part of the al-Jafr basin, the existence of a large "cistern" going back to the middle or late PPNB, which is said to have provided drinking-water, and of a small weir for "basin-irrigated" agriculture, the construction of that latter being attributed to the nomadic population of the middle PPNB. In the al-Jafr basin other hydraulic structures have been pointed out as well, which are attributed to the Neolithic, especially at Wadi Ruweishid al-Sharqi (Fujii 2007) and Wadi Quweir 106 (Fujii *et al.* 2011): at either site two basin-irrigation dams have been discovered. Also in east Jordan, at the so-called Wisad Pools, G.O. Rollefson, Y.M. Rowan and A. Wasse (Rollefson *et al.* 2010: 2-4) have discovered an important necropolis, crossed in the centre by the Wadi Wisad. In the bed of the wadi nine pools follow upon each other, natural basins dug into the local basalt rock and retaining the runoff caused by the winter rains. The two largest pools, at either end of the system, show no traces of alteration, whereas several of the smaller pools in-between have been retouched in order to heighten their retaining capacity. The system could not be dated precisely. The artefacts that were gathered over all of the site go back to different periods: early and middle Epipaleolithic (ca. 20,000 – 13,000 BC), PPNB (ca. 8500 – 6900 BC), Late Neolithic (ca. 6400 – 4700 BC), Chalcolithic and Early Bronze (ca. 4700 – 2200 BC).

19 A third pond, lying a bit uphill from the two here mentioned, had also been interpreted as an installation in a thalweg (Geyer and Coqueugniot 2013). A more detailed analysis by M.-L. Chambrade (Chambrade 2012: 394, 396) has put it among the water derivations on slope.

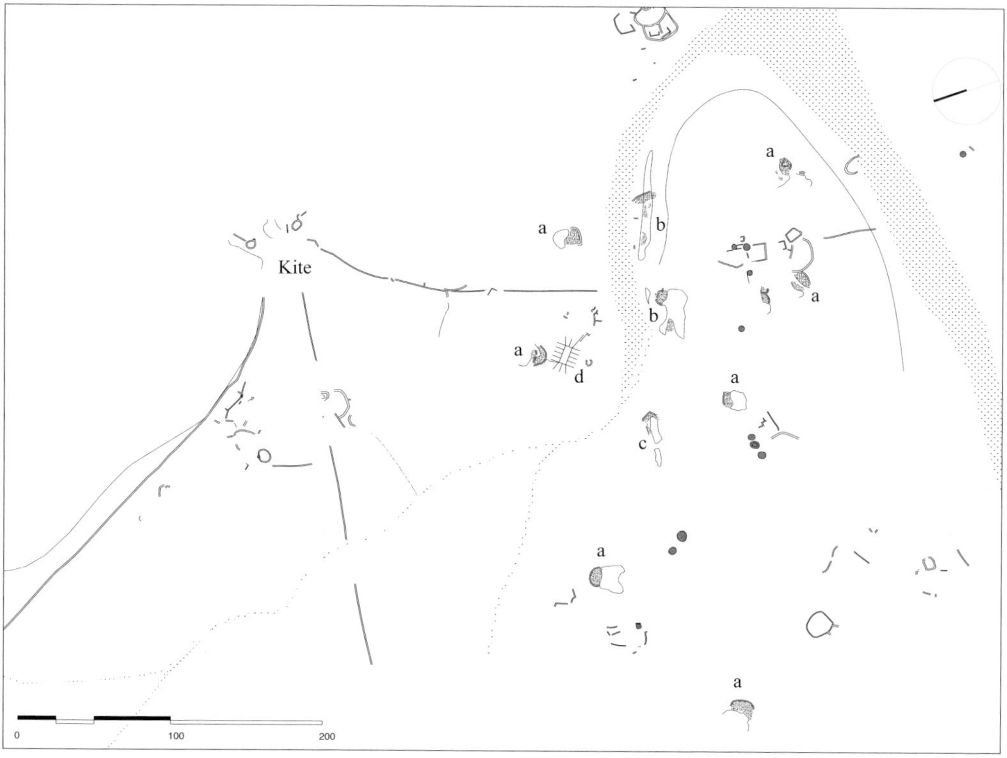

Fig. 6: The site of Tulūl al- Ḥafyān 5 (november 2000).

Legend: a = "artificial pond" on the slope; b = "artificial pond" on thalweg; c = derivation of the slope; d = beduin tent.

Ruğum Zbayda

The sites of Ruğum Zbayda are also situated on high ground, but in the 'Iṯriyā's uplands. The chain where they are located lies along a WNW-ESE axis extending over ca. 6 km and is framed by two *wadis*, that start off in the uplands to feed the vast embayment of the Sahl 'Iṯriyā. The NNE slope of the chain opens onto a wide and flat *fayda* (Fig. 11) making a very suitable pastureland,[20] in a context resembling that of Tulūl al-Ḥafyān: a denuded rocky hillock on the margins of a *fayda* suitable as a pasture, in an environment of arid limestone.

In all, nine sites have been determined in the sector and, by the lithic artefacts, been dated to the Paleolithic (3, of which 1 to the lower Paleolithic and 2 to the middle Paleolithic), to the PPNB (6, of which 5 to the late/final PPNB), to the Chalcolithic/Bronze (3) and, by the ceramics, to Early Bronze IV (2, with little ceramic sherds). Here the potential for development is about equivalent to that already described for Tulūl al-Ḥafyān, there prevail the same constraints of access to water and the same solutions are implemented to mitigate that lack. At four of the six PPNB sites structures of the *kamenitsas* and/or "artificial ponds" type have been found, especially at Ruğum Zbayda 1 and 6, in association with lithic artefacts going back, by far the most often, to the late/final PPNB.

20 Due to evident overgrazing this plain bears today only little palatable vegetation, but the attractivity of the place for animal-breeders is emphasized by that fact, alone.

Fig. 7: "Pond" upstream from the site of Tulūl al-Ḥafyān 5
(© mission Marges arides, negative B. Geyer).

Fig. 8: *Kamenitsa* turned into a "pond" at the site of Tulūl al-Ḥafyān 5; on the right, part of the basin for rainwater (© mission Marges arides, negative B. Geyer).

On a west-looking slope of Ruğum Zbayda 1, that is, exposed to the main winds, an installation on the hillslope has been discovered (Figs. 12, 13), that, just as at Tulūl al-Ḥafyān 5, allows the harvesting of runoff water on the open-lying limestone slope. A small wall, almost 50 m long, guides the water towards an "artificial pond". The material gathered around the installation belongs exclusively to the late/final PPNB. The top of the hillslope is punctuated by several *kamenitsas* (Fig. 14).

Fig. 9: "Artificial pond" localized in a thalweg, at the site of Tulūl al-Ḥafyān 5 (© mission Marges arides, negative B. Geyer).

The evaluation of the great number of artefacts suggests the presence of an encampment mainly dedicated to domestic activities. But in the absence of any other structure in this place the installation might be associated with the site Ruğum Zbayda 4, which lies less than 300 m away, is exclusively dated to the late/final PPNB as well and features numerous circular and rectangular structures with scanty walls. Also the abundant material collected at that site goes in the direction of a dwelling-site, where hunting played a considerable role among the activities ensuring subsistence, as is in addition suggested by the find of 11 shafted arrows (Byblos points on twisted blades). The Ruğum Zbayda sites dating to the late/final PPNB have in common a similar location: high up on the slope below the outcrop at the summit. With the exception of Ruğum Zbayda 1 and 4, those sites show also tools dating to the Paleolithic, to the end of the Chalcolithic or to the Early Bronze (Canaanean debitage is scarce at Ruğum Zbayda 3, tabular scrapers are rare at Ruğum Zbayda 2 and 6), artefacts belonging to the PPNB remain dominant. They are characterized by the presence of diverse structures, often partly buried and therefore difficult to interpret.

"Artificial ponds" could at any rate be identified. For example at Ruğum Zbayda 6, in an amphitheatre at the top of a network, below the crest rising from the Oligocene limestone rocks, partly buried enclosures and/or shelters are associated with artefacts dating to the periods mentioned above but mainly to the late/final PPNB. The inhabitants found their water supply on a hillock west of the site, where several *kamenitsas* have been discovered (Fig. 15), two of which had been made over into "ponds" (Fig. 16). Finally, *kamenitsas* may also be found on the cornice dominating Ruğum Zbayda 3; the great number of artefacts dating to the late/final PPNB permit to propose the existence of an encampment where diverse domestic activities were exercised (a variety of tools, few arrowheads).

Micro-regional Potentials: Territories and Strategies of Subsistence

Sometimes mobility appears to be a necessary adaptation to the constraints of climatic and edaphic aridity, a mobility that is likely never to have been given up in spite of the sedentariness developing in the course of the PPN. The communities made the best of this way of life, that was adapted to the arid zones: they occupied the steppic plateaus and mountains and exploited the various seasonal and/or permanent resources. By their location and sometimes also by their lithic industry the Tulūl al-Ḥafyān and Ruğum Zbayda sites could be interpreted as camps for temporary/seasonal activities, in some cases characterized by considerable amounts of artefacts to do with hunting.

A second sign of adaptation to arid surroundings, where the success of planta-

Fig. 10: Water derivation on a slope at the site Tulūl al-Ḥafyān 5 (negative M.-L. Chambrade).

Fig. 11: Ruğum Zbayda: in the foreground the limestone hillock, the fayda is visible in the background (© mission Marges arides, negative J. Besançon).

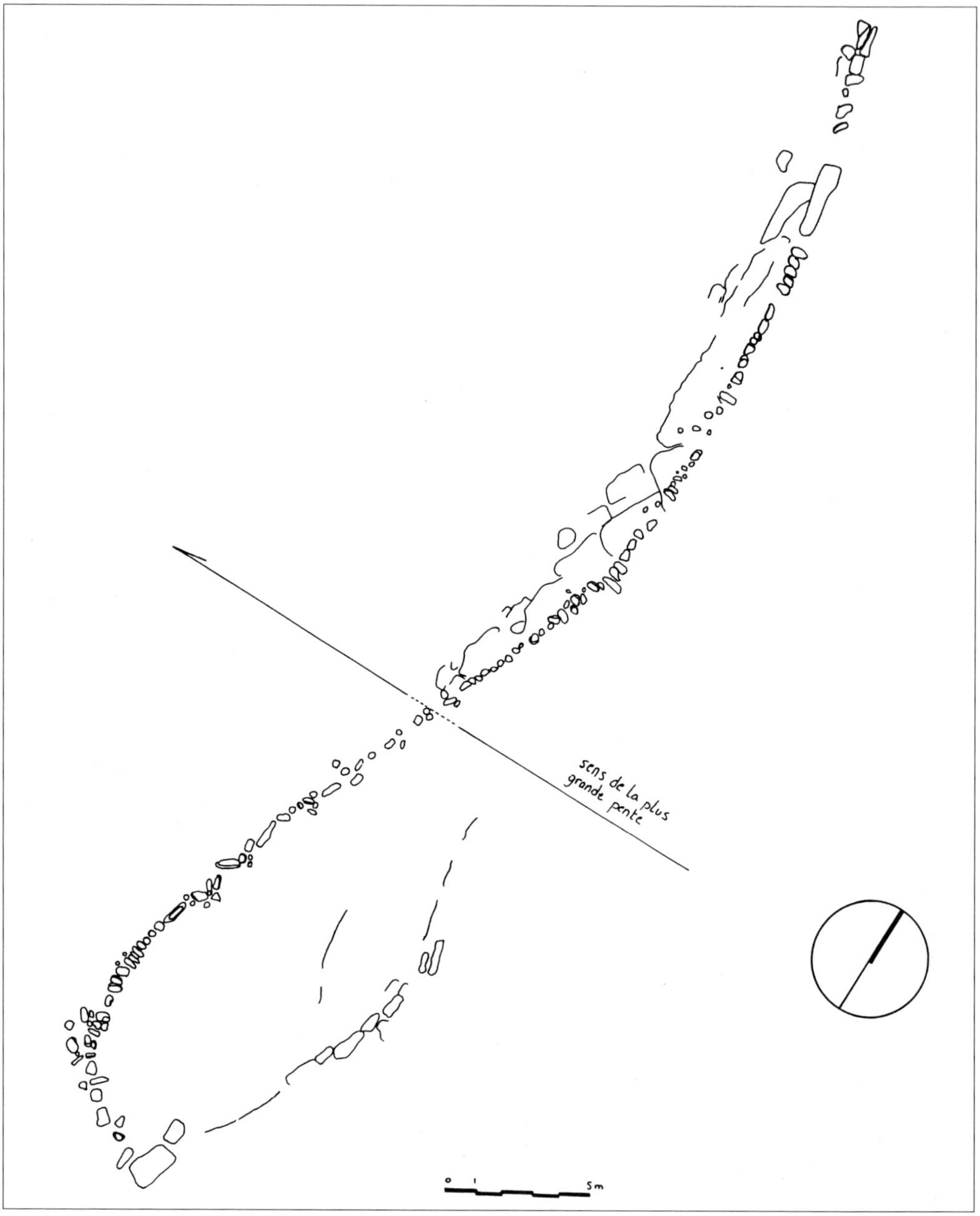

Fig. 12: Water derivation on a slope at the site of Ruğum Zbayda 1, in 1998 (© mission Marges arides, copy J.-C. Moncel and B. Chalich).

tions is rather uncertain, may be seen in the predominant exploitation of meat, of wild and/or domestic animals. As the final PPNB villages were pure farming communities, it is possible that the subsistence economy of the groups that had frequented Tulūl al-Ḥafyān and Ruğum Zbayda had not only rested on game-hunting. In the Late Neolithic, mobile groups that very

Fig. 13: Water derivation on a slope at the site of Ruğum Zbayda 1
(© mission Marges arides, negative B. Geyer).

Fig. 14: *Kamenitsas* on the upper surface dominating Ruğum Zbayda 1
(© mission Marges arides, negative J. Besançon).

Fig. 15: *Kamenitsas* at Ruğum Zbayda 6 (© mission Marges arides, negative B. Geyer).

probably practiced a combined hunting-breeding economy wandered through the arid zones of eastern Jordan (Baird 1993; Betts and Cropper 2013) and may have appeared in the Ğabal al-Bal'as since the second half of the PPNB (Abbès 2007, 2009; Chambrade 2012: 383). At Qdeir 1, in the basin of El-Kowm, regarded as a camp of semi-nomadic herdsmen and dated to the final PPNB (Cauvin 1990; Stordeur 1993), arrowheads constitute quite a considerable part of the lithic assemblage (Aurenche and Cauvin 1982). And the importance of hunting activities suggested by the study of the material is confirmed by archaeozoological investigations: in the various phases of occupation the remains of gazelles make up between a third to almost half of the fauna (Gourichon 2004: 380).

One could imagine these two forms of subsistence economy – based on hunting and livestock husbandry or on pastoral nomadism – for the late/final PPNB camps of the arid margins of northern Syria, and especially so in the cases of Tulūl al-Ḥafyān and Ruğum Zbayda. As other archaeological elements that might support or gainsay this proposal are missing, a micro-regional environmental investigation[21] may already help to be more

21 Such micro-regional studies are founded upon a model of the economic territory of the sites based on the daily exploitation of the resources (cf. next footnote), the analysis of the potentials and the constraints of the environments making up the territory and the identification of the available resources. They follow the method of Site Exploitation Territory proposed by E.S. Higgs and C. Vita-Finzi (Vita-Finzi and Higgs 1970; Higgs and Vita-Finzi 1972, 1975), here further developed thanks to GIS and adjusted to the issues under study and to the available facts. Once the territories are defined, the geosystems that they are composed of will be mapped and their potentials and their constraints evaluated – on the basis of the terrain's qualities – in maps and bibliographies. For a complete overview of the theoretical foundations of that method see Chambrade 2012: 489-501 and, forthcoming, for a more detailed study of this case.

Fig. 16: "Artificial pond" on *kamenitsa* at Ruğum Zbayda 6 (© mission Marges arides, negative B. Geyer).

precise as to whether the occupation of these sites is suitable for herding practices and whether a seasonal occupation is possible, i.e. whether the available water resources and the techniques applied for their acquirement prove sufficient to satisfy relatively long-term needs. Based on the available archaeological data one site has been selected for either sector: Tulūl al-Ḥafyān 9 and Ruğum Zbayda 4. They are taken to be representative of either group of sites, and their territorial environment to mirror the general situation of either sector.

The case of Tulūl al-Ḥafyān 9

The site of Tulūl al-Ḥafyān 9 has yielded the largest number of artefacts, exclusively dating to the late/final PPNB. The circular structures at the site are the clearest indications of the existence of shelters, which in this sector may go back to that period. It is situated in the southernmost part of the relief, on a narrow, west-east running interfluve. The interfluve separates a small shallow valley from a U-shaped valley in the north that cuts out a semicircle from the relief. The dominant position of the site makes for a viewpoint of 360°, towards the heart of this hilly sector in the west and the northwest and towards part of the Sahl 'Iṯriyā in the other directions, until the Ǧabal al-Bal'as that appears on the horizon towards the south. This location, protected from floods and from secondary runoff supports the hypothesis of a residential character that has already been attributed to Tulūl al-Ḥafyān 9 because of the diversity of domestic tools.

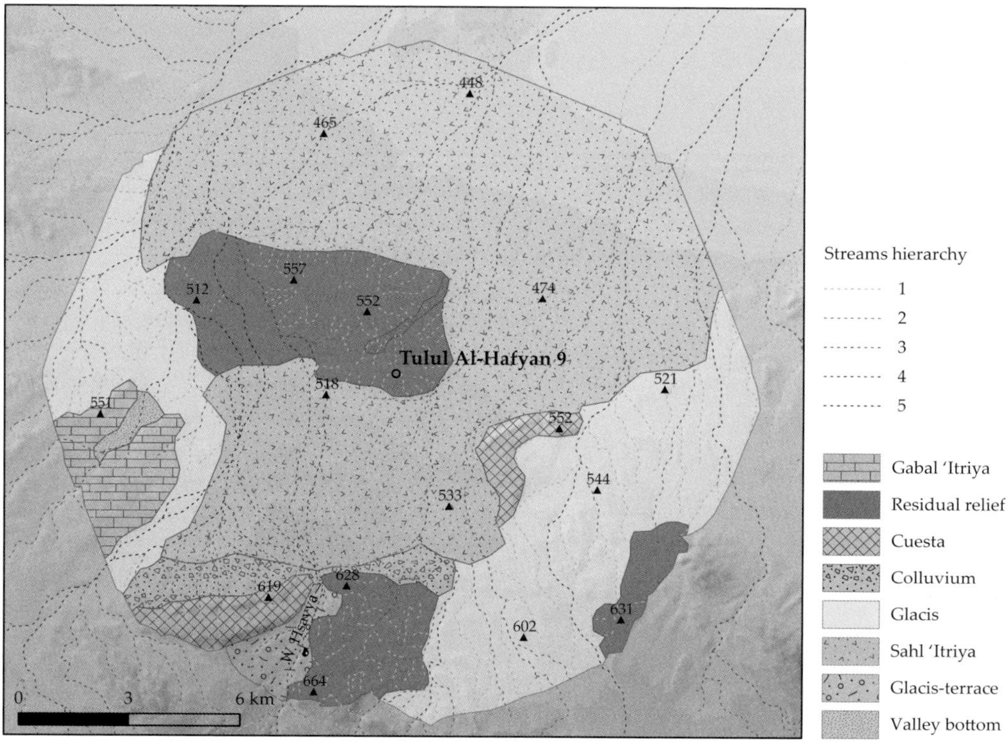

Fig. 17: Geosystems of the economic territory of Tulūl al-Ḥafyān 9 (execution M.-L. Chambrade).

Half of its economic territory[22] (49.5 %) is taken up by the Sahl 'Iṯriyā (Fig. 17). The glacis account for an important percentage (25.5 %), followed by the reliefs (21 %). Some glacis-terraces and valley bottoms complete the surroundings of the site but cover an insignificant part within the territory (1.5 and 2.5 %).

The potential for cultivation is rather limited in the micro-region of Tulūl al-Ḥafyān 9 (Figs. 18c, d). The relief where the site is located appears like an island surrounded by the Sahl 'Iṯriyā, except in the west. Given its fine alluvial sediments and a supply with rainwater in the winter season, that is, in the growing period of cereals and pulses exploited in the PPNB, and also due to the numerous channels that cross it and feed underflow water[23] providing a more lasting flow, the *fayda* might be a vast exploitable area. But its potential for cultivation is much limited by the saltiness of its soils containing gypsum, and the risk of floods destroying the crops (Fig. 18b).

22 The limits of the economic territories have been determined on the basis of ethnographic studies (see Chambrade 2012: 494-496 for more details and for the works consulted). The maximum limits of the territories are put at 2 walking-hours, corresponding to the area of gathering. For cultivated territories the limits are put at 30 minutes to one hour of walking; often those territories do not exceed a distance longer than may be covered in 30 minutes on an average, but the distances may be twice that long depending on the distribution and the availability of arable ground. Obviously for similar reasons, the limits for pastures have been put at 30 minutes and two hours. They have been worked out with the help of a GIS software taking into account the constraints lying in the topography. The limits are represented by isochronous lines on figures 18 and 20.

23 Underflow water are flows under the surface in the bed of a watercourse. They are active during the humid season and may survive for all or part of the dry season, even when the surface runoff has dried up, provided they are fed by groundwater not too far from the surface. They are of great importance in the arid zone.

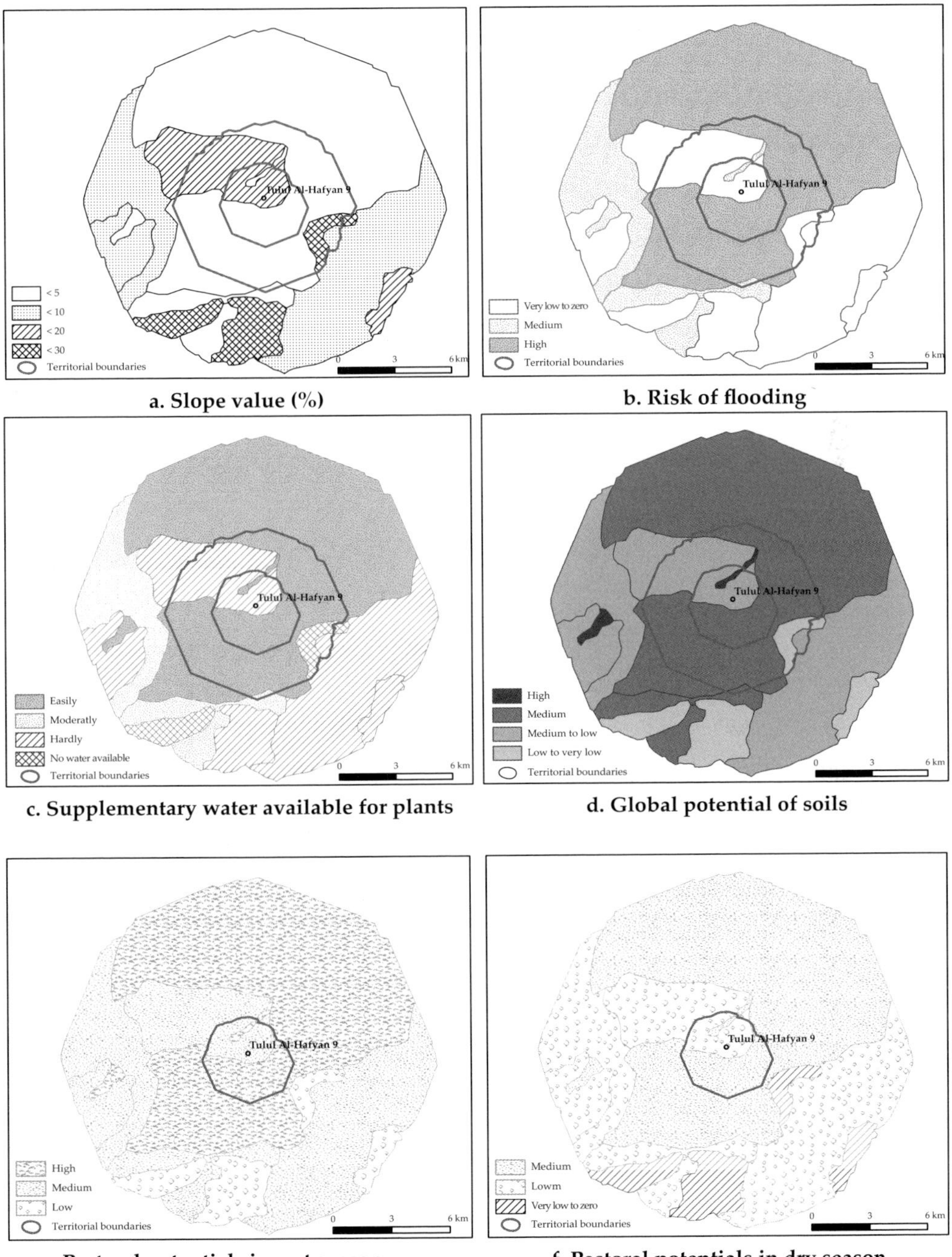

Fig. 18: Evaluation of the agro-pastoral potentials of the economic territory of Tulūl al- Ḥafyān 9 (execution M.-L. Chambrade).

The other geosystems of the micro-region are hardly more suitable for cultivation. The reliefs, and especially the slopes – sometimes quite steep and eroded (Fig. 18a) – are the least attractive sectors. Possibilities for cultivation do exist but are restricted to the valley floor and, perhaps, to certain glacis (Figs. 18c, d), to sectors impregnated by the floods of the watercourses and thus benefiting by some additional moisture that may compensate – at least minimally – for the meagre precipitations. But those areas are always limited, probably they just allow the sowing of say, barley, which needs less water than wheat or pulse: the success of the cultivation and the volume of the yield remain nevertheless risky because of the varying amounts of annual rainfall and the already mentioned danger of flooding. In addition, most of them are more than an hour's walk away from the site.

The potential for grazing is greater but essentially limited to the humid season (Figs. 18e, f). Except for the slopes and certain residuary reliefs with low to no potential at all for grazing, the different geo-systems provide diversified pasture grounds, quite considerable on the alluvial layers that retain moisture beyond the occasional rainfalls.

The Sahl 'Iṯriyā, above all, harbours a relatively dense vegetation of grasses and shrubs, especially on the few low dune-like elevations and in the beds of watercourses. Vegetation on the glacis is low and more sporadic, but by their vast extension – combined to that of the Sahl 'Iṯriyā – they offer quite important grazing possibilities on almost 85 % of the area. Nevertheless, for a large part of the year the severe aridity and the summer dryness reduce that potential for grazing considerably. Besides the possibilities for grazing due to the vegetation in the humid season, the micro-region appears to be attractive for the wild fauna of the steppe. In addition, the dominant location of the site made it an ideal observation post from where to spot the game crossing the *fayda* and the glacis. The scarceness of arrowheads suggests on the other hand that hunting played only a secondary part for the inhabitants of Tulūl al-Ḥafyān 9 and for several other sites in the sector, with the exception of Tulūl al-Ḥafyān 5, where domestic activities, hunting and the dressing of flint were combined.

With the exception of the source of the Wādī Ḥsayya, which lies at a distance of almost two walking hours from the site (Fig. 17), the access to water in the sector of Tulūl al-Ḥafyān is determined, for humans as well as for wild or domestic animals, by the seasonal climate. In the humid season, water is available for a few days after a rainfall at a few points in the *kamenitsas* on the reliefs and at the bottom of the watercourses. A more lasting access to water may be achieved by digging in the loose alluvial layers down to the underflow water bodies,[24] the volume of which depend on the thickness of the sediments, the yield of the floods and the extension of the catchment basins (Besançon and Geyer 2006: 41). However, the watercourses in the immediate vicinity of Tulūl al-Ḥafyān 9 are only fed by the precipitation falling on the rather low Ǧabal Hrībǧa, or ooze out of the gypso-saline alluvial layers. This situation emphasizes the real utility of the "artificial ponds", where the rainwater falling rarely and at strictly circumscribed spots could be hoarded: one might even think of a temporary – and not seasonal – occupation of the sector.

24 R. Jaubert *et al.* (2002) have underlined the importance of these underflows, today as well as in the past. They have indeed been exploited since Antiquity by means of hydraulic installations such as wells and underground draining galleries (qanats), providing water for the households and for irrigation. Water supply from those reserves may be assumed since the PPNB, because in Cyprus the boring of wells is attested since the early PPNB (see especially Guilaine *et al.* 1999, 2011; Peltenburg *et al.* 2000) and on the continent from the late PPNB on (Nishiaki 2009; Galili and Nir 1993).

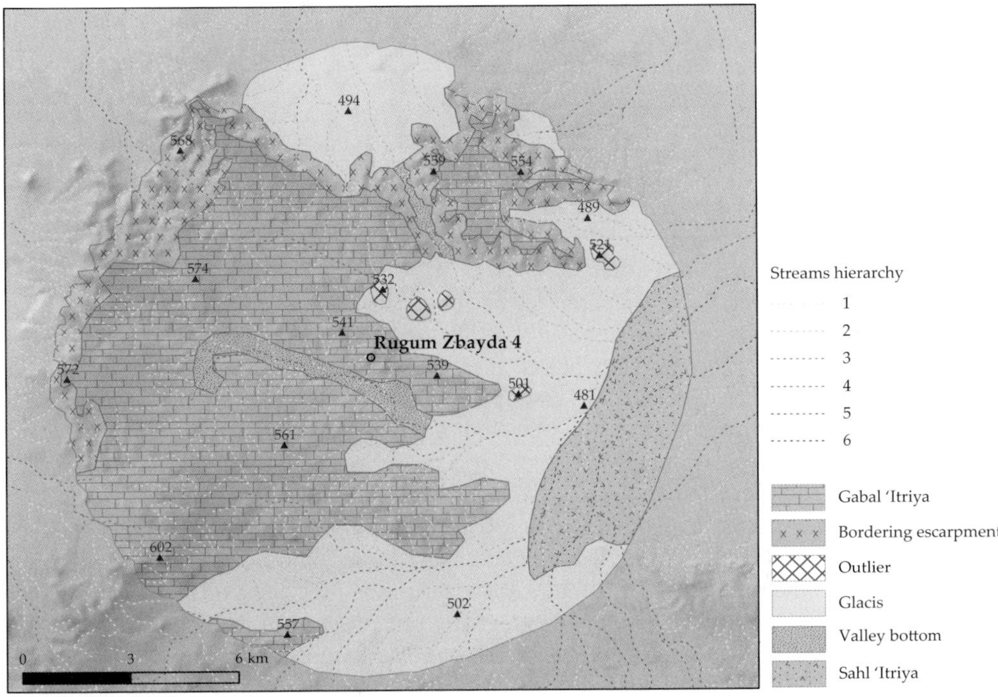

Fig. 19: Geosystems of the economic territory of Ruğum Zbayda 4 (execution M.-L. Chambrade).

The case of Ruğum Zbayda 4

Ruğum Zbayda 4 is among the sites that have yielded the greatest amount of artefacts in this small eastern chain of the Ǧabal 'Iṯriyā; the association of the material with the documented structures is reliable because the site dates exclusively to the late/final PPNB. Ruğum Zbayda 4 lies on the southwest-orientated slope of the chain, below the outcrop at the summit. All sites of the sector share this sunny location, protected from the north and east winds prevailing in the cold season; the place towards the top of the slope was a protection against the floods of the wadi in the south (Fig. 19) as well as against the secondary runoff on the very slope.

The economic territory of Ruğum Zbayda 4 is made up of three main geosystems (Fig. 19): the Ǧabal 'Iṯriyā, its piedmont glacis and the Sahl 'Iṯriyā. With 55 % of the territory, the Ǧabal 'Iṯriyā definitely holds an important place, the glacis take up 37 % and the Sahl 'Iṯriyā the remaining 8 %.

As we have seen in the case of Tulūl al-Ḥafyān 9, the Ǧabal 'Iṯriyā is not really suitable for cultivation, except locally, and that more so in the southeast than in the north-northwest. Its bordering escarpment looks like a cliff with pretty steep slopes (Fig. 20a). In its eastern part the valley floors with slightly embedded watercourses are more attractive (Figs. 20c, d), as they are covered by alluvial sediments and supplied by underflow water, especially the principal valley situated at the foot of Ruğum Zbayda 4. The interfluves are much eroded, on their summits the rock lies open at places – a phenomenon that may at any rate further the formation of *kamenitsas* the utility of which has been demonstrated above.

The connecting slopes with the glacis are a bit more favoured because they are covered with colluvium, but essentially in the downhill section, the uphill part is generally covered with pebbles.

The glacis themselves provide some potential for cultivation, especially due to a rather dense hydrographic network. Finally, the constraints which the Sahl 'Iṯriyā presents to cultivation have already been mentioned. The areas somewhat favourable to cultivation – meaning in this context, above all able to compensate, at least to a minimum, the aridity of the climate – are more extensive than in the case of Tulūl al-Ḥafyān 9, but probably just allowed to grow plants that needed little water.

As for Tulūl al-Ḥafyān 9, the potential of the economic territory of Ruǧum Zbayda 4 lies above all in pastures. The areas are just as diversified in this sector, but also as strictly reduced in the dry season (Figs. 20e, f). They are as attractive for the wild fauna of the steppe, the small chains and buttes provide as many observation posts for spotting game.[25] An abundance of arrowheads at this site shows the importance of hunting activities, but we are not looking at a simple hunting camp, as domestic tools (scrapers, burins…) are also quite frequent. The potentials of the economic territories may be similar, the activities apparently not.

The problems connected with the water supply resemble those at Tulūl al-Ḥafyān 9. On the summit of the chain there exist *kamenitsas*, and let us bear in mind that it has been proposed that the installation on the slope of Ruǧum Zbayda 1 might – because of its proximity – be associated with the site of Ruǧum Zbayda 4. At any rate, the bed of the southern wadi may have protected exploitable underflows, which were probably more lasting than in the case of the valley close to Tulūl al-Ḥafyān 9, but it is not certain whether they could be maintained throughout the humid season.

Looking at the ensemble of those different elements touching the micro-regional territories of Tulūl al-Ḥafyān 9 and of Ruǧum Zbayda 4 one tends to opt for the possibility of occupation during the humid season and subsistence activities that have mainly to do with the exploitation of meat products, be they wild and/or domestic. Due to the difficulty of providing water for a long term in spite of the hydraulic techniques applied, those settlements were probably temporary ones. This hypothesis is debatable in the case of Ruǧum Zbayda 4, but the arguments in favour of a seasonal installation, *i.e.* the presence of a watercourse that could have delivered a lasting underground water reserve, are weak and insecure. Due to the variety of imaginable possibilities the exact nature of those sites remains uncertain as well: it could only be clarified by an excavation. They may be encampments depending on a village or on a camp of semi-nomadic herdsmen, or just as well encampments of groups of hunter-breeders or also of nomadic herdsmen.

Conclusion

The arid margins of northern Syria, until a short time ago next to unknown, especially with regard to the prehistoric periods, proved a very interesting field of investigation. The large surface that has been surveyed – about 7000 sq. kilometres – and the good state of preservation of the vestiges have enhanced our understanding of the modalities of land occupation in a region suffering from considerable constraints due to aridity. At the outset, the existence of artesian springs in the centre of the surveyed region played an important role in the localiza-

25 For example, the hunting camp of Ǧabal al-ʿAlūšiyya (cf. *supra*, Diverse installations) is situated on a outlier northeast of Ruǧum Zbayda 4, on the outer limit of the micro-region.

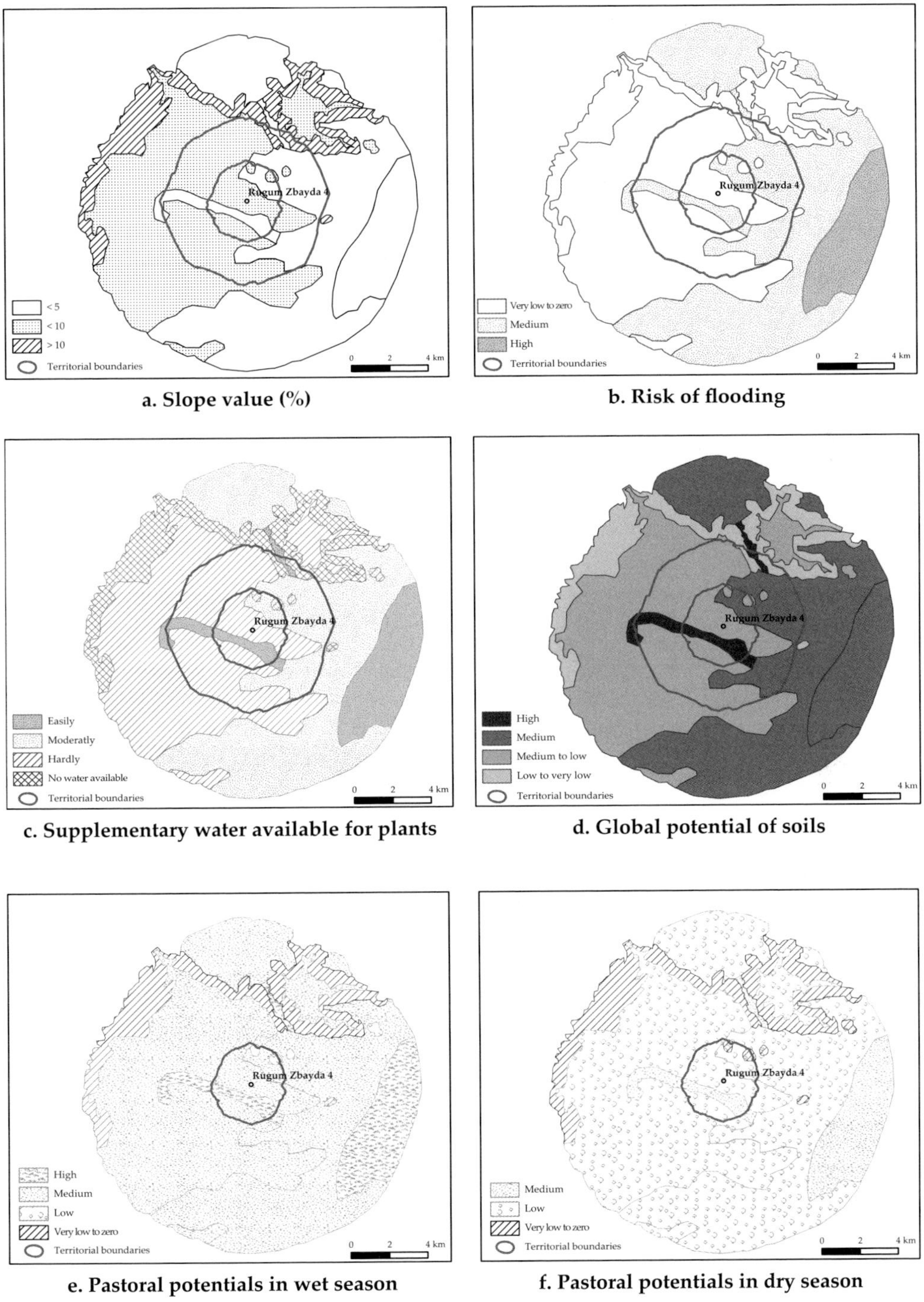

Fig. 20: Evaluation of the agro-pastoral potentials of the economic territory of Ruğum Zbayda 4 (execution M.-L. Chambrade).

tion of human settlements. But the wish to profit from the resources, especially from game but probably also from raw materials, in sectors where the daily water supply was not sure, made the hunter-gatherers – perhaps since the Kebaran – look for means to mitigate the lack in water: first of all they are likely to have made use of the installation provided by nature in the way of the *kamenitsas*, natural basins receiving runoff water, later – with certainty since the late/final PPNB – they arranged, after the pattern of those *kamenitsas*, "artificial ponds" and had thus the possibility to explore and, above all, to settle temporarily in sectors that would otherwise have been particularly forbidding.

Those settlements had diverse functions, sometimes specific ones such as the debitage of flint or hunting but mostly habitation; they were founded in micro-regions with diversified potentials, essentially relating to the exploitation of the fauna in the mountainous south-eastern part of the region. The adaptation to the climatic, hydrological and edaphic constraints in this sector achieved in the late/final PPNB was based on regular removals according to the seasonal availability of resources, a subsistence economy relying mainly on the exploitation of meat products and the use of innovative hydraulic techniques that were able to compensate for the difficult access to water.

Bibliography

Abbès, F.

2007 *Bal'as. L'occupation des zones arides durant la néolithisation,* Scientific report of the french-syrian archaeological mission, unpublished.

2008 Wadi Tumbaq 1: A Khiamian Occupation in the Bal'as Mountains, *Neo-Lithics* 1/08: 3-9.

2009 *Bal'as. L'occupation des zones arides durant la néolithisation*, Scientific report of the french-syrian archaeological mission, unpublished.

Aurenche, O. and M.-C. Cauvin

1982 Qdeir 1, Campagne 1980, Une installation néolithique du VIIème millénaire, *Cahiers de l'Euphrate* 3: 51-77.

Baird, D.

1993 *Neolithic chipped stone assemblages from the Azraq basin, Jordan and the significance of the Neolithic of the arid zones of the Southern Levant*, thesis, University of Edinburgh.

Besançon, J. and B. Geyer

2006 Contraintes écogéographiques et modes d'occupation du sol. In R. Jaubert and B. Geyer (eds.), *Les marges arides du Croissant fertile. Peuplements, exploitation et contrôle des ressources en Syrie du Nord,* TMO 43, series "Conquête de la steppe" no. 2, Maison de l'Orient, Lyon: 11-53.

Betts, A.V.G.

1988 Excavations at Dhuweila, eastern Jordan: a preliminary report, *Levant* 20: 7-21.

Betts, A.V.G. and S.W. Helms

1989 A water harvesting and storage system at Ibn el-Ghazzi in Eastern Jordan: a preliminary report, *Levant* 21: 3-11.

Betts, A.V.G. and D.N. Cropper

2013 Chapter 7, The Eastern Badia. In A.V.G. Betts, D.N. Cropper, L. Martin and C. McCartney (eds.), *The Later Prehistory of the Badia: Excavations and Surveys in Eastern Jordan,* Vol. 2, Levant supplementary series 11, Oxbow Books, Oxford.

Bocquet-Appel, J.-P. and O. Bar-Yosef (eds.)
2008 *The Neolithic Demographic Transition and its Consequences,* Springer, Dordrecht.
Cauvin, J.
1990 Les origines préhistoriques du nomadisme pastoral dans les pays du Levant: le cas de l'oasis d'El Kowm (Syrie). In H.P. Francfort (ed.), *Nomades et sédentaires en Asie centrale : apports de l'archéologie et de l'ethnologie*, CNRS Editions, Paris: 67-78.
Chambrade, M.-L.
2012 *Dynamique spatio-temporelle et environnement des sites néolithiques pré-céramiques de Syrie intérieure*, thesis, Université de Franche-Comté, Besançon.
forthcoming Étude microrégionale du site PPNB récent/final de Ḏahrat al- Ǧdayda (Syrie): contexte environnemental, modes d'implantation et activités de subsistance. In M.-O. Rousset, B. Geyer and P.-L. Gatier (eds.), *Habitat et environnement. Prospections dans les Marges arides de la Syrie du Nord*, T.M.O., series "Conquête de la steppe" no. 4, Maison de l'Orient, Lyon.
Chatty, D. and R. Jaubert (eds.)
2002 Alternative Perceptions of Authority and Control: The Desert and the Ma'moura of Syria, *The Arab World Geographer* 5/2, Special Issue: 71-140.
Coqueugniot, É.
2011 Apport des industries lithiques à la connaissance de l'occupation des Marges arides (Syrie) du Paléolithique supérieur au Bronze ancien (premières observations), *Syria* 88: 23-30.
Fujii, S.
2007 PPNB Barrage Systems at Wadi Abu Ṭulayḥa and Wadi ar-Ruweishid ash-Sharqi: A Preliminary Report of the 2006 Spring Season of the Jafr Basin Prehistoric Project, Phase 2, *ADAJ* 51: 403-427.
2009 A Half-Buried Cistern at Wadi Abu Ṭulayḥa: A Key to Tracing the Pastoral Nomadization in the Jafr Basin, Southern Jordan. In B. Finlayson and G. Rollefson (eds.), *Jordan's Prehistory: Past and Future Research,* Department of Antiquities of Jordan, Amman.
Fujii, S., T. Adachi, L.A. Quintero and P.J. Wilke
2011 Wādī Ghuwayr 106: A Neolithic Barrage System in the north-eastern Al-Jafr Basin, *ADAJ* 55: 189-211.
Galili, E. and Y. Nir
1993 The Submerged Pre-Pottery Neolithic Well of Atlit-Yam, Northern Israel, and its Palaeoenvironmental Implications, *The Holocene* 3: 265-270.
Gatier, P.-L., B. Geyer and M.-O. Rousset (eds.)
2010 *Entre nomades et sédentaires. Prospections en Syrie du Nord et en Jordanie du Sud,* TMO 55, series "Conquête de la steppe" no. 3, Maison de l'Orient, Lyon.
Geyer, B.
2000 Des fermes byzantines aux palais omayyades, ou l'ingénieuse mise en valeur des plaines steppiques de Chalcidique (Syrie), *Aux origines de l'Archéologie aérienne. A. Poidebard (1878-1955),* Presses de l'USJ, Beirut: 109-122.
2009 Pratiques d'acquisition de l'eau et modalités de peuplement dans les Marges arides de la Syrie du Nord. In M. al-Dbiyat and M. Mouton (eds.), *Stratégies d'acquisition de l'eau et société au Moyen-Orient depuis l'Antiquité: études de cas*, B.A.H. 186, Institut Français du Proche-Orient, Beirut: 25-43.

2011 The Steppe: Human Occupation and Potentiality. The example of Northern Syria's Arid Margins, *Syria* 88: 7-22.

Geyer, B. (ed.)

2001 *Conquête de la steppe et appropriation des terres sur les marges arides du Croissant fertile,* T.M.O. 36, series "Conquête de la steppe" no. 1, Maison de l'Orient, Lyon.

Geyer, B. and É. Coqueugniot

2013 Occupation du sol et conquête territoriale durant le Néolithique dans les Marges arides de Syrie du Nord, *Syria* 90: 159-176.

Geyer, B, M.-O. Rousset, M. Al-Dbiyat, N. Awad, O. Barge, J. Besançon, Y. Calvet, P.-L Gatier and R. Jaubert

2004-2005 Les Marges arides de la Syrie du Nord : première synthèse d'une prospection géoarchéologique, *Annales Archéologiques Arabes Syriennes* 47-48: 17-34.

Gourichon, L.

2004 *Faune et saisonnalité. L'organisation temporelle des activités de subsistance dans l'Epipaléolithique et le Néolithique précéramique du Levant Nord (Syrie),* thesis, Université Lyon 2.

Guilaine, J., F. Briois, I. Carrere, J. Coularou, E. Crubezy, C. Manen, T. Perrin and J.-D. Vigne

1999 Shillourokambos (Chypre), *Bulletin de Correspondance Hellénique* 123/2: 541-544.

Guilaine, J., F. Briois and J.-D. Vigne (eds.)

2011 *Shillourokambos: Un établissement néolithique pré-céramique à Chypre. Les fouilles du secteur 1*, Errance, École française d'Athènes, Paris, Athènes.

Higgs, E.S. and C. Vita-Finzi

1972 Prehistoric economies: a territorial approach. In E.S. Higgs (ed.), *Papers in economic Prehistory*, University Press, Cambridge: 27-36.

1975 Appendix A. Site catchment analysis: A concise guide to field methods. In E.S. Higgs (ed.), *Palaeoeconomy*, 2nd. Vol. of *Papers in economic Prehistory*, University Press, Cambridge: 223-224.

Jaubert, R., F. Debaine, J. Besançon, M. Al-Dbiyat, B. Geyer, G. Gintzburger and M. Traboulsi

1999 *Land use and vegetation cover in the semi-arid and arid areas of Aleppo and Hama provinces (Syria),* Cahiers du GREMMO, hors série, Lyon.

Jaubert, R., B. Geyer, F. Debaine and M. Al-Dbiyat

2002 L'exploitation des eaux de subsurface des marges arides de Syrie : dégradation ou transformation ?, *Sécheresse* 13/1: 43-50.

Jaubert, R. and B. Geyer (eds.)

2006 *Les marges arides du Croissant fertile. Peuplements, exploitation et contrôle des ressources en Syrie du Nord,* TMO 43, series "Conquête de la steppe" no. 2, Maison de l'Orient, Lyon.

Nishiaki, Y.

2009 Excavation of a 9 000-year-old water well in Mesopotamia, *Kakenhi News* 1: 5.

Peltenburg, E., S Colledge, P. Croft, A. Jackson, C. McCartney and M.A. Murray

2000 Agro-pastoralist colonization of Cyprus in the 10th millennium BP: initial assessments, *Antiquity* 74: 844-853.

Rollefson, G.O., Y.M. Rowan and A. Wasse

2010 Desert Monuments to the Dead: Views of the Early Pastoral Lifeways in Jordan's Eastern Desert, *ACOR Newsletter* 22/1: 1-5.

Seymour, D.J.
2009 Distinctive places, suitable spaces: Conceptualizing mobile group occupational duration and landscape use, *IJHA* 13: 255-281.

Stordeur, D.
1993 Sédentaires et nomades du PPNB final dans le désert de Palmyre (Syrie), *Paléorient* 19/1: 187-204.

Traboulsi, M.
2004 *Les précipitations au Proche-Orient : variabilité spatio-temporelle et relations avec la dynamique de l'atmosphère (1960-1961/1989-1990),* thesis, Université de Bourgogne, Dijon.
2010 Les précipitations dans les Marges arides de la Syrie du Nord. In P.-L. Gatier, B. Geyer and M.-O. Rousset (eds.), *Entre nomades et sédentaires. Prospections en Syrie du Nord et en Jordanie du Sud,* TMO 55, series "Conquête de la steppe" no. 3, Maison de l'Orient, Lyon: 73-95.

Vita-Finzi, C. and E.S. Higgs
1970 Prehistoric economy in the Mount Carmel area of Palestine: Site catchment analysis, *Proceedings of the Prehistoric Society* 36:1-37.

DANIELE MORANDI BONACOSSI
University of Udine

Desert-kites in an Aridifying Environment. Specialised Hunter Communities in the Palmyra Steppe during the Middle and Late Holocene

Recent archaeological and palaeoenvironmental studies in the Palmyra steppe attest to a major change in settlement and the exploitation patterns of natural resources on a regional scale during the final Pre-Pottery Neolithic – early Pottery Neolithic. The geoarchaeological survey conducted by the joint Italo-Syrian Project in the Palmyra steppe has shown a significant concentration of Palaeolithic, Epipalaeolithic and Pre-Pottery Neolithic open-air sites around the margins of the ancient lake basins of Palmyra and Abu Fawares, possibly hunter-gatherers' temporary camps (Cremaschi and Zerboni 2012; Morandi Bonacossi and Iamoni 2012: 33-34 and figs. 2-3). These site are associated with substantial and widespread lacustrine deposits, evidence of the existence of a large lake exceeding the boundaries of the present-day seasonally-flooded, saline marsh of the Sabkhat al-Mouah found to the south of the Palmyra Oasis (together with a smaller one in the Abu Fawares depression), suggesting that the climate was wetter during the Upper Pleistocene and Early Holocene.[1] Several sites have also been recorded in the area of the mountain saddle connecting both lake basins, and in the surrounding steppe, which date to the period characterised by higher rainfall.

The survey evidence concerning the following drier periods of the Pottery Neolithic, Chalcolithic and Bronze Age – which witnessed a drop in the Palmyra and Abu Fawares lake levels, the Palmyra and Qaryatein Oasis nucleation and dry-steppe formation – is rather problematic and has been discussed elsewhere (Morandi Bonacossi and Iamoni 2012; Cremaschi and Zerboni 2012). What emerges from the archaeological record, however, is a remarkable shift in the regional settlement landscape. With respect to an Upper Pleistocene and Early Holocene settlement horizon characterized by numerous open-air sites, possibly seasonal hunting-camps around the lakes, and the occupation of caves, rock-shelters and other mountain sites (often associated with or close to flint quarries marked by the presence of lithic workshops), a new occupation pattern emerges during the Middle and Late Holocene. Settlements become less widely distributed throughout the landscape and strongly polarized towards oasis settlement on one hand, and the specialized occupation of the dry-steppe by pastoral and seasonal hunter groups on the other.

In this paper, discussion will focus on the Palmyrena desert-kites, large dry-stone structures composed of two long guiding walls which converge, funnel-like, on a large circular or polygonal enclosure equipped with cells around the perimeter. The latter are stone rings, closed in on all sides. Some are circular, whilst others form elongated ovals with lengths of 10 m or more and widths of 2-3 m (Fig. 1). Their use as blinds for hunters or pit-like traps is still de-

1 For a full presentation and discussion of the palaeoenvironmental evidence, see Cremaschi and Zerboni 2012.

Studia Chaburensia 4 (2014), pp. 33–47.

bated (see below). Together with about 900 tumuli identified through satellite imagery analysis and survey in the region (see Iamoni this volume), desert-kites are the most widespread and distinctive structure type characterizing the region's archaeological landscape. They bear witness to a major adaptive shift in settlement patterns and economic strategies linked to increasing aridity and the parallel emergence of specialized pastoral and hunting communities.

In recent decades, and particularly during the last few years, field research has gained a strong impetus in Israel, Jordan, Syria, the Arabian Peninsula and lately also in Armenia, aimed at better defining the elusive desert-kite phenomenon in geographical, functional and, not least, chronological terms (Bar-Oz *et al.* 2011a and b; Betts 1998; Betts and Helms 1986; Betts and Yagodin 2000; Brunner 2008; Échallier and Braemer 1995; Eddy and Wendorf 1999; Gasparyan *et al.* 2013; Helms and Betts 1987; Holzer *et al.* 2010; Kempe and Al-Malabeh 2013; Kennedy 2011 and 2012; Meshel 2000; Morandi Bonacossi 2013; Morandi Bonacossi and Iamoni 2012; Nadel *et al.* 2010; Nadel *et al.* 2013; Perevolotsky and Baharav 1991; Rosen and Perevolotsky 1998; Van Berg *et al.* 2004; Zeder *et al.* 2013).

As a result of these new investigations huge numbers of desert-kites have been found to be scattered across the steppe and desert regions of the Near East from the northwestern fringes of the Ararat Depression in Armenia and the south of the Arabian Peninsula in Yemen to Syria, Jordan, the Negev and Sinai deserts, with large concentrations especially in central and southern Syria, eastern Jordan and central and northern Saudi Arabia.

The long and complex discussion on the function of desert-kites, which was started by the pioneering articles published in the journal *Antiquity* by Maitland (1927), Rees (1929) and Crawford (1929) and revived more recently by Échallier and Braemer's essay in *Paléorient* (1995) and the subsequent debate, has by now reached firm and widely accepted conclusions.

The hypothesis that desert-kites were game traps for the mass-killing of gazelle, and possibly also other large ungulates such as oryx and onager (and perhaps also ostrich), which would have been driven into the large terminal enclosures by means of the long convergent walls, has been put forth by many scholars (Bar-Oz et al. 2011a; Betts and Yagodin 2000; Fowden 1999; Gasparyan *et al.* 2013; Helms and Betts 1987; Kennedy 2011 and 2012; Legge and Rowley-Conwy 1987; Morandi Bonacossi 2013; Morandi Bonacossi and Iamoni 2012; Nadel *et al.* 2010; Van Berg *et al.* 2004). The main problem, however, was that until very recently there was no primary archaeological and faunal evidence which made it possible to directly associate kites with contemporary faunal remains originating from mass gazelle-hunting operations using kites.

However, two seminal articles published recently by Bar-Oz *et al.* (2011b) and Zeder *et al.* (2013) have described a remarkable well-stratified deposit of bones of Persian Gazelle (Gazella subgutturosa) from the small site of Tell Kuran in the Khabur River Basin of northeastern Syria, which is situated in close proximity to a number of desert-kites in use during the fourth and third millennia BC identified by Van Berg *et al.* (2004). This faunal assemblage, which was C14-dated to between 3600 and 3100 BC, provides for the first time direct and compelling evidence of the use of kites for hunting gazelle in post-Neolithic times, since it contained a dense mass of gazelle bones with butchery marks representing a minimum of 93 individuals. The shallowness and limited horizontal extent of the deposit suggest that all of the remains were of animals killed at one time, dumped in a single butchery event (Bar-Oz *et al.* 2011b: 7346; Zeder *et al.* 2013: 117).

The extraordinary and unique Kuran gazelle assemblage was dominated by juveniles and prime-age adults, with fewer old adults and with males and females approximately equally

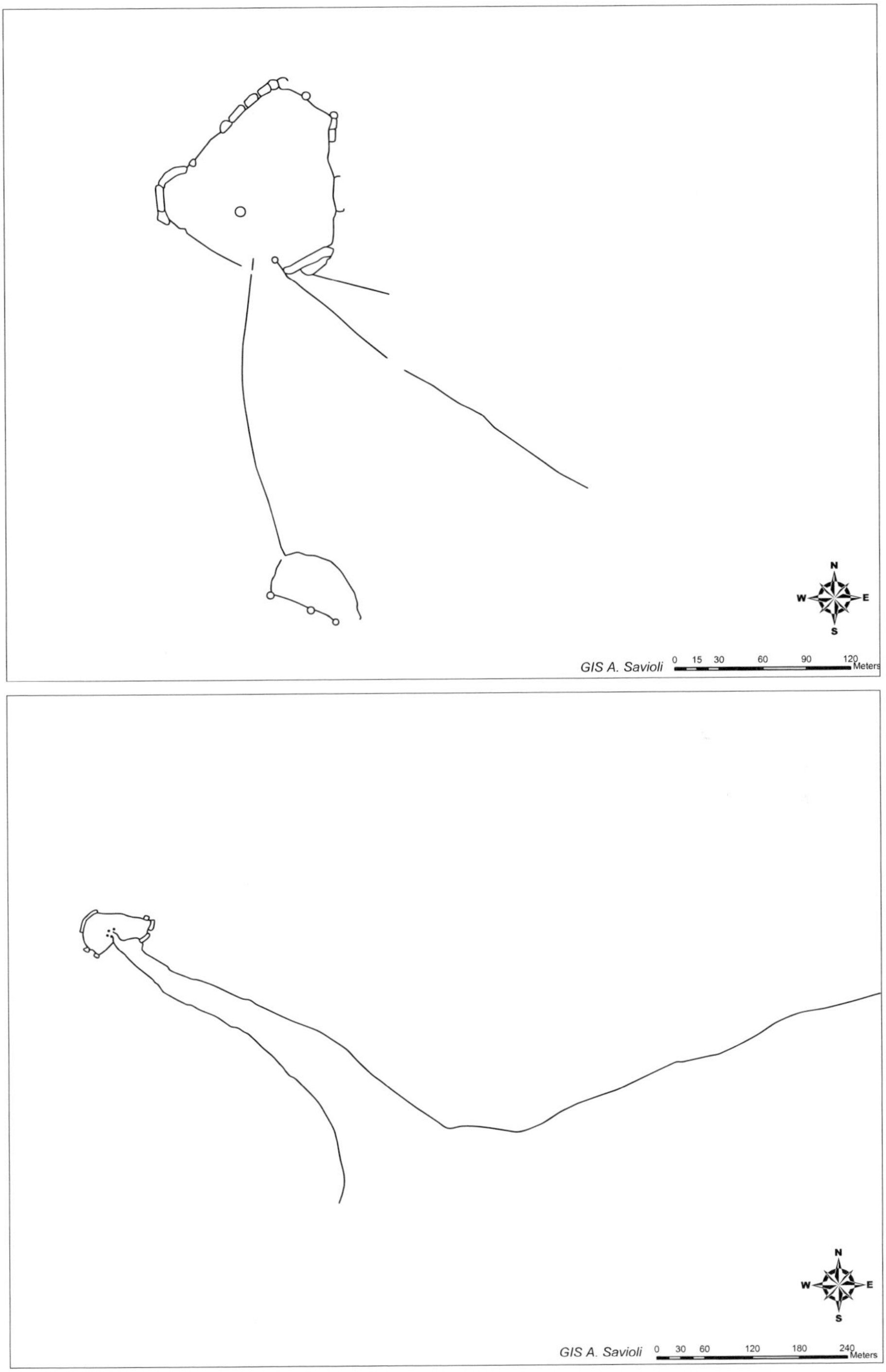

Fig. 1: Examples of desert-kites in the Palmyra region with circular and elongated cells around the perimeter (survey and drawing A. Savioli).

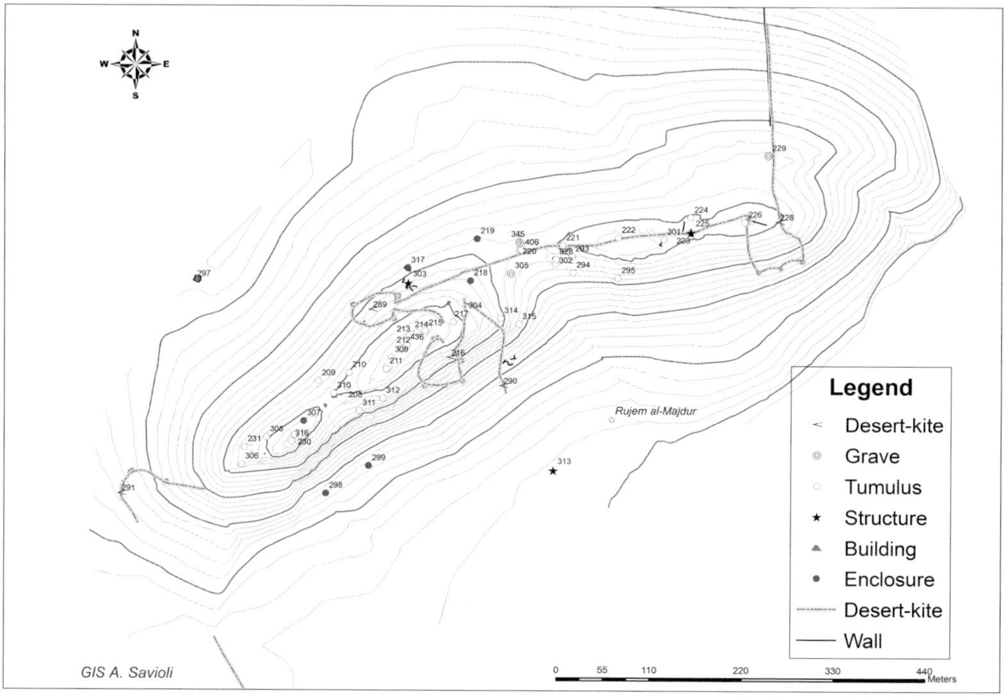

Fig. 2: Surveyed structures at Rujem al-Majdur. Note the hilltop tumulus positions (GIS and map preparation A. Savioli).

represented. The age and sex profiles of the hunted Tell Kuran gazelle resemble a catastrophic demographic structure, indicating that the hunt, which probably occurred in mid- to late summer (most likely August), was targeted at the mass-killing of an entire, probably migrating, mixed herd of males and females of all ages – and that the nearby Hemma Plateau desert-kites were most likely the traps used to hunt these animals (Bar-Oz *et al.* 2011b: 7348; Zeder *et al.* 2013: 117-119).

The current state of knowledge, however, does not allow us to exclude the possibility that at least some desert-kites were multifunctional structures, perhaps used – not necessarily contemporarily – both for the entrapment of wild game animals (mainly in mass-kills of gazelle) and, as suggested by Échallier and Braemer (1995), as corrals employed in the managing of ovicaprids that grazed freely on the steppe.

Recent field research has not only clarified the kites' function, but also gathered sufficient archaeological evidence to demonstrate that the widespread use of kites in Armenia, Syria, Jordan and in the Negev and Sinai Deserts was concentrated in the fourth through the third millennium BC,[2] although a few kites in the Dhuweila region of eastern Jordan have been dated to the Late PPNB on the basis of their association with Late PPNB structures and Beidha and Byblos arrowheads recovered from the surface (Betts 1998: 191-205; Betts and Russel 2000; Helms and Betts 1987: 47, 50, figs. 5-6).

2 For a fresh review and critical discussion of the available data, see in particular Échallier and Braemer 1995: 54-55; Morandi Bonacossi and Iamoni 2012: 36-38 and Zeder *et al.* 2013: 111-117, in particular Table 1.

However, these associations of architectural features and surface lithics with the Dhuweila kites have been contested due to the fact that both the interpretation of the structures as kites (rather than complex enclosures) and their relation to the Late PPNB architectural features and the surface projectile points are uncertain (Échallier and Braemer 1995: 54-55).

Current evidence for an early use of the desert-kites is thus still meagre and, as we have seen, disputed. However, the data from the joint Italo-Syrian Project in the Palmyra steppe provide new evidence in support of the use of kites and mass-killing strategies for wild game animals already in the Neolithic.

Attempts at the absolute dating of desert-kites may be supported by relative chronological information obtainable from the archaeological record. The large necropolis of Rujem al-Majdur (Fig. 2), a burial ground made up of more than thirty tumuli located on the summit of a flat-topped limestone ridge which runs eastwards from the Jebel al-Abtar range and dominates the mouth of the Wadi al-Hallabat (Morandi Bonacossi and Iamoni 2012: 46-52; Iamoni this volume), has been investigated by the joint Italo-Syrian Project.

Eight tumuli, two satellite burials associated with the tumuli and an isolated grave were subjected to archaeological excavation in 2009 and 2010. Radiocarbon determination of the age of two tumuli made it possible to date the cairn field to the Early Pottery Neolithic (Iamoni this volume). The charcoal samples retrieved from the floors of cairns T 213 and T 302[3] yielded radiocarbon dates of 6066-5921 cal. BC and 5876-5658 BC respectively (2σ; Iamoni this volume). The two tumuli can therefore be dated to the beginning of the sixth millennium BC.

Fig. 3: View of tumulus T 221 built on top of the northern guide wall of desert-kite 289.

The large tumulus T 221 (Morandi Bonacossi and Iamoni 2012: 48-49) was built on top of the northern guide wall of desert-kite 289 (Fig. 3), one of the four kites built along the slopes of the Rujem al-Majdur ridge. This provides us with a *terminus ante quem* for the construction of the Rujem al-Majdur kite, which is referable to or earlier than the beginning of the sixth millennium BC. Such an early date for kite 289 is further reinforced by the state of preservation of this and the other Rujem al-Majdur kites (216, 228 and 291). Whilst several of the almost 300 desert-kites identified on the ground or in satellite imagery in the Palmyrena mountainous region are quite well preserved, with guide walls standing up to 0.5-0.7 m high and of the circular cell walls standing to heights of 2 m and more, the guide walls of kite 289 are very low (only one course of stones) and extremely heavily eroded.

3 For a discussion of the results of the cairns' excavation, see Morandi Bonacossi and Iamoni 2012: 49-52.

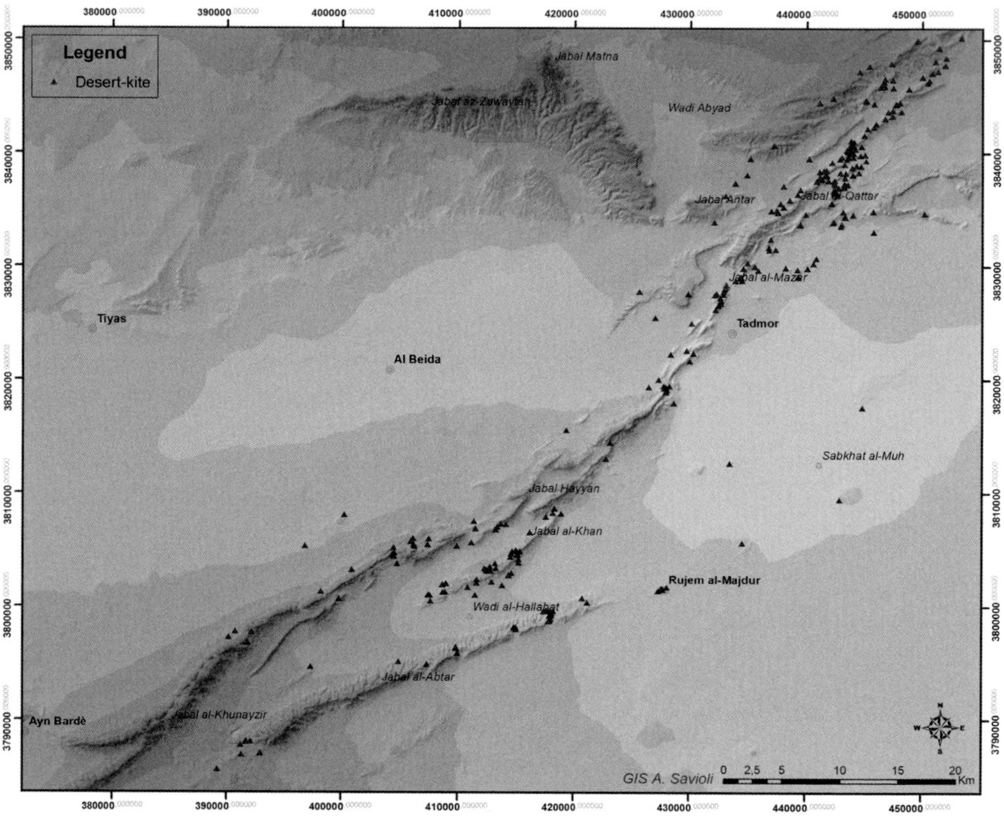

Fig. 4: Distribution map of desert-kites in the Palmyra dry-steppe (GIS and map preparation A. Savioli).

Thus the archaeological and radiocarbon evidence from the Rujem al-Majdur cairn field offers new and reliable support for dating the beginning of the kite phenomenon – and indeed of specialized mass-hunting techniques – in the Palmyra dry-steppe to at least the early sixth millennium BC.

Analysis of remotely-sensed images (Ikonos, Landsat, Corona and Aster) and geoarchaeological survey work (founded on an extensive stratified sample) have led to the identification of 289 kites in the Palmyra hinterland (Fig. 4). These are clustered in two large groups located along the mountain chains of the Jebel Hayan, Jebel al-Khan and Jebel al-Abtar and other high ground in the plain south of the Sebkhat al-Muah, as well as on the Jebel Mazar and Qattar mountain ranges to the northeast of the great Palmyra Oasis. Unfortunately the latter kite cluster could not be ground truthed due to the outbreak of the Syrian crisis and was analysed only through satellite imagery.

Most kites were isolated and apparently functionally independent, but in several cases they were interconnected, forming sets of long kite chains, complex and articulated systems of up to six structures with the same orientation often and with the outer ends of guiding walls touching, which allowed large herds of animals to be intercepted (Fig. 5). This phenomenon has also been recorded in the Hemma Plateau, the north Jordan desert and in Saudi Arabia (Betts and Yagodin 2000: 40; Kempe and Al-Malabeh 2013; Van Berg *et al.* 2004: 93).

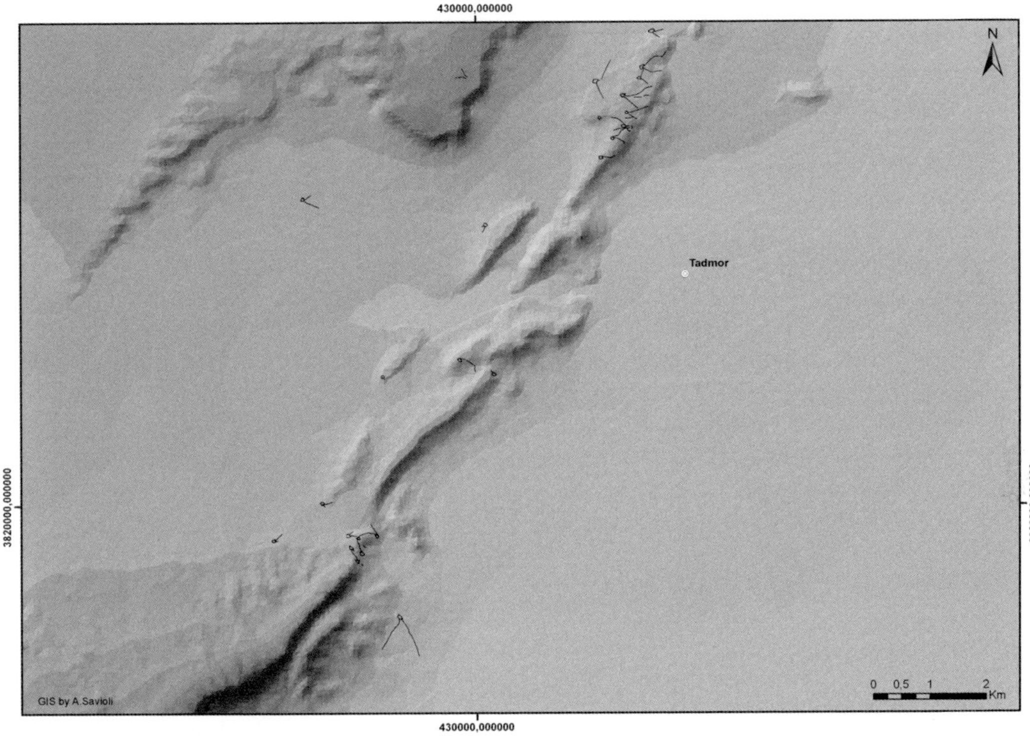

Fig. 5: A chain of six kites with the same orientation and the outer ends of guiding walls touching (top centre; GIS and map preparation A. Savioli).

The typological classification of desert-kites already devised by Echallier and Braemer (1995) for the regions comprised between the Jebel al-ʿArab massif and the southern Safa and Qaryatein and Dumeir, in southern and central Syria respectively, could be usefully applied to the Palmyrena record (Fig. 6). Circular (Fig. 7), arrow-shaped and trapezoidal kites are most common,[4] whilst 'axe'-shaped (Fig. 8) and star-shaped kites (Fig. 9) are much less widespread.

I have already discussed the distribution of kites in the Palmyra region in a previous publication (Morandi Bonacossi and Iamoni 2012), emphasizing how an aspect of considerable interest in the study of these game traps – normally built astride mountain saddles in

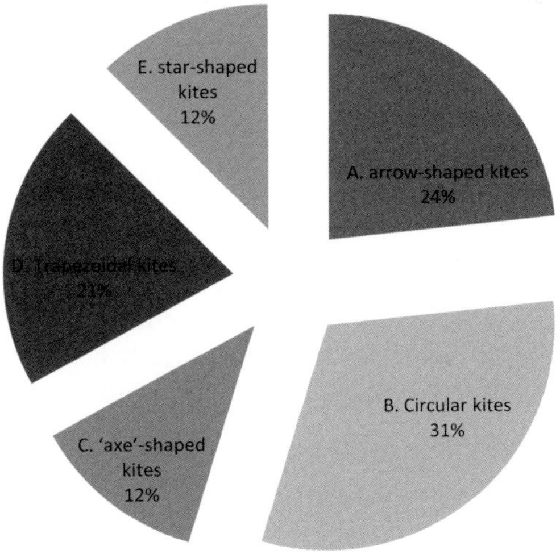

Fig. 6: Typological classification chart of the Palmyrena desert-kites.

4 For photographs of trapezoid and arrow-shaped desert-kites, see Morandi Bonacossi and Iamoni 2012, figs. 8-9.

Fig. 7: A Group B (circular) desert-kite.

order to hide the slaughtering enclosures beyond watersheds, whereas the walls continued for hundreds of metres on the other side of the mountain ridges – regards their preferential orientation. The 273 desert-kites identified so far along the Palmyrena mountain ranges, of which it is possible to determine the orientation with certainty, exhibit a significant pattern in the arrangement of their guide walls (Fig. 10). As I have already remarked (Morandi Bonacossi and Iamoni 2012: 41-42), most of the kites spread along the eastern surveyed part of the Jebel al-Abtar and the Jebel al-Khan ranges have divergent walls pointing southwards (37% of the total), while the traps located on the mountain slopes of the Jebel Hayan are mostly aligned towards east and southeast (29%), although a not insignificant number of kites (which appear to be concentrated on the southern slopes of the Jebel Hayan above the Wadi Surra valley) point northwards (28%). Very few kites, all in the Jebel al-Khan and Jebel al-Abtar, are aligned towards the west (4%). The situation changes remarkably along the mountains of the Jebel Mazar and Jebel Qattar to the northeast of Palmyra (Fig. 11), where the number of westward-oriented kites sharply increases due to the presence of the eastern Jebel az-Zuwaytah slopes that close the way to gazelle herds migrating from the Middle Euphrates spring/early summer breeding grounds to the fall/winter pastures of eastern Jordan, thus forcing them to turn east to find gaps in the Jebel Mazar and Jebel Qattar ranges.

In this northern kite concentration the game traps aligned towards south and east are also numerous, whereas kites pointing northwards are much less frequent. In general, of the 273 kites the orientation of which could be established through field survey or satellite imagery analysis (in 16 cases the kites' state of preservation did not allow for the determination of their alignment), 88 (32.2%) had divergent walls pointing eastwards, 96 (35.1%) southwards and only 42 (15.4%) and 47 (17.2%) westwards and northwards respectively.

These topographic orientations of the kites are certainly partly determined by the general NE-SW orientation of the Palmyrena mountain ranges, but (as mentioned above) are

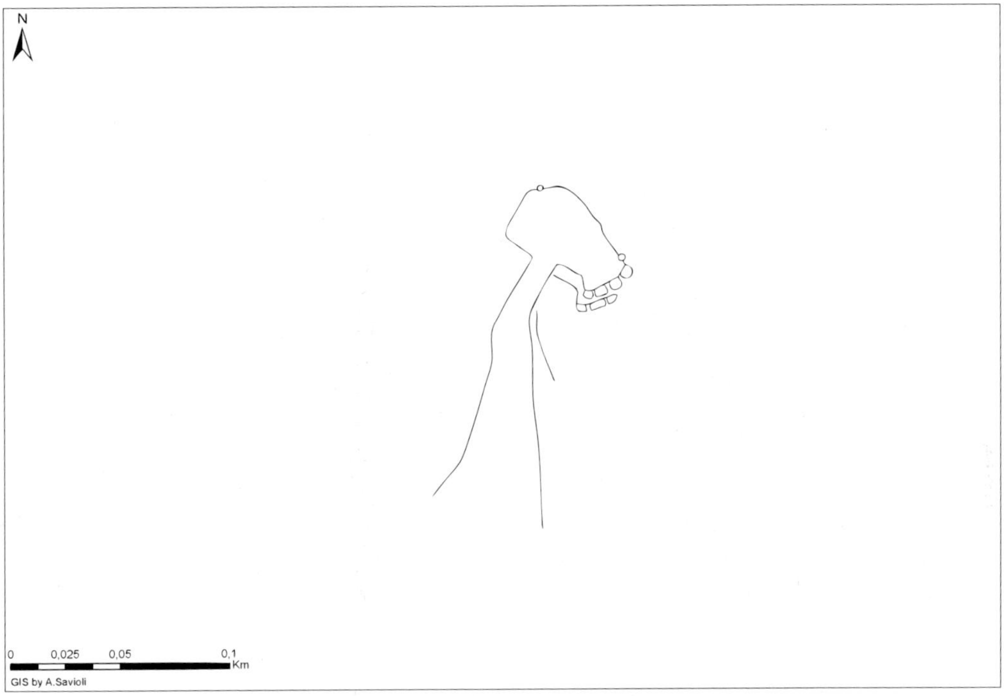

Fig. 8: A Group C ('axe'-shaped) desert-kite (survey and drawing A. Savioli).

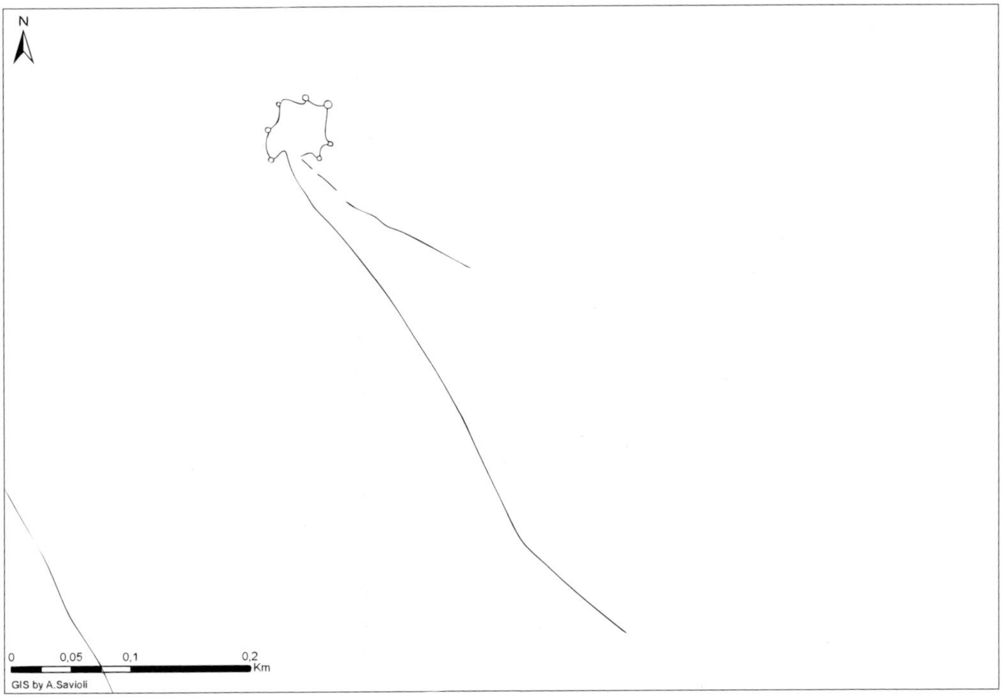

Fig. 9: A Group E (star-shaped) desert-kite (survey and drawing A. Savioli).

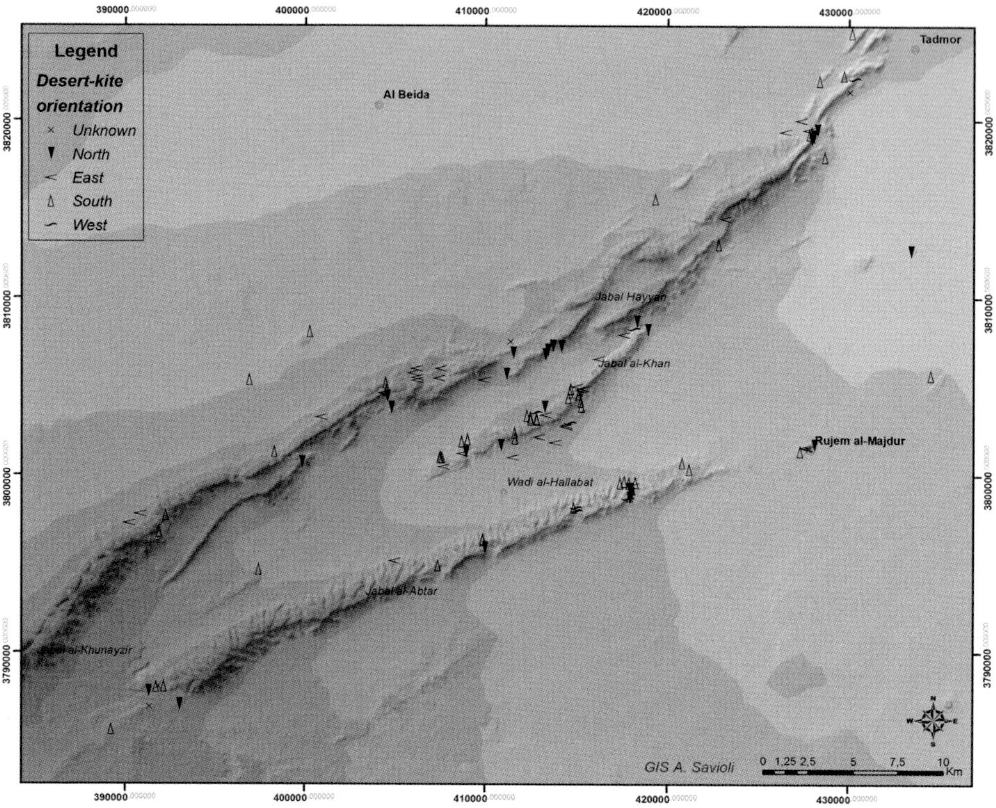

Fig. 10: The orientation of the desert-kites in the mountains west of Palmyra (GIS and map preparation A. Savioli).

also influenced by local geographical features, such as for instance the obstacle to gazelle herds migrating south constituted by the Jebel az-Zuwaytah massif within our northern kite cluster.

However, comparison with the migration routes recorded in the literature on modern gazelle behavioural ecology[5] would suggest that the north-facing traps might have been used to capture animals arriving from the Middle Euphrates during the late summer, when the gazelle herds were largest after the birth of the young in the late spring/early summer.

The majority of the Palmyrena kites, which were aligned to the south and east/southeast (67%), might have been designed to intercept the smaller groups of animals which, during the fall/winter, moved back from southern Syria and eastern Jordan to reach the Middle Euphrates breeding grounds. This would help to explain the clear disproportion between east and south/southeast facing kites and those which opened northwards (17%) and westwards (15%) that were possibly meant to hunt gazelles moving southwards during the dry season, when gazelle herds reached their maximum size.

5 For a recent critical review of the available zoological and archaeological literature, see Martin 2000; Morandi Bonacossi and Iamoni 2012: 38-40; Zeder *et al.* 2013: 111-114.

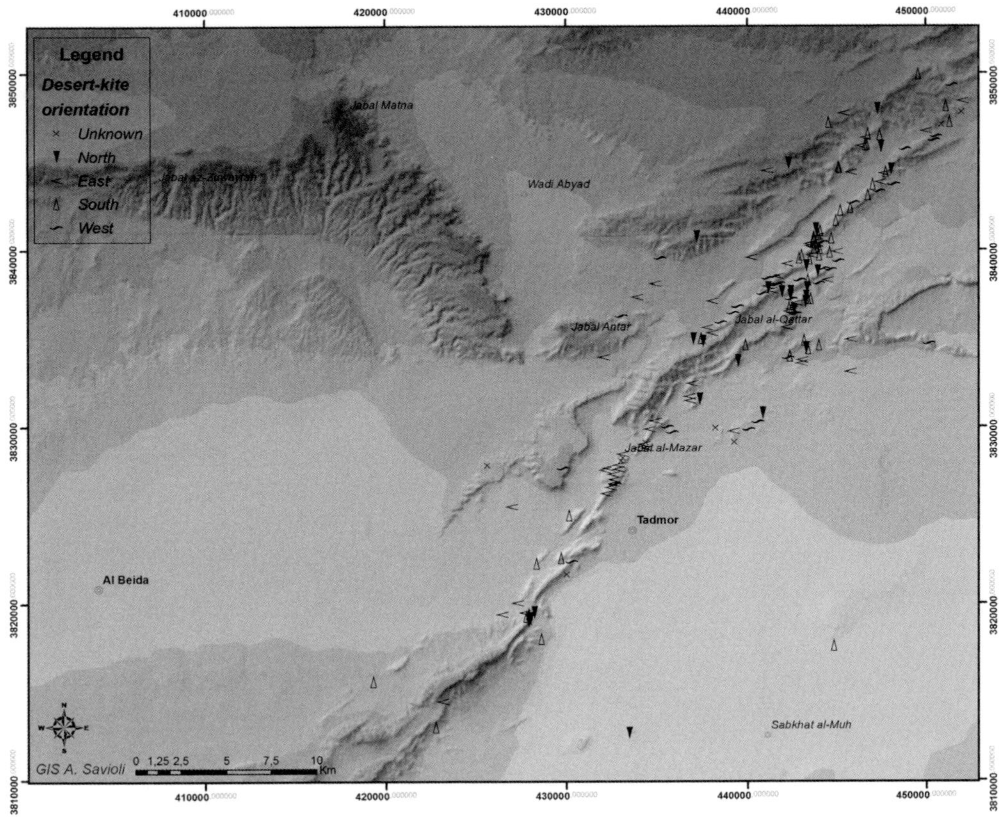

Fig. 11: The orientation of the desert-kites in the mountains northeast of Palmyra (GIS and map preparation A. Savioli).

The hypothesis that this seasonal gazelle mass-hunting strategy based on a larger number of traps involved during the fall/winter hunt and fewer during the spring/summer season was the result of a conscious attempt at seasonal culling the migrating gazelle herds calibrated so as to prevent the extinction (or serious reduction) of this primary source of animal protein seems somewhat improbable, even if it cannot be entirely excluded. However, the evidence of the age and sex profile of the hunted Tell Kuran gazelle, which resembles a catastrophic demographic structure, suggests that the hunt was targeted at the mass-killing of an entire migrating herd, not at culling individual gazelle (Bar-Oz *et al.* 2011b: 7345).

Finally, the apparent overlapping of kite and tumulus distribution in both the investigated mountain regions to the northeast and west of Palmyra seems to be significant (Fig. 12). Cairns are much more numerous than kites in the Palmyrena (about 900 tumuli have been identified in the remotely-sensed images and during the survey) and are often interspersed with desert-kites. Although the presence of numerous cairns adjacent to kites can hardly be considered surprising, due to the ubiquitous presence of tumuli in the Palmyra mountain ranges, a possible functional association between these two landmark features characterizing the landscape and natural resource exploitation patterns in the dry Palmyra steppe after the occurrence of aridification cannot be ruled out.

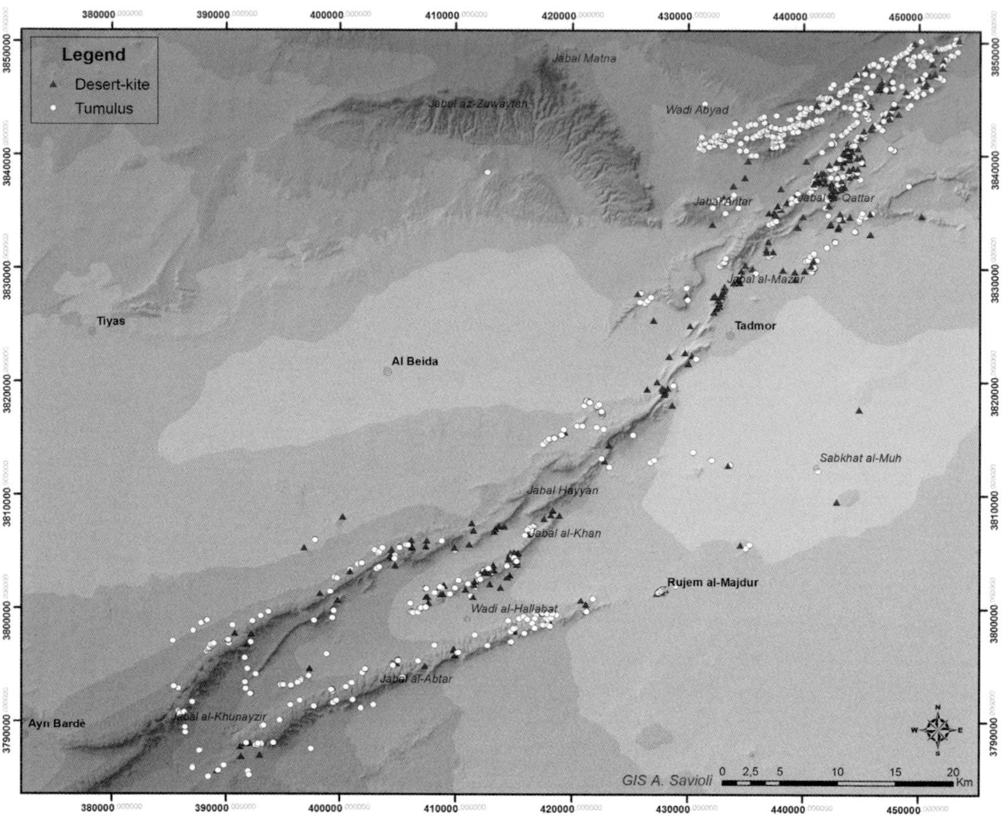

Fig. 12: Distribution map of tumuli and desert-kites in the Palmyra region (GIS and map preparation A. Savioli).

The same apparent spatial association of cairns and kites has been recently observed in the fringes of the Ararat Depression of Armenia and the Negev (Gasparyan *et al.* 2013; Rosen 2009: 61).

Unfortunately, the huge tragedy that Syria is undergoing in these years has determined the interruption of archaeological research in the Palmyra region, preventing us from intensifying the exploration of these fascinating wild game animal traps by digging trial trenches in selected kites and performing radiocarbon and IRSL (Infrared Stimulated Luminescence) age determinations. Only the excavation and multidisciplinary investigation of these structures will make it possible to better define their chronology and understand their wider social and environmental context, as well as the complex pattern of interaction between specialized mobile pastoral and hunting communities in the dry-steppes of Syria and throughout the Levant that is reflected by features such as desert-kites and tumuli.

Bibliography

Bar-Oz G., D. Nadel, U. Avner, and D. Malkinson

2011a Mass Hunting Game Traps in the Southern Levant: the Negev and Arava 'Desert Kites'. *Near Eastern Archaeology* 74: 208-215.

Bar-Oz G., M. Zeder., and F. Hole

2011b Role of mass-kill hunting strategies in the extirpation of Persian gazelle (Gazella subgutturosa) in the northern Levant, *PNAS* 108/18: 7345-7350.

Betts A.V.G.

1998 *The Harra and the Hamad: Excavations and Surveys in Eastern Jordan*, Sheffield.

Betts A.V.G. and S. Helms

1986 Rock art in Eastern Jordan: kite carvings?, *Paléorient* 12/1: 67-72.

Betts A.V.G. and K.V. Russel

2000 Prehistoric and Historic Pastoral Strategies in the Syrian Steppe. In M. Mundy, B. Musallam (eds.), *The Transformation of Nomadic Society in the Arab East*, Cambridge: 24-32.

Betts A.V.G. and V. Yagodin

2000 A new look at 'desert-kites'. In L. Stager, J. Greene, and M.D. Coogan (eds.), *The Archaeology of Jordan and Beyond. Essays in Honor of James A. Sauer*, Winona Lake: 31-44.

Brunner U.

2008 Les pièges de chasse antiques au Yémen, *Chroniques Yéménites* 15: 29-34.

Crawford O.G.S.

1929 Note by editor, *Antiquity* 3: 400-401.

Cremaschi M. and A. Zerboni

2012 Adapting to Increasing Aridity. The Cuvette of Palmyra (Central Syria) from Late Pleistocene to Early Holocene. In F. Bertoncello, J.-F. Berger, L. Carozza, F. Braemer, and P. Dapaepe (eds.), *XXXIIes Rencontres internationales d'archéologie et d'histoire d'Antibes: Variabilités environnementales, mutations sociales: nature, intensités, échelles et temporalités des changements*, Antibes: 37-52.

Echallier J.C. and F. Braemer

1995 Nature et fonctions de 'desert-kites': données et hypothèses nouvelles, *Paléorient* 21/1: 35-63.

Eddy F.W. and F. Wendorf

1999 *An Archaeological Investigation of the Central Sinai, Egypt,* Boulder.

Fowden G.

1999 Desert-kites': ethnography, archaeology and art. In J.H. Humphrey (ed.), *The Roman and Byzantine Near East: Some Recent Archaeological Research, vol. 2*, Journal of Roman Archaeology. Supplementary Series 3: 107-136.

Gasparyan B., A. Khechoyan, G. Bar-Oz, D. Malkinson, A. Nachmias, and D. Nadel

2013 The northernmost kites in south-west Asia: the fringes of the Ararat Depression (Armenia) Project, *Antiquity* 87. http://antiquity.ac.uk/projgall/bar-oz336/

Helms S. and A.V.G. Betts

1987 The desert 'kites' of the Badiyat esh-Sham and North Arabia, *Paléorient* 13/1: 41-67.

Holzer A., U. Avner, N. Porat, and L. Kolska Horwitz
2010 Desert-kites in the Negev desert and northeast Sinai: Their function, chronology and ecology, *Journal of Arid Environments* 74/7: 806-817.
Kempe S. and H. Al-Malabeh
2013 Desert kites in Jordan and Saudi Arabia: Structure, statistics and function, a Google Earth study, *Quaternary International* 297: 126-146.
Kennedy D.L.
2011 The "Works of the Old Men" in Arabia: remote sensing in interior Arabia, *Journal of Archaeological Science 38:* 3185-3203.
Kennedy D.L.
2012 Kites – new discoveries and a new type, *Arabian Archaeology and Epigraphy* 23: 145-155.
Legge A.J. and P.A. Rowley-Conwy,
1987 Gazelle-killing in Stone Age Syria, *Scientific American* 275/2: 76-83.
Maitland R.A.
1927 The 'works of the old man' in Arabia, *Antiquity* 1: 197-203.
Martin L.
2000 Gazelle (Gazella spp.) behavioural ecology: predicting animal behaviour for prehistoric environments in south-west Asia, *Journal of the Zoological Society of London* 250: 13-30.
Meshel Z.
2000 Desert kites in Sinai and southern Negev. In Z. Meshel (ed.), *Sinai: Excavations and Studies*, British Archaeological Research International Series, Oxford: 121-142.
Morandi Bonacossi D.
2013 I 'desert kite' del deserto di Palmira (Siria). Le catene montuose della Palmirena occidentale come spazi di approvvigionamento di risorse e aree di connessione in epoca pre-classica. In S. Magnani (ed.), *Le aree montane come frontiere. Spazi d'interazione e connettività*. Roma: 415-440.
Morandi Bonacossi, D. and M. Iamoni
2012 The Early History of the Western Palmyra Desert Region. The Change in the Settlement Patterns and the Adaptation of Subsistence Strategies to Encroaching Aridity: A First Assessment of the Desert-kite and Tumulus Cultural Horizons. *Syria* 89: 31-58.
Nadel D., G. Bar-Oz, U. Avner, E. Boaretto, and D. Malkinson
2010 Walls, ramps and pits: the construction of the Samar Desert-kites, southern Negev, Israel, *Antiquity* 84: 976-992.
Nadel D., G. Bar-Oz, U. Avner, D. Malkinson, and E. Boaretto
2013 Ramparts and walls: Building techniques of kites in the Negev Highland, *Quaternary International* 297: 147-154.
Perevolotsky A. and D. Baharav
1991 The distribution of desert-kites in eastern Sinai and sub-regional carrying capacity: an ecological perspective, *Journal of Arid Environments* 20: 239-249.
Rees L.W.B.
1929 The Trans-Jordan Deserts, *Antiquity* 3: 389-407.

Rosen B. and A. Perevolotsky
1998 The function of 'desert-kites': hunting or livestock husbandry?, *Paléorient* 24/1: 107-111.

Rosen S.A.
2009 History Does Not Repeat Itself: Ciclicity and Particularism in Nomad-Sedentary Relations in the Negev in the Long Term. In J. Szuchman (ed.), *Nomads, Tribes and the State in the Ancient Near East. Cross Disciplinary Perspectives.* Chicago: 57-86.

Van Berg P.-L. , M. Vander Linden, S. Lemaitre, N. Cauwe , and V. Picalause
2004 Desert-kites of the Hemma plateau (Hassake, Syria), *Paléorient* 30/1: 89-100.

Zeder M.A., G. Bar-Oz, S.J. Rufolo, and F. Hole
2013 New perspectives on the use of kites in mass-kills of Levantine gazelle: A view from northeastern Syria, *Quaternary International* 297: 110-125.

MARCO IAMONI
University of Udine

Late Neolithic Funerary Evidence from Palmyra:

the Rujem al-Majdur necropolis and the "desert-kite and tumulus cultural horizons" in Central Syria

Abstract

Recent discoveries have shown that the region around Palmyra in the pre-classical age was densely occupied, in agreement with historical sources and archaeological data from Palmyra itself, which attest to the existence of a Bronze Age settlement on the Palmyra oasis. Yet the nature of the occupation in the surrounding region has been always elusive, due to its nomadic character. Furthermore, despite a few exceptions such as El Kowm, little is known of prehistoric occupation. Recent research, however, has started to provide significant information, shedding light on the pre-classical archaeological horizon of the Palmyra region. The following is an account of the funerary evidence excavated and surveyed to date by the Syro-Italian mission (DGAMS and Universities of Milan and Udine), adding more details to the evidence already published and updating the discussion of the results.

Introduction

During the 2008-2010 campaigns a significant part of the resources of the Syro-Italian Mission in Palmyra (for previous activities see Al-Maqdissi, Kanhouche, Cremaschi and Morandi Bonacossi 2009; Cremaschi and Zerboni 2012; Morandi Bonacossi and Iamoni 2012) were focussed on the survey of tumuli/cairns in the region and the excavation of those considered to be potentially the best preserved. In particular, the hills of Rujem al Majdur contain a significant number of tumuli, which, although apparently all looted *ab antiquo*, together constitute a valuable body of evidence regarding funerary rituals and traditions in the region. Part of this evidence has been already published (Morandi Bonacossi and Iamoni 2012) and the following study aims at completing the picture, with the hope that in the future it will be possible to continue the investigation of the area.[1]

1 The excavations and survey were conducted during the 2009 and 2010 seasons: I take this opportunity to thank Yasmin Khanouche, Eveline Samaan and Muhadden Hamoudeh, with whom I had the pleasure to work at Rujem al Majdur. Investigation in the area stopped in 2011 due to the climate of instability that progressively overtook Syria. It is our hope that the political situation will soon improve, so as to permit the Syrian people to live in peace and democracy and allow foreign missions to continue their research together with Syrian teams in the field.

Studia Chaburensia 4 (2014), pp. 49–61.

Archaeological evidence: new data

Chronology

Two C14 determinations have been obtained from tumuli T213 and T302,[2] where significant evidence of burning activity was found (most likely related to the construction of the funerary structures). The two charcoal samples (probably small twigs used to set the fire) were collected from the ash layers of two fireplaces located in the northeastern areas of the cists of both tumuli. The results (Fig. 1) have been re-calibrated by the author on the basis of the recently published new calibration curve (Reimer *et al.* 2013).

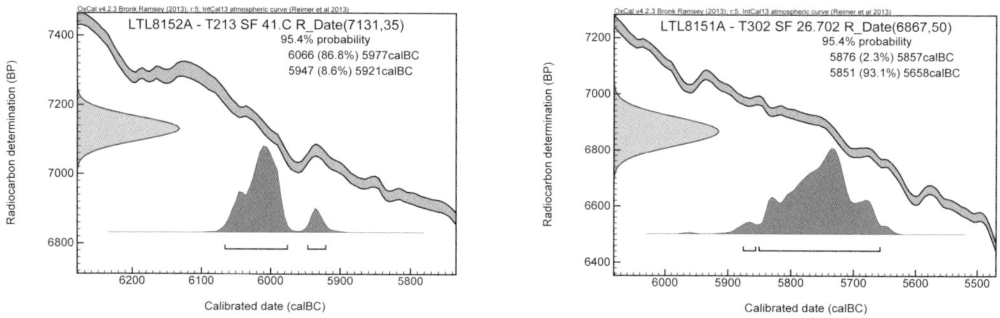

Fig. 1: calibration of C14 determination with OxCal 4.2 (© Ramsey).

Both determinations point to a very early (and somewhat unusual) chronology for the tumuli, with T213 dating to the very end of the 7th or beginning of the 6th millennium BC (6066-5921 BC) and T302 to the first half of the 6th millennium (5876-5658 BC, both datings 2σ). Generally, tumuli and cairns in the Levant, as well as in the Jebel Bishri and Euphrates areas, have been dated to the Late Chalcolithic-Early Bronze Age (Bradbury and Philip 2011: 175; Falb, Krasnik and Meyer 2005; Haiman 1992; Lönnqvist 2010; Numoto and Kume 2010; Steimer-Herbet 2004: 20) or even the beginning of the 2nd millennium (Fuji and Adachi 2010; Kepinski 2006: 103–104). The Rujem al-Majdur tumuli provide evidence in support of a much earlier phase, the late Pottery Neolithic, that has been found already in the southern Levant and eastern Jordan (Avner and Carmi 2001; Avner *et al.* 1994; Fuji 2013: 70-73; Rollefson 2013; Rowan, Rollefson and Kersel 2011), and thus suggest a much earlier beginning for the use of such burial practices in the area.

Tumuli

In previous work, attention was focused on the type of tumuli situated on Rujem al Majdur and their position on the hilltop (Fig. 2). Three tumuli with significant features were discussed (221, 213 and 302): the first (T221) was the largest excavated in Rujem al Majdur. Although it had been almost completely emptied (only a few intrusive sherds occurred in the upper layers of the tumulus), T221 encompassed a circular area with a diameter of more than 10 m (compared to an average of about 5 m for the other tumuli). T213 and 302, although plundered as well (a few intrusive sherds occurred also in their upper layers), still showed

2 C14 samples were processed by the CEDAD laboratory of the University of Salento.

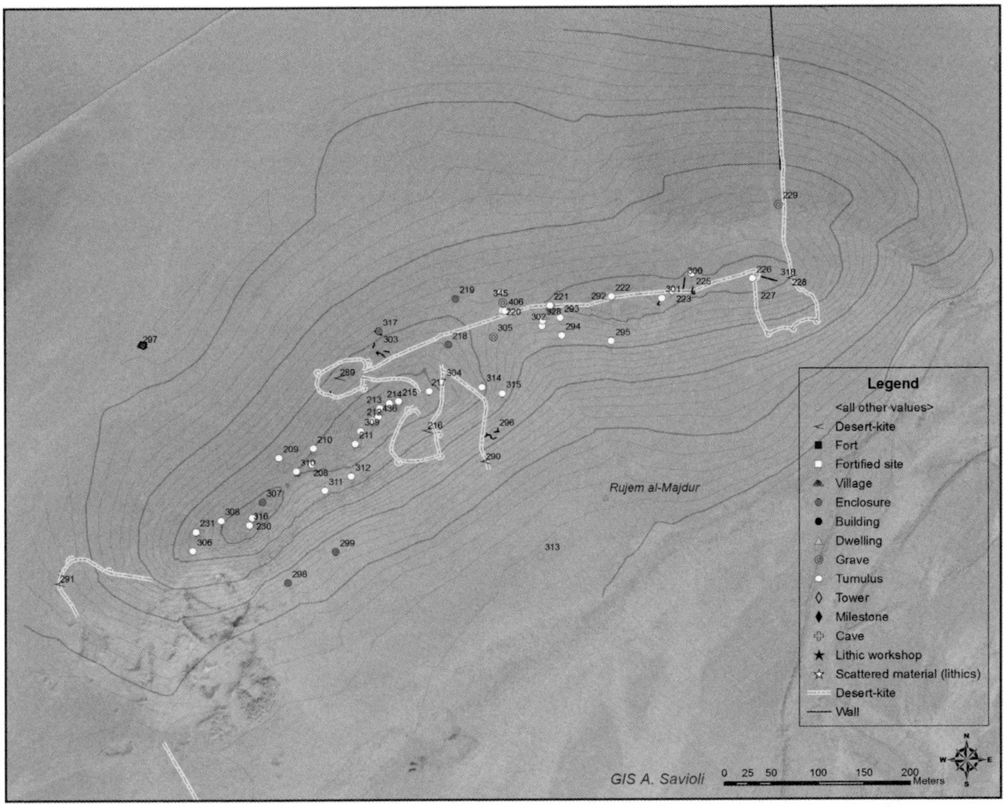

Fig. 2: map of tumuli and structures on the top of Rujem al Majdur.

traces of burning activity (perhaps interpretable as evidence of ritual fires, that may have involved offerings to the dead) which provided a rare opportunity to study the funerary rituals of tumulus/cairn horizons in Central Syria (Morandi Bonacossi and Iamoni 2012: 49).

Further interesting data have also come from the excavations of two other tumuli. Firstly T220, which features adjacent grave T406 on its eastern side (Figs. 3 and 4), is located on the top of Rujem al Majdur, more or less in the central part, in correspondence to a slight depression in the hillcrest.

It is semicircular with a diameter of about 3.3 m; including the adjacent grave T406, the entire tumulus is oval in shape with a length of 5.8 m along the (E-W oriented) major axis, whereas the shorter (N-S) axis measures about 3.3 m.

The tumulus is made up of medium and large-sized stones and at the time of the excavation reached a height of about 1 metre above the surrounding surface. The internal chambers of T220 and T406 differ slightly in dimensions and shapes: the larger is that in T220 and is quite irregular, although very roughly circular, while the smaller T406 chamber is narrow and oval; both are about 1 metre deep. The main cist (T220) was prepared with vertical stone slabs that are still visible, especially along the eastern side, whereas the other walls of the tumulus are made of large stones. The cist was in origin likely sealed with flat stones that were still present in the northern and western portions of the chamber: their absence in the southern and central part of the tumulus suggests that T220 had been looted, perhaps in ancient times.

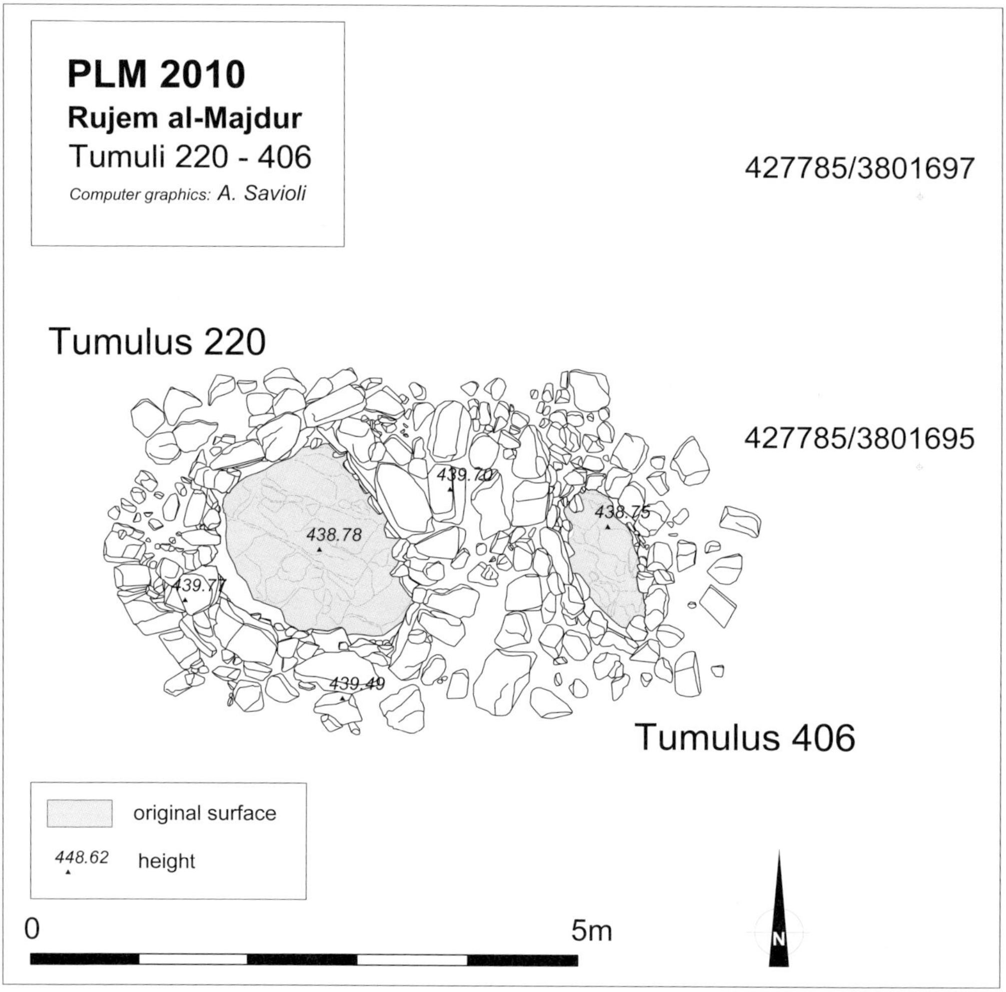

Fig. 3: T220 and its adjacent satellite grave T406.

Fig. 4: T220; during (left) and at the end of the excavation (right).

Fig. 5: lithic artefact (scraper ?) from tumulus T220.

The excavation revealed the presence of a hardened surface which in the northern sector was directly covered by the stone slabs; it seems likely that this was the tumulus floor (though one cannot rule out the possibility the surface was simply the result of the pressure of the stones on the cist fill).

On the eastern side traces of burning were visible, suggesting also in this case the occurrence of some kind of ritual connected with the tumulus (as in T302 and T213 – though in T220 the evidence is much less compelling – see Morandi Bonacossi and Iamoni 2012). Clearly recognizable human remains have not been retrieved: only a very few bone fragments were found during the excavation, but they were so small and their state of preservation so bad that it has not proved possible to reach any conclusions about their nature. Nothing was discovered inside or on the floor of the chamber that might have been part of an original assemblage.

However, a lithic artefact was found in the upper layer of the tumulus fill (Fig. 5). It is possibly a denticulate scraper that[3] measures about 12 cm long and about 6 cm wide. A preliminary evaluation suggests a wide possible date range, from the Palaeolithic to Neolithic/ Bronze Age (the latter seems most likely).[4] In the absence of a more refined and precise analysis, it is impossible to go further, but it should be noted that the Neolithic date agrees very well with the C14 determinations discussed above.

3 A more careful re-examination of this – and of all of the other artefacts excavated and collected during the survey at Rujem al Majdur – was planned for a subsequent study season. Due to the tragic events that have hit Syria, it has been impossible to complete this (as well as other detailed analyses). The classification of the artefact has consequently been done on the basis of a photograph taken during the excavation season; I wish to thank C. Conati for her precious help in providing this preliminary classification.

4 I am most grateful to F. Abbés for dating this artefact.

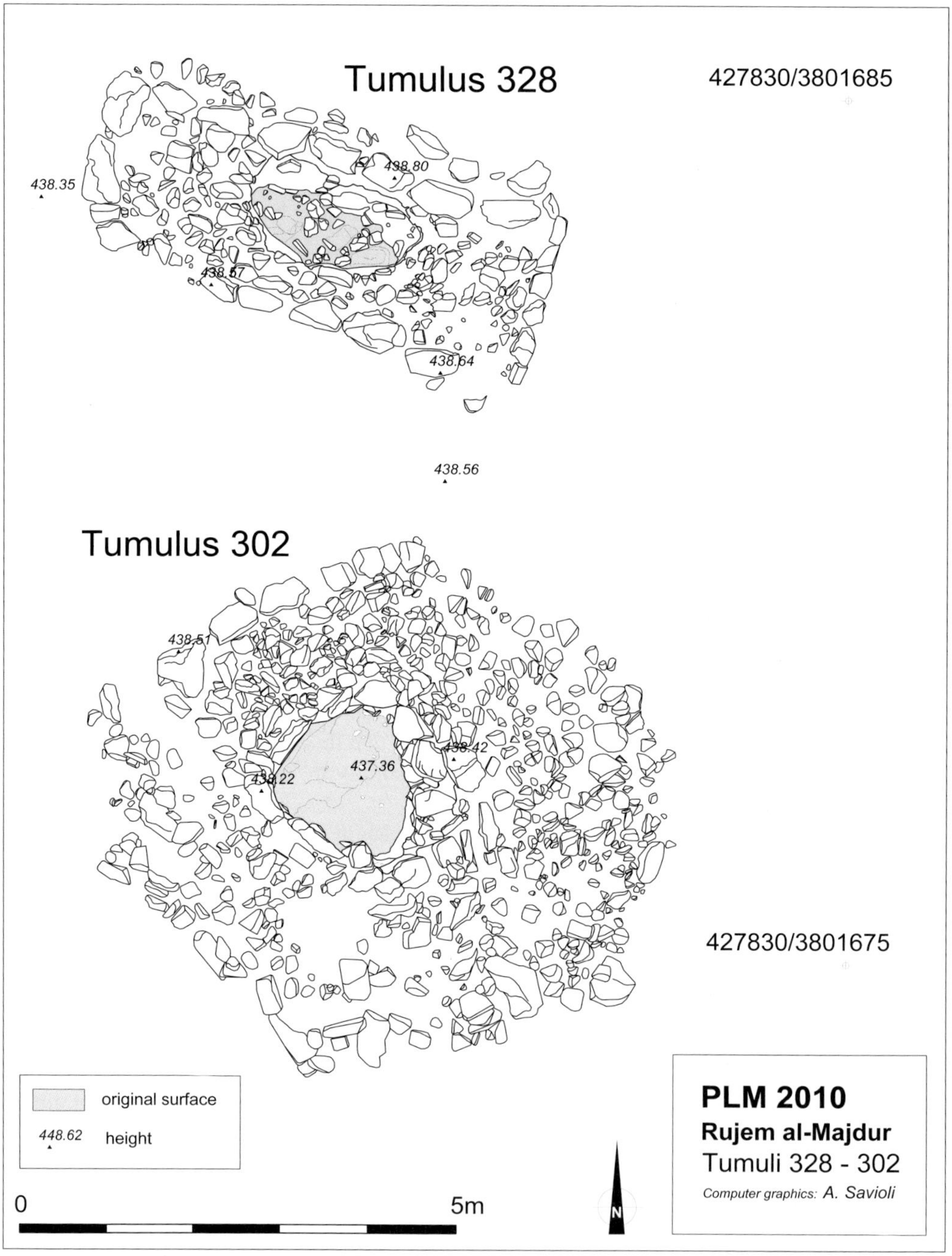

Fig. 6: tumuli T302 and T328.

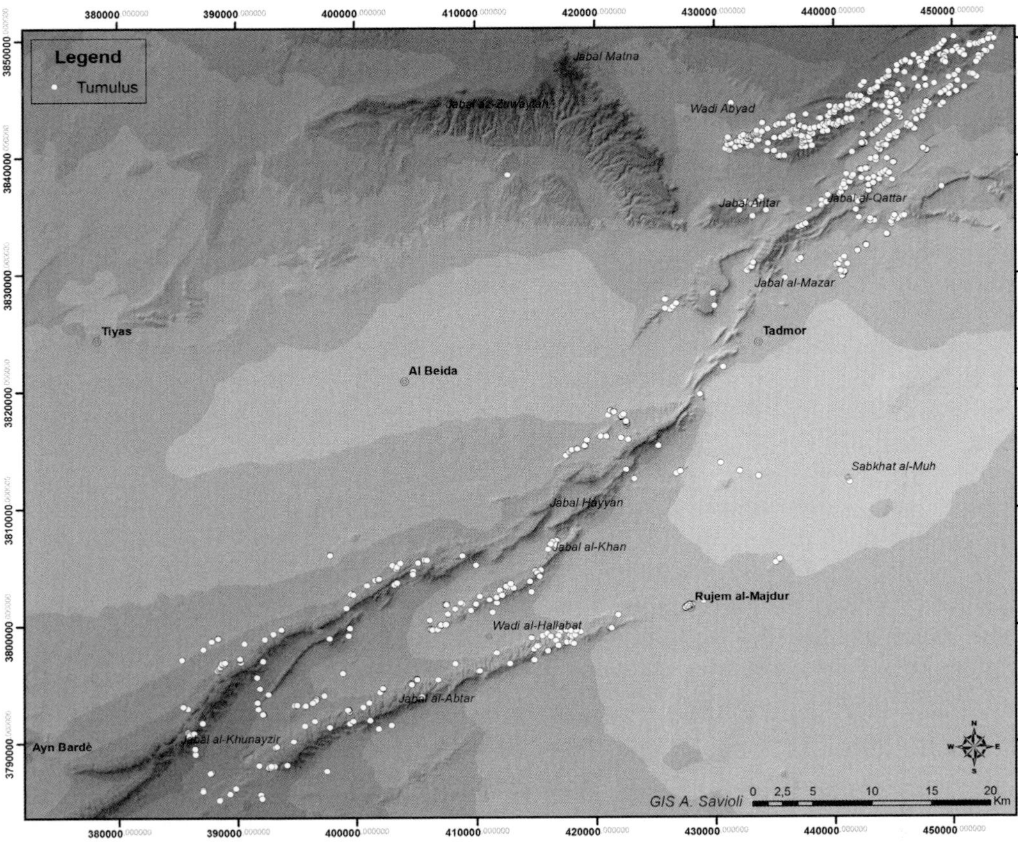

Fig. 7: tumulus distribution in the Palmyra Plain area (map by A. Savioli).

The second grave of some importance is T328, which lies next to T302 (the latter has been already discussed in Morandi Bonacossi and Iamoni 2012). Its shape is rather unusual, rectangular with rounded corners; the structure is long (about 5 metres in length and about 2 metres wide) and oriented NW-SE (Fig. 6). This suggests that T328 does not belong to the usual tumulus/cairn typology, but is instead a different (funerary?) structure somehow connected with (in the light of its proximity to) the main tumulus T302. Given the likely early date of the tomb (if one assumes a close relationship with T302, which has been C14-dated to the very beginning of the 6th millennium BC), it is difficult to look for parallels. Some similarities may be seen with respect to tombs in the necropolis of Abu Hamed (Falb, Krasnik and Meyer 2005: 32, fig. 50), but it must be kept in mind that these are definitely much later, that is EB III-IVA (Meyer 2010: 161). T328 was also looted in the past and consequently neither grave goods nor significant human remains were found in its fill. Nevertheless, some very small bone fragments were retrieved during the excavation of the upper layers of T328's fill. The recovery of a non-diagnostic potsherd suggests that the looting probably occurred in antiquity, though the quality of the ceramic find does not permit any greater precision. The importance of T328 lies in its proximity to T302, and the consequent possibility of a connection between them (perhaps of kinship; there is also the probable "centrality" of T302, as in the case of T213 –which has a satellite grave as well).

With reference to the presence of tumuli in the region of Palmyra, the continuation of research has permitted the building of a more ample tumulus database using information obtained from both field survey and the study of satellite images.[5]

The number of tumuli identified amounts now to 889. The most significant (and evident) pattern concerns their position atop the ridges of hills and mountains that cross the Palmyra Plain from SW to NE (Fig. 7). This confirms the tentative conclusions presented in the preliminary report (Morandi Bonacossi and Iamoni, 2012: 46) and serves to underline once more the importance that these funerary structures had as landmarks and territorial reference points, as well as significant elements of the cultural landscape (Bradbury and Philip 2011: 176–178; Porter 2002).

Tumuli were thus crucial visual guides for (semi) nomadic tribes crossing the Palmyra Plain. Their position on mountaintops must have made them useful markers, perhaps indicating paths that crossed the Palmyra oasis. In the light of the chronology discussed above, however, the significance of this pattern needs to be reconsidered, since the role of tumuli as reference points might extend further back into prehistory. This suggests that occupation features that became common during the 4th-3rd millennium BC in the region had already emerged from 6th millennium onwards. It is true, of course, that the overall number of tumuli detected in the region includes different chronological phases as well as different burial customs; at present we have not sufficient data to identify specific chrono-cultural clusters. Yet it is also true that the homogeneity of tumulus distribution must surely reflect a common cultural background, which has continued to be influential over a very long period.

Conclusion

The evidence discussed in this article, especially if taken together with that already presented in previous publications, adds interesting information regarding the nature of the prehistoric occupation of the Palmyra region. It further emphasizes the strong correspondence between tumuli and desert kites: the map discussed in this volume (see Morandi Bonacossi this volume, fig. 13) highlights quite clearly the overlapping distribution of these two structure types and suggests a strong correlation between them. Rujem al Majdur in this respect provides a further confirmation, although it seems clear that some of the tumuli postdate the desert kites (in particular those that, because of their positions, would have interfered with the function of certain desert kites: T221, 217, 293, 294, 314 and 315, and quite likely 302 and 328). The tumulus evidence thus suggests the abandonment of the desert kite of fig. 2 at some point. The remaining tumuli (in particular those on the eastern ridges of the hills) do not show a clear relationship that indicates their chronology with respect to the desert kites. Nevertheless, they would not have interfered with the desert kites' function, and so they might in theory have been contemporary or even antecedent. If this holds true, the different C14 determinations of T302 and T213 (significantly different at a 95% probability level, see the Chi Square test in Figure 8) may constitute crucial reference points, since they could be taken as *terminus post quem* (T213) and *ante quem* (T302) to establish the beginning and end of use of the desert kite (see Morandi Bonacossi this volume for a more detailed discussion of desert kite chronology). In the absence of excavation data from the desert kites, however, it is impossible to verify this point, which remains a question for future research.

5 The study was conducted by Alberto Savioli using satellite images provided by Google Earth.

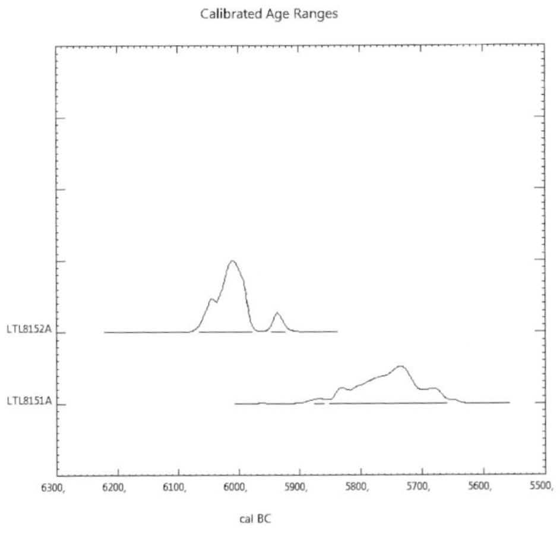

Test Statistic T	Chi Square value at 0.5	Degrees of freedom
18.71034	3.84	1

Fig. 8: C14 determinations in single plot and Chi Square analysis, showing the statistically significant difference between the two datings (using Calib, © Stuiver and Reimer).

The Rujem al Majdur necropolis would have thus been in use at least from the very end of the 7th millennium BC and this could be taken as a preliminary chronological reference point for what has been defined as the "desert kite and tumulus cultural horizon", i.e. the material culture evidence relating to the semi-nomadic/nomadic lifestyle (and related sustenance economy) in the semi-arid region of Palmyra.

With respect to the role of tumuli as landmarks, they have been considered as visual points that somehow marked the routes of paths crossing the Palmyra area (Morandi Bonacossi and Iamoni 2012: 53–54). In this respect, the very early C14 dates suggest extreme caution – and the necessity of more data, since the second millennium historical sources that mention crossroads at Palmyra (Astour 1995; Durand 1987; Joannès 1996; 1997: 397–399; Van Koppen 2007) are of little help in this respect.

The general alignment of the tumuli in the region does not seem to follow specific paths, but rather seems simply to follow the crests of the hills. The tumuli's association (or at least that of some of them) with the desert kites is indeed a key factor for an understanding of their position, since the burials may mirror the movement of groups dedicated to intensive specialized and cooperative hunting (Morandi Bonacossi and Iamoni 2012: 42–43).

However, the large number of tumuli and their widespread occurrence throughout the Palmyra region suggest the existence of multiple patterns that are impossible to single out without further work in the field. For example the mountains to the immediate southwest of Palmyra contain only a few tumuli on their crests; most are located instead in the adjacent plains, which suggest that reasons other than visibility are needed to explain their positions.

An interesting point concerns tumulus distribution across the area under investigation: it is noteworthy that there is a greater concentration of tumuli in the northeastern sector, compared to the southwest. At a very general level, this could be due to the existence of two distinct burial traditions, one of which was possibly more prevalent in the Bishri area (and thus perhaps much later in date than that excavated at Rujem al Majdur).

Some of the tumuli (Fig. 9) located on top of the northeastern mountains of Palmyra exhibit an unusual structure, with the main burial mound surrounded by a circular stone wall. This funerary construction bears a strong resemblance to the tumuli excavated at Wadi Hedaja (Fuji and Adachi 2010: 67) on the northwestern flank of Jebel Bishri.

Fig. 9: comparison of the tumuli excavated in the Bishri area and in the mountains NW of Palmyra: On the left the tumulus BC-10 excavated at Wadi Hedaja (after Fuji and Adachi 2010, 66, Fig. 10); on the right three tumuli with surrounding stone walls in-dentified via examination of the satellite images of Google Earth (source and date of photo: Google Earth, 9/8/2010; image © 2014 DigitalGlobe).

These date to the Middle Bronze Age and have been convincingly attributed to the Amorite cultural horizon (Fuji and Adachi 2010: 73). If this association is confirmed, their presence in the northeastern area of Palmyra might imply a burial tradition which is the direct expression of (or was influenced by) an Amorite nomadic horizon, thus extending the Amorites' area of origin from the Jebel Bishri to the entire northeastern region of Palmyra. At the same time, the absence of this tumulus type in the area southwest of Palmyra means that there was probably a boundary here to the diffusion of this burial tradition.[6]

A last comment concerns the symbolism and rituals associated with the tumuli. The evidence from the Rujem al Majdur excavations offers interesting insights into the feasting and mortuary activities conducted at the tumuli, as well as their symbolic value for the community. With regard to the first aspect, the discovery of ash layers in at least two of the largest tumuli in the area (T213 and T302) points to the development of an elaborate ritual for the burial of the dead. We cannot say whether this implied offerings, but it must have involved special ceremonies which were not common to all the tumuli. At the same time the presence in T213 and T302 of satellite burials (T406 and 328) highlights the importance of stronger kinship relationships as well as of common ancestors whose role was emphasised by the community. The burial of people in proximity to a major tumulus reflects at least the voluntary (if not necessarily obligatory) display of kinship linkages. It ultimately underlines group identity and membership and confirms the importance of family ties (a trait particularly accentuated in nomadic societies (Porter 2012: 63) from a period much earlier than that conventionally proposed in the study of similar funerary evidence. Clusters of tumuli located towards the southwestern edge of the survey area (but here on the plain, not in the hills) may point in this direction too, reflecting the occurrence of even stronger family/tribal linkages.

Both lines of evidence (rituals and the symbolic role of specific tumuli), when considered together with the concomitant occurrence of desert kites, denote the emergence of processes of increasing social complexity, whose origin must be seen in the environmental changes that characterised the region of Palmyra during the PPNB (Cremaschi and Zerboni 2012). The encroaching aridity that affected the area and whose peak perhaps coincided with the "8.2 cal Ka BP event" (Weninger *et al.* 2006), forced human communities to look for new communal subsistence strategies, as well as for different modes of organization. The latter, in turn, implied (and generated) much stronger social cohesion, strengthening genealogy/the community of ancestors (Porter 2012: 57–58), visually fixed and memorized by the living community through the construction of monumental burials. The continuation of research on the body of data collected will permit further investigation of this point, as well as into the occurrence of different burial traditions in the Palmyra area, in order to improve our understanding of the "desert-kite and tumulus cultural horizon".

6 A more detailed article on this subject is in preparation.

Bibliography

Al-Maqdissi, M., Y. Kanhouche, M. Cremaschi and D. Morandi Bonacossi

2009 Présentation sommaire des travaux archéologiques de la mission syro-italienne: Les fouilles archéologiques à Mishirfeh – Qatna et la prospection de la Palmyrène occidentale, *Studia Orontica* I: 5-20.

Astour, M. C.

1995 Overland Trade Routes in Ancient Western Asia. In J. Sasson (ed.), *Civilizations of the Ancient Near East*, New York (Vol. III): 1401-1420.

Avner, U., and I. Carmi

2001 Settlement patterns in the Southern Levant deserts during the 6th–3rd millennia BC: a revision based on 14C dating, *Radiocarbon* 43: 1203-1216.

Avner, U., I. Carmi and D. Segal

1994 Neolithic to Bronze Age settlement of the Negev and Sinai in light of radiocarbon dating: a view from the southern Negev. In R. Bar-Yosef and O. Kra (eds.), *Late Quaternary chronology and paleoclimates of the eastern Mediterranean*, Tucson: 265-300.

Bradbury, J. and G. Philip

2011 The world beyond tells: preclassical activity in the basalt landscape of the Homs region, Syria. In T. Steimer-Herbet (ed.), *Pierres levées, stèles anthropomorphes et dolmens = Standing stones, anthropomorphic stelae and dolmens*, Lyon/Oxford: 169-180.

Cremaschi, M. and A. Zerboni

2012 Adapting to increasing aridity: The cuvette of Palmyra (central Syria) from late Pleistocene to early Holocene. In F. Braemer, F. Bertoncello (eds.), *Variabilités environnementales, mutations sociales Nature, intensités, échelles et temporalités des changements,* Antibes: Éditions APDCA: 37-52.

Durand, J.-M.

1987 Documents pour l'histoire du royaume de Haute-Mésopotamie, I/C: Les trois routes de l'Euphrate à Qatna à travers le désert. *Mari, Annales de Recherches Interdisciplinaires* 5: 159-167.

Falb, C., K. Krasnik, J.-W. Meyer and E. Vila

2005 *Gräber des 3. Jahrtausend v.Chr. im syrischen Euphrattal – 4. Der Friedhof von Abu Hamed,* Schrifet zur vorderasiatischen Archäologie, Band 8, Saarwellingen: Saarländische Druckerei & Verlag.

Fuji, S.

2013 Chronology of the Jafr Prehistory and Protohistory: a Key to the Process of Pastoral Nomadization in the Southern Levant, *Syria* 90: 49-125.

Fuji, S. and T. Adachi

2010 Archaeological Investigations of Bronze Age Cairn Fields on the Northwestern Flank of Mt. Bishri. In Ohnuma *et al.* (eds.), *Formation of Tribal Communities: Integrated Research in the Middle Euphrates, Syria (Al-Rafidan - Special Issue)*, Tokio: 61-77.

Haiman, M.

1992 Cairn Burials and Cairn Fields in the Negev, *Bulletin of the American Schools of Oriental Research* 287: 25-45.

Joannès, F.
1996 Routes et voies de communication dans les archives de Mari. In J.-M. Durand (ed.), *Amurru 1. Mari, Ébla et les Hourrites, dix ans de travaux*, Paris: 323-361.

Joannès, F.
1997 Palmyre et les routes du désert au début du deuxième millénaire av. J.-C, *Mari, Annales de Recherches Interdisciplinaires* 8: 393-415.

Kepinski, C.
2006 Mémoires d'Euphrate et d'Arabie, les tombes à tumulus, marquer territoriaux de communautés en voie de sédentarisation. In C. Kepinski, O. Lecomte and A. Tenu (eds.), *Studia Euphratica. Le moyen Euphrate iraquien révélé par les fouilles préventives de Haditha*, Paris: 87-128.

Lönnqvist, M.
2010 Tracing tribal implications among the Bronze Age tomb types in the region of Jebel Bishri in Syria. In Ohnuma *et al.* (eds.), *Formation of Tribal Communities: Integrated Research in the Middle Euphrates (Al-Rafidan - Special Issue)*, Tokio: 165-173.

Meyer, J.-W.
2010 The Cemetery of Abu Hamad: a Burial Place of Pastoral Groups? Formation of Tribal Communities: Integrated Research in the Middle Euphrates. In Ohnuma *et al.* (eds.), *Formation of Tribal Communities: Integrated Research in the Middle Euphrates, Syria (Al-Rafidan - Special Issue)*, Tokio: 155-164.

Morandi Bonacossi, D. and M. Iamoni
2012 The early history of the Western Palmyra Desert Region. The change in the settlement patterns and the adaptation of subsistence strategies to ecncroaching aridity: a first assessment of the desert-kite and tumulus cultural horizons, *Syria* 89: 31-58.

Numoto, H. and S. Kume
2010 Surveys and sondages at the cemeteries near the site of Tell Ghanem al-'Ali. Formation of Tribal Communities: Integrated Research in the Middle Euphrates. In Ohnuma *et al.* (eds.), *Formation of Tribal Communities: Integrated Research in the Middle Euphrates, Syria (Al-Rafidan - Special Issue)*, Tokio: 49-60.

Porter, A.
2002 The Dynamics of Death: Ancestors, Pastoralism, and the Origins of a Third-Millennium City in Syria, *Bulletin of the American Schools of Oriental Research* 325: 1-36.
2012 *Mobile pastoralism and the formation of near eastern civilizations,* New York: Cambridge University Press.

Reimer, P. J., E. Bard, A. Bayliss, J. W. Beck, P. G. Blackwell, C. Bronk Ramsey, C. E. Buck, H. Cheng, R. L. Edwards, M. Friedrich, P. M. Grootes, T. P. Guilderson, H. Haflidason, I. Hajdas, C. Hatté, T. J. Heaton, D. L. Hoffmann, A. G. Hogg, K. A. Hughen, K. F. Kaiser, B. Kromer, S. W. Manning, M. Niu, R. W. Reimel, D. A. Richard, E. M. Scot, J. R. Southon, R. A. Staf, C. S. M. Turney and J. van der Plicht.
2013 IntCal 13 and marine 13 radiocarbon age calibration curves - 50,000 years Cal BP, *Radiocarbon* 55(4): 1869-1887.

Rollefson, G. O.
2013 Late Prehistoric aggregation patterns in Jordan's Eastern Badia, *Syria* 90: 211-230.

Rowan, Y. M., G. O. Rollefson and M. M. Kersel
2011 Maitland's "Mesa" reassessed: a late prehistoric cemetery in the eastern Badia, Jordan, http://www.antiquity.ac.uk/projgall/rowan327/

Steimer-Herbet, T.
2004 *Classification des sépultures à superstructure lithique dans le Levant et l'Arabie Occidentale (IVe et IIIe millénaires avant J.-C.)*, BAR International Series, Oxford.

Van Koppen, F.
2007 Syrian Trade Routes of the Mari Age and MB II Hazor. In M. Bietak and E. Czerny (eds.), *The Synchronization of Civilizations of the Eastern Mediterranean in the second millennium BC,* Vol. III ,Vienna: 367-374.

Weninger, B., E. Alram-Stern, E. Bauer, L. Clare, U. Danzeglocke, O. Jöris, C. Kubatzki, G. Rollefson, H. Todorova and T. Van Andel.
2006 Climate forcing due to the 8200 cal yr BP event observed at Early Neolithic sites in the eastern Mediterranean, *Quaternary Research* 66: 401-420.

SUMIO FUJII
Kanazawa University

Chronology of the Bishri Pastoral Prehistory and Protohistory

A Cross-check against the Jafr Chronology in Southern Jordan

A top-priority issue of the late prehistoric archaeology of dryland Syria is to establish a local chronology for seriating various features dotted in the steppe and desert and, at the same time, locating them within the general framework of the Levantine Badia archaeology. This is no easy task, however, because reliable data sets, especially those on post-Neolithic sites, are still patchy and less synthesized (Akkermans and Schwartz 2003; Al-Maqdissi *et al.* 2011; Anfinset 2008; Braemer *et al.* 2004; Cauvin 1994; Cauvin and Stordueur1977; Geyer 2002; Geyer and Monchambert 2003; Jaubert and Geyer 2006; Lönnqvist 2008; Lönnqvist and Lönnqvist 2011; Molist and Cauvin 1990; Ohnuma *et al.* 2010; and Zarins1989a, 1989b, 1990, 1992a, 1992b). The problem is the scarcity of basic information backed with full-fledged investigations involving excavations. It appears that the deficiency of excavated data results in the imperfection of the local chronology, which in turn causes further isolation of the Syrian dryland archaeology.

Our research project in the northwestern piedmont of Mt. Bishri, central Syria, may provide a breakthrough in the vicious circle. The initial stage of the project, undertaken from 2007 to 2010, excavated several dozen Middle Bronze Age burial cairns (Fujii and Adachi 2010). Subsequent investigations have slightly extended the research area westwards and examined several sites including a Khiamian encampment (Wadi al-Hajana 1), two Pre-Pottery Neolithic B (hereafter PPNB) flint workshops (Wadi al-Hajana 1 and 2), a Late Neolithic open sanctuary (Fakat Bidewy 1), and a few Chalcolithic cairn fields (Fakat Bidewy 2 and 3, and Jabal Gala). The discovery of such a wide variety of sites has not only attested to the archaeological potential of the Bishri Hills but also opened a way for incorporating the Bishri pastoral prehistory and protohistory into the general scheme of the Levantine Badia archaeology. This paper briefly reviews the investigation results and offers a chronological perspective of the Bishri sites through a cross-check against contemporary sites in the Jafr Basin in southern Jordan, another research field of ours.

Research Field

The Bishri region has been only poorly investigated despite its supposed archaeological potential especially in the Bronze Age (Buccellati 1966, 1992, 1997; Charpin 2004, 2010; Charpin and Ziegler 2003; Durand 1987, 2004; Joannès 1997; Kupper 1957; Matthews 2003; Nilhamn 2008; Whiting 1995). Even general surveys, to say nothing of excavations, are far outnumbered by the surrounding drylands such as the el-Kowm and Palymra Basins (Buccellati and Kelly-Buccellati 1967; Poidebard 1934). Although the situation has been improved to some extent in recent years (Lönnqvist 2006; Lönnqvist 2008; Lönnqvist and Lönnqvist 2011), there still remain many archaeological blanks in the area.

Studia Chaburensia 4 (2014), pp. 63–92.

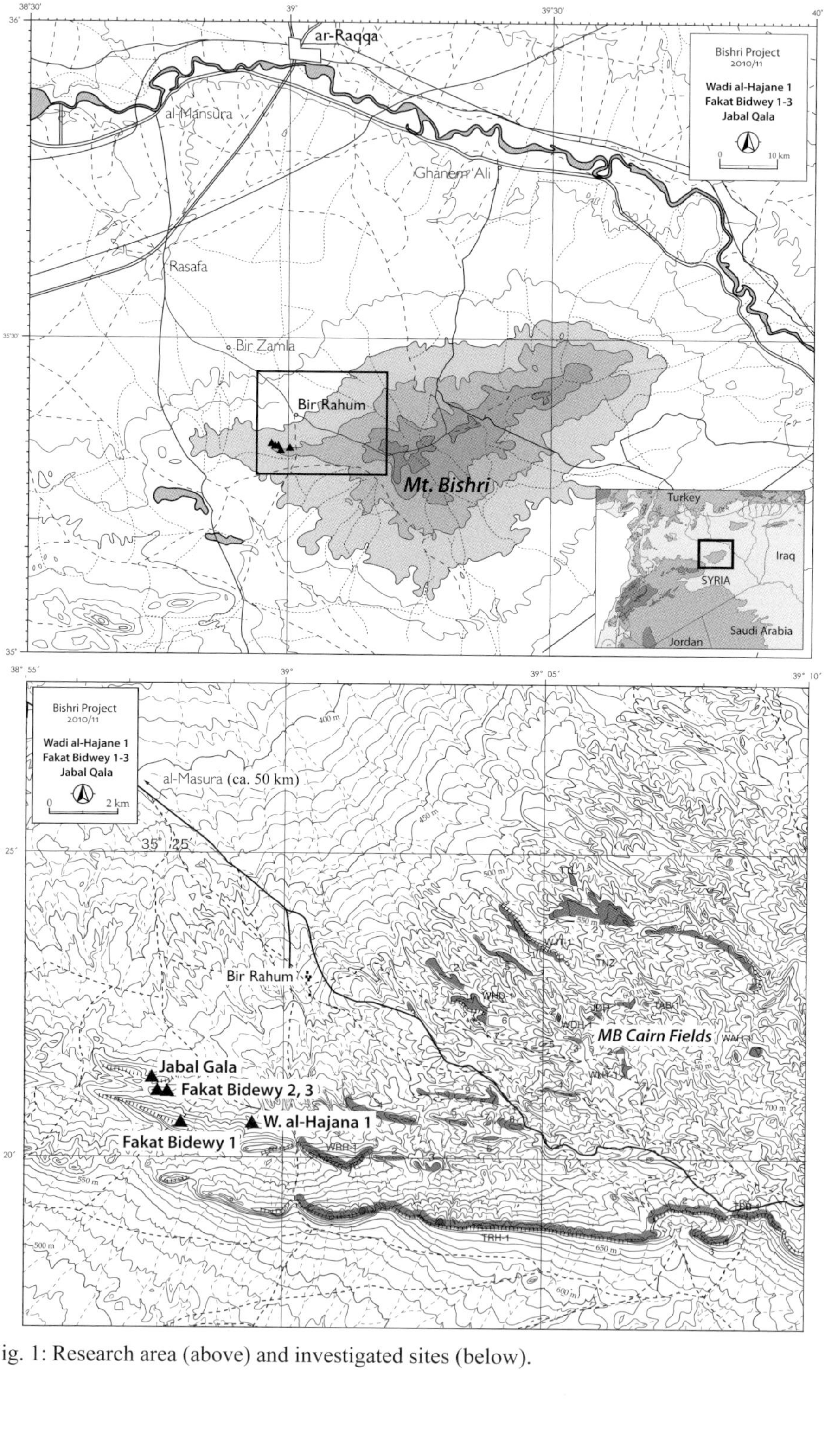

Fig. 1: Research area (above) and investigated sites (below).

Our research project has been focused on one of the blanks, the northwestern piedmont zone of Mt. Bishri. The present research field covers an area of ca. 350 square kilometers around the small village of Bir Rahum (Fig. 1). Topographically, the area presents a plateau-like landscape with an elevation of ca. 400-850 m and gradually ascends toward the escarpment that fringes the southern edge of the limestone hills. Following the general topography, local drainage systems run north- or northwestward to converge on Wadi Rasafa and eventually flow into the middle stream of the Euphrates ca. 60 km north. Thus the area, though seemingly isolated in the remote wilderness, is in fact closely connected to the sedentary cultural sphere in the Middle Euphrates River Basin. Not only that, it has a direct connection with the neighboring el-Kowm and Palmyra Basins as well through gentle slopes at the western edge of the escarpment. The location of the Bishri sites should be understood in such a broader perspective.

The natural environment of the research area is characterized by an arid climate and its consequent poor vegetation (Akashi and Tanno 2009). Since no perennial natural water sources are available, no traditional settlements exist in the area. The only exception to this is Bir Rahum referred to above, but even this small village was founded merely a few decades ago in the course of a government-led permanent residence project. This is not to say, however, that the prehistoric and protohistoric Bishri was entirely deserted. As described below, our investigation results are suggestive of a rather continuous land-use history during the early Holocene when northern Syria and its surrounding areas underwent serious climatic fluctuations (Issar and Zohar 2007; Staubwasser and Weiss 1986; Weiss 2013; Widell 2007; Wossnik 2009).

Wadi al-Hajana 1

The investigation results shall be briefly reviewed in a chronological order. First comes the site of Wadi al-Hajane 1. Being located on the north-facing hillside of an elongated foothill ca. 5 km south of Bir Rahum, it occupies the lower end of a gentle slope sandwiched between two small gullies (Fig. 2). The site elevation is ca. 530-540 m, and the site area is estimated at ca. 0.5-1 ha. The site was excavated in the spring field season of 2011 and briefly reported elsewhere (Fujii *et al.* 2011; Fujii and Adachi 2013). The excavation revealed two stratified cultural entities: a Khiamian indoor flint workshop belonging to Layer 2, and a PPNB open-air flint workshop included in Layer 1.

Structure A

Structure A or the Khiamian flint workshop was found at the northeastern corner of Area I, near the confluence of the two gullies. This small semi-subterranean structure was dug from the upper surface of Layer 2, measuring ca. 4-4.5 m in diameter and ca. 0.3-0.5 m in floor depth (Figs. 3 and 4). The indoor space was roughly round and fringed with a single row and course of undressed or partly dressed upright limestone slabs. The construction materials were put slightly inclining outwards and formed a retaining wall to cope with sideways soil pressure. As is usual with such a simple semi-subterranean structure, no clear evidence for a superstructure was confirmed. Thus, thatched or tent-like roofing is conceivable, although neither postholes nor pillar bases were left on the floor.

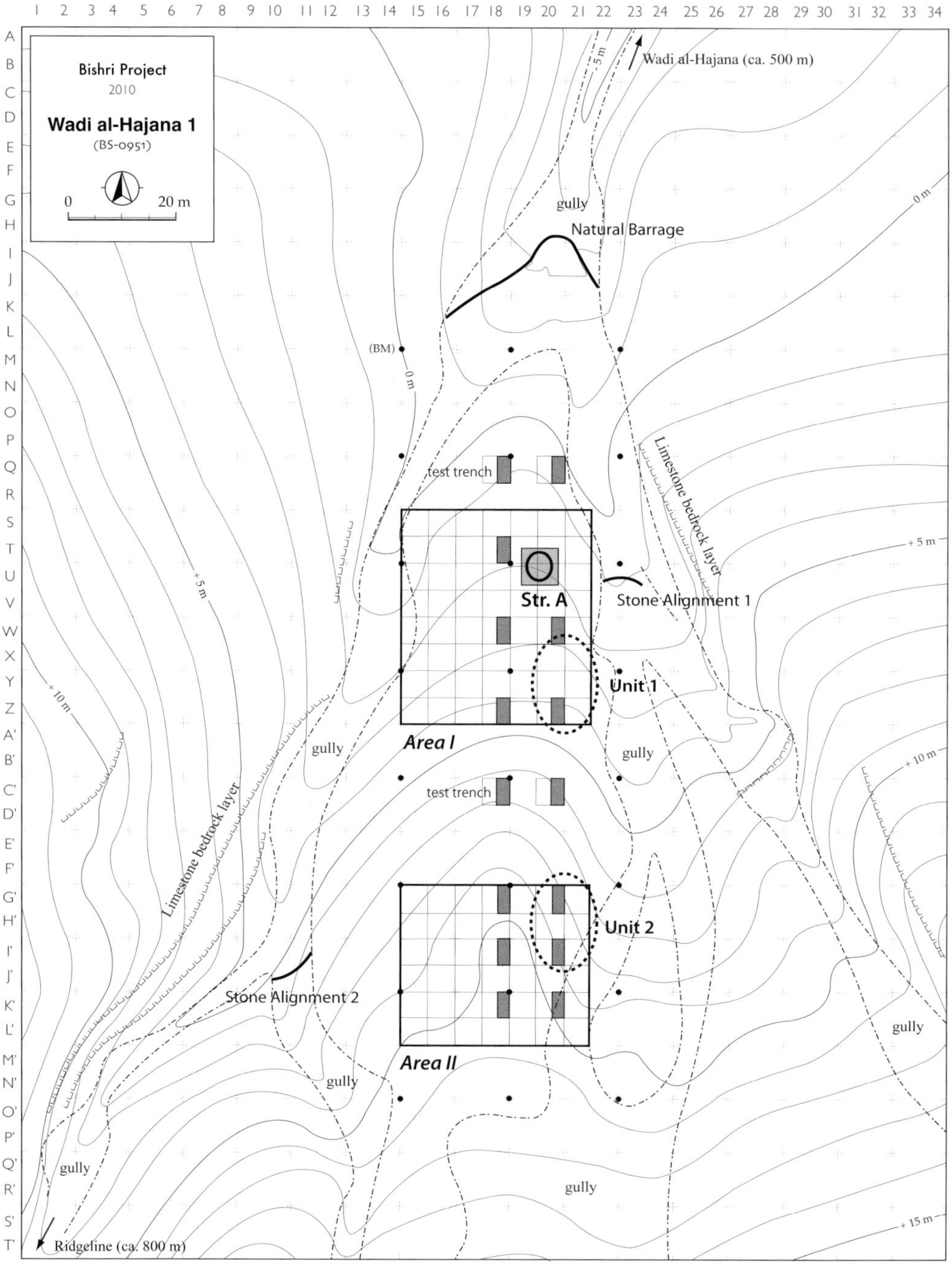

Fig. 2: Wadi al-Hajana 1: site plan and operation areas.

Fig. 3: Wadi al-Hajana 1: Structure A (looking N).

Though not clearly defined, it appears that the entrance occupied the eastern corner where smaller slabs were used. No special treatment was confirmed on the floor, but a large hearth, ca. 1.2 m in diameter and ca. 0.5-0.7 m in wall height, was incorporated into its northwestern corner. This feature contained a large amount of black ash and soot-covered angular limestone rubble ca. 10-15 cm long. However, neither carbonized seeds nor burned animal bones, to say nothing of flint artifacts, were recovered despite 2 mm-mesh dry sieving and water flotation. The specific function of this unique feature is still unknown, but it should be noted that it has much in common with *foyers creusés en cuvette* in Phases II-III of Tell Mureybet (van Loon 1968: fig. 4, pl. IIIB; Cauvin 1977: fig. 10; Molist 2008: 95-96, fig. 1), the type site of the Euphrates Khiamian.

Another remarkable indoor feature was a partition-like wall that diagonally traversed the center of the floor. It was directly connected with the northern wall to form a small room ca. 3 m by ca. 2 m in floor area. This means that the structure was reconstructed in the course of its use. In addition, five fallen slabs were found *in situ* between the partition-like wall and the southeastern wall. They probably represent a remnant of another wall that ran across the southern half of the floor. It would follow that the structure was reconstructed at least twice (Fig. 4). The structure was not only reduced in size but also decreased in quality on every reconstruction, suggesting that the dwellers gradually became smaller in group size and even more mobile in behavior pattern. It is possible that this episode went hand-in-hand with the decline of the Khiamian culture itself.

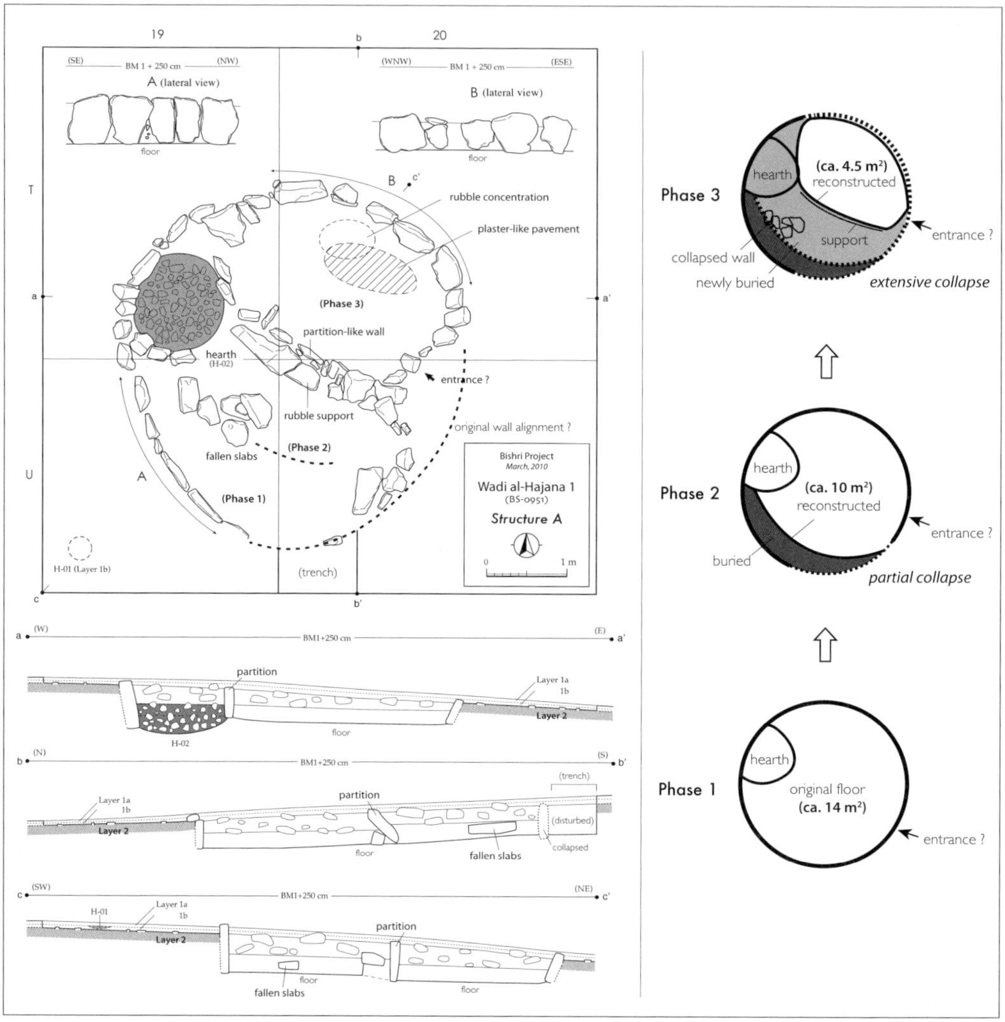

Fig. 4: Wadi al-Hajana 1: plan and sections of Structure A (left) and its reconstruction history (right).

Small finds from Structure A

Structure A yielded a total of 3,258 flint artifacts. Other finds were limited to a small shell pendant only (Fig. 5: 29). Neither faunal nor floral remains were recovered despite the dry sieving and water flotation of heath fills and floor deposits. Thus the structure can probably be regarded as a workshop specializing in flint production. The artifacts were focused on the indoor space, suggesting that the structure served as a roofed workshop. Since the contents of the flint assemblage have been described elsewhere in some detail (Fujii *et al.* 2011; Fujii and Adachi 2013), we will only give an outline below.

The flint artifacts were produced with light to dark brown, slightly mat, fine-textured flint, and no obsidian products were included. Single-platform blade/bladelet cores and their related debitage class specimens amounted to 95.0 % of the grand total (Fig. 5: 1-2). This, coupled with the oc-

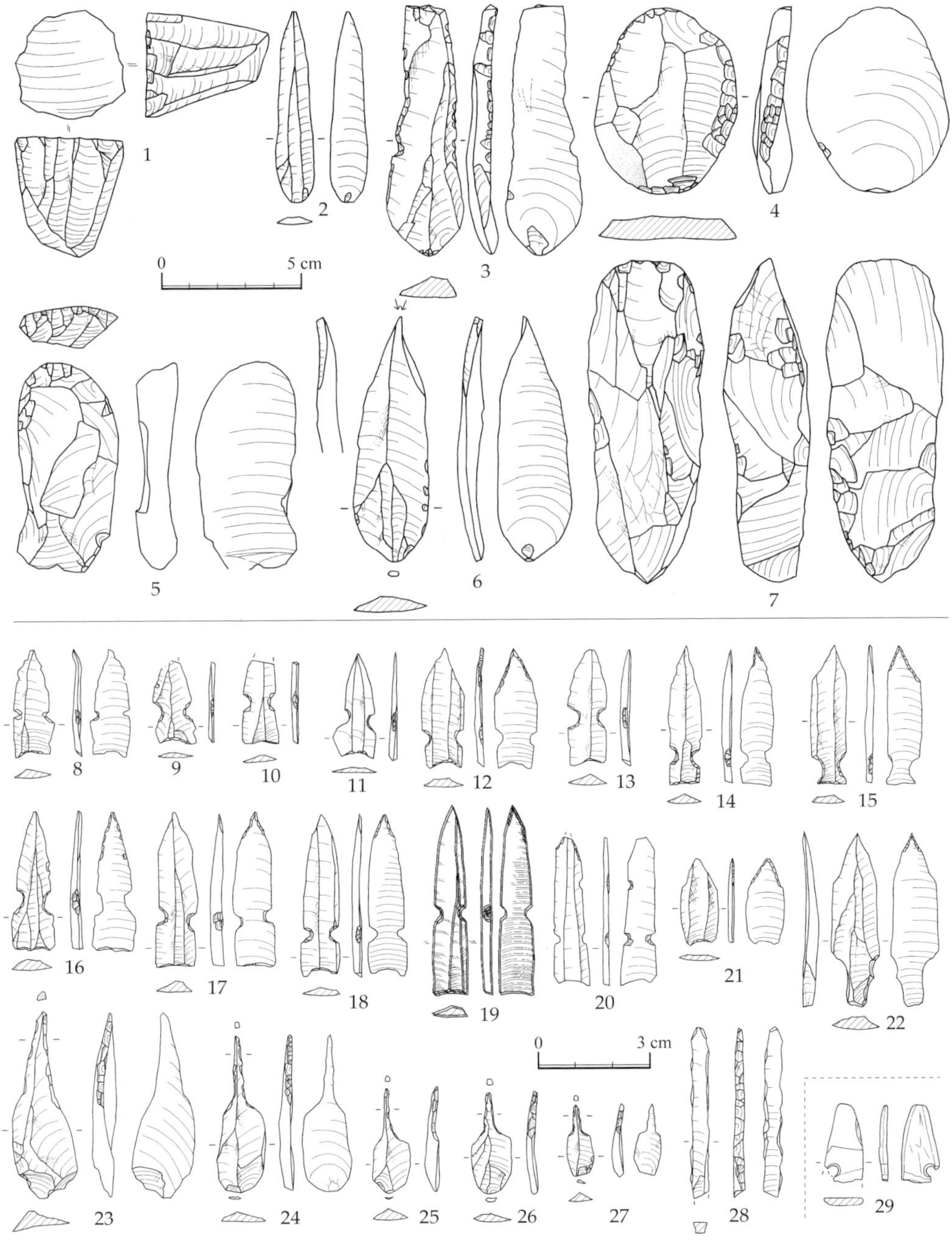

Fig. 5: Wadi al-Hajana 1: small finds from Structure A.

currence of several hammerstones, attests to the on-site production of the flint assemblage. However, cortical nodules and primary elements were nearly absent, suggesting that the initial core preparation took place elsewhere, probably around flint outcrops dotted below the escarpment.

Retouched tools accounted for only 3.1 % of the grand total, but the predominance of el-Khiam type points enabled us to date the assemblage to the Khiamian period (Fig. 5: 8-20). The tool kit also included microdrills (Fig. 5: 23-27), retouched blades/bladelets/flakes (Fig. 5: 3), sidescrapers (Fig. 5: 4), and endscarpers (Fig. 5: 5) in small numbers. In addition, a dihedral burin (Fig. 5: 6), an elongated Mureybetian-like adze (Fig. 5: 7), a heavy-duty digging tool, and a needle-like rod (Fig. 5: 28) occurred sporadically. Noticeable is the fact that many of the el-Khiam type points are equipped with a pair of lateral notches at a relatively higher position, which seems to have something to do with the absence of Hagdud truncations. Both traits are shared between Wadi al-Hajana 1 and Tell Mureybet, and can be regarded as the norm of the Euphrates Khiamian. Also of significance is the absence of geometric microliths including lunates, which suggests that the assemblage falls into Mureybet Phase IIB or the final phase of the Khiamian.

The question is the origin of the flint knappers who repeatedly visited the site. Seeing that few traces of actual life other than flint artifacts are left in the workshop, it is unlikely that they represent an indigenous group in the Bishri Hills. As a matter of fact, no contemporary sites including simple flint scatters have been located in the neighboring el-Kowm and Palmyra Basins as well as our research area. Thus it seems more reasonable to assume that they were a small, task-specific group derived from the Middle Euphrates River Basin where contemporary settlements were located. Suggestive in this regard is the fact that their flint production activities were focused on delicate tools such as el-Khiam type points and microdrills. Both of them necessitate high quality flint rarely available in the parent settlement area, suggesting that the expedition was intended for the procurement of raw material.

Surface collection

Prior to the excavation, a systematic surface survey took place in the two operation areas and recovered a total of 7,497 chipped flint artifacts. The predominance of naviform core-and-blade components suggested a PPNB date for the surface collection. Unexpected was the substantial absence of Khiamian artifacts. This is probably because they were entirely sealed in the buried indoor workshop.

Two concentrations were found along the eastern slope: Unit 1 at the southeastern corner of Area I, and Unit 2 at northeastern corner of Area II (Fig. 3). Both units were equally dominated by core and debitage class specimens, which included naviform cores (Fig. 6: 1-2), crest blades (Fig. 6: 4), unmodified blades/flakes, and chips/chunks as major elements. Here again, cortical flint nodules and primary elements were scarce, suggesting that the initial core preparation took place elsewhere. Although the analysis is still underway, it appears that the assemblage has no clear evidence for the off-set bi-directional strategy. It would follow that it bears resemblance to the el-Kowm type of flint assemblage rather than the Palmyra/Douara assemblage (Borrel 2010; Borrel *et al.* 2011, 2013) - a reasonable interpretation in view of the geographical vicinity to the former. On the other hand, retouched tools accounted for only ca. 0.9 % of the grand total and included adzes drills, notches, and endscrapers. In addition, small retouched blades/flakes (Fig. 6: 8-10) and scaled/splintered pieces (Fig. 6: 11) also occurred in a limited number, but they might possibly represent later contaminations. No points/arrowheads were included.

Both the scarcity of standardized tools and the absence of structural remains suggest that the PPNB flint workshop was even more ephemeral in nature than that of the Khiamian. The question is what phase of the PPNB the assemblage belongs to, but the absence of C-14 data

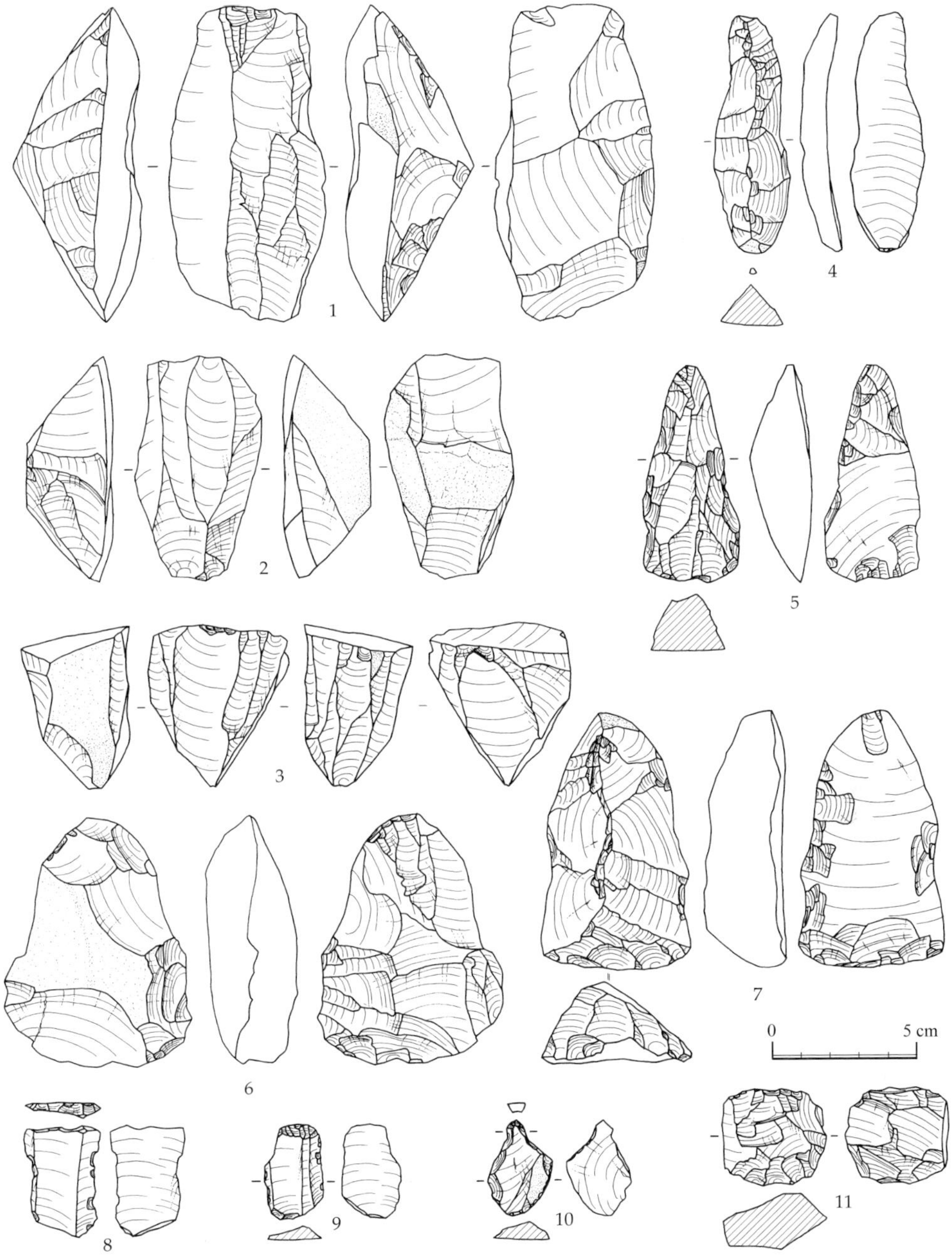

Fig. 6: Wadi al-Hajana 1: surface collection.

and diagnostic finds such as points makes the precise dating difficult. However, in view of the general trend in the neighboring drylands (Borrel *et al*. 2011; Cauvin 1981, 1990, 1994: 247-259; Stordeur 1993, 2000; Nishiaki 2000), it is highly likely that the assemblage represents

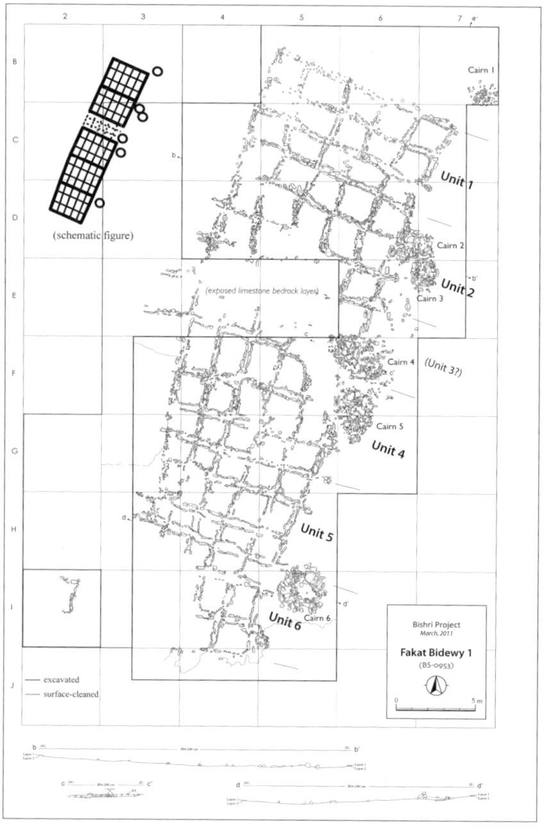

Fig. 7: Fakat Bidewy 1: plan and elevations of the structural complex.

ad hoc flint exploitation activities in the course of seasonal pasturing by the Final PPNB pastoral nomads. Given this, it would follow that the initial pastoral nomadism began to penetrate into the Bishri Hills in the first half of the 6th millennium uncal. BC. Incidentally, a similar assemblage has been found at a nearby site (Wadi al-Hajana 2) as well. The existence of the two adjacent flint scatter sites suggests that such temporary land use was the norm of the Bishri Final PPNB.

Fakat Bidewy 1

Being located ca. 2 km west of Wadi al-Hajana 1, the site of Fakat Bidewy 1 occupies a narrow saddle ca. 570 m in elevation that crosses the ridge of the elongated foothill (Fig. 1). It was found for the first time during our general survey in 2009 and excavated in the spring field season of 2011 (Fujii, Adachi and Yamafuji 2013). The excavation revealed a unique structural complex as yet unknown in Syria.

Structural complex

The excavated structural complex, measuring ca. 35 m wide, ca. 9-12 m deep, and ca. 10-30 cm in preserved height, was extended in the NE-SW direction along the major axis of the saddle (Figs. 7, 8). It was a ground type structure, being constructed by means of a two-rowed upright slab wall technique unique to the Neolithic Badia (Bar-Yosef 1982; Fujii 2001; Fujii, Adachi, Endo and Yamafuji in press; Fujii, Yamafuji and Nagaya forthcoming; Garrard *et al.* 1994; Goring-Morris 1993). The stony ground surface of those days (i.e. the upper surface of Layer 2) was used as a natural floor without any special treatment. No clear evidence for a superstructure was confirmed.

Though poorly preserved at the southwestern corner, it appears that the complex consisted originally of a total of 105 (= 5 by 17 rows) square to rectangular cells. Of interest is the fact that a slightly wider row intervenes at regular intervals. This phenomenon enables us to divide the elongated complex into six homogeneous units with a wider row being the center (the only exception to this is Unit 3, which consists only of two rows of cells. This is probably because the existence of an exposed limestone bedrock layer made it difficult to follow the standard layout). Close examinations of junctions between any two adjacent units suggested that the complex developed from the NE toward the SW. It would follow that the complex started with the largest Unit 1 and, through the irregular Unit 3, gradually tapered to end eventually with the poorly preserved Unit 6.

Aside from Unit 3, every unit consisted of 15 (= 5 by 3 rows) empty cells without any small features such as hearths. Although no clear evidence for an entrance was found, the southeastern wall can probably be regarded as the front of every unit in view of its higher construction quality. Most of the units were associated with a small cairn along their front wall, but the combination way of the two components was not always regular. Unit 1, for example, was associated with an isolated cairn around its northeastern corner, whereas Unit 2 incorporated two cairns into the middle part of its front wall. We excavated two of the six cairns, both of which proved to be a simple stone concentration without any clearly-defined internal structure. Neither human skeletal remains nor burial gifts were included, but this is not uncommon to cairn tombs of prehistoric nomads in the Levantine arid peripheries (Haiman 1992).

Fig. 8: Fakat Bidewy 1: general view of the structural complex (above, looking NW) and a close-up view of the northern half (below, looking WNW).

The function of the unique complex admits of various interpretations. It is most unlikely, however, that the complex was constructed for a residential purpose, because there are no clear traces of actual life such as heaths and *in situ* finds. The isolated location at the saddle poor in water supply also casts doubt on the residence hypothesis. In view of the absence of no contemporary settlements around it, the use as an extramural warehouse also seems doubtful. Thus some impractical use is more likely. Suggestive in this regard is the series of unique traits that characterize the complex: the combination of a two-rowed upright slab wall structure without any trace of real life and a small cairn again empty in contents, the lateral connection of such homogeneous units, and its consequent formation of an elongated empty complex. As discussed below, all of these remind us of Late Neolithic open sanctuaries in southern Jordan.

Small finds

The site yielded some 1,500 flint artifacts. The other finds were limited to a dozen wheel-made pottery sherds including a few glazed ware sherds, several iron nails, two animal bone fragments, and a glass sherd. These sporadic finds belong largely to the surface collection and, therefore, can be regarded as stray finds that have nothing to do with the complex. This is

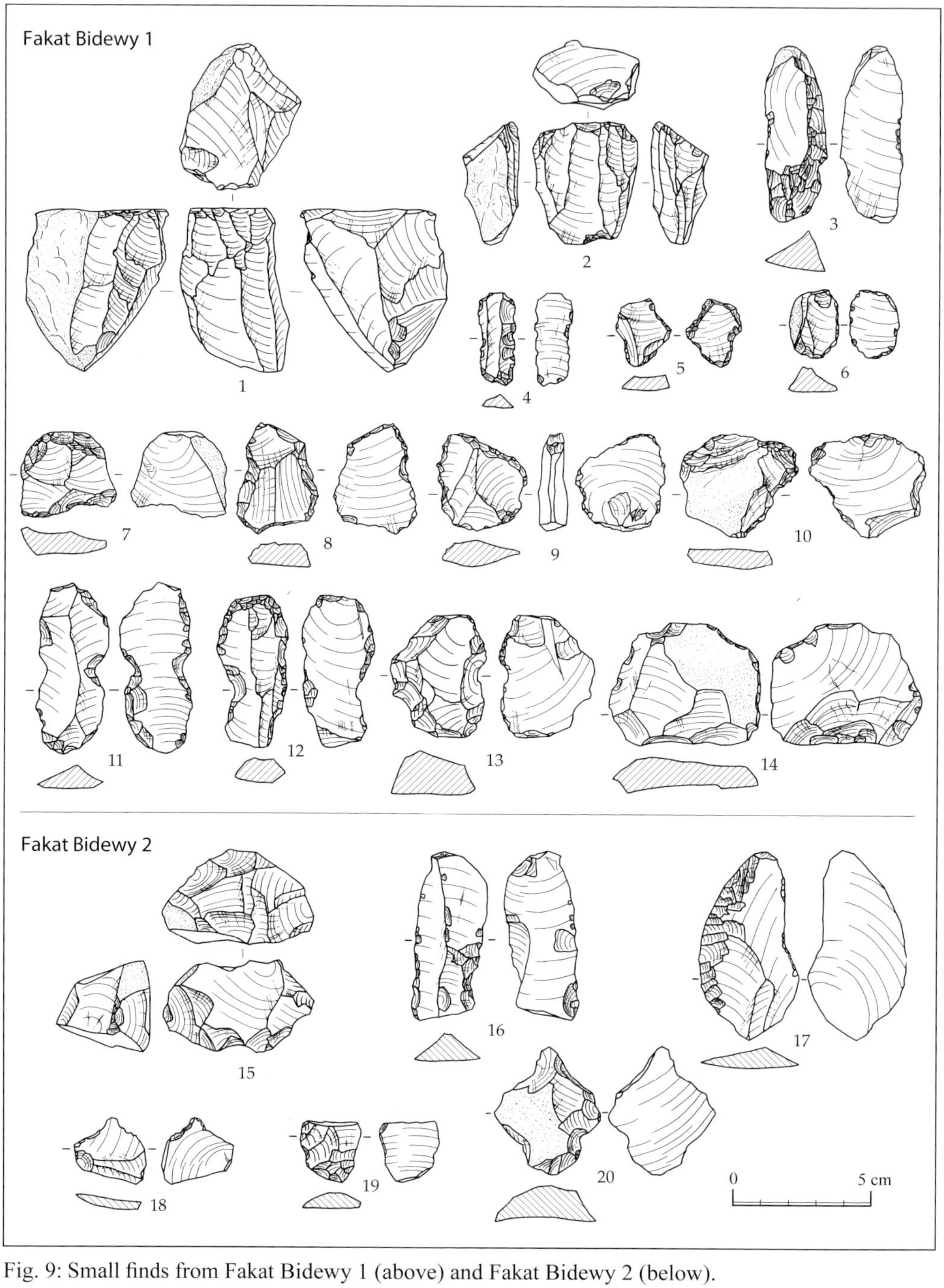

Fig. 9: Small finds from Fakat Bidewy 1 (above) and Fakat Bidewy 2 (below).

not to say, however, that the flint artifacts are directly connected to the complex. As discussed below, the contextual correlation between the two are also somewhat debatable.

The flint artifacts were produced with raw materials similar to those used for the Khiamian and PPNB assemblages at Wadi al-Hajana 1. The frequency of core and debitage class specimens (ca. 77 % of the grand total) is suggestive of on-site production, but, here again, cortical nodules and primary elements both related to the initial core preparation are scarce. Unlike the Wadi al-Hajana 1 assemblage, tool class specimens account for nearly 20 % of the ground total, possibly suggesting that the assemblage derived from everyday life contexts rather than a task-specific expedition.

Naviform cores were replaced by a small number of single-platform blade cores (Fig. 9: 1-2). However, the assemblage was rather flake-oriented, and tool class specimens included amorphous retouched flakes (Fig. 9: 5-10), retouched blades (Fig. 9: 4), end- or sidescrapers, notches/denticulates (Fig. 9: 11-12), and scaled/splintered pieces (Fig. 9: 13-14). In view of the absence of PPNB Naviform core-and blade components, on one hand, and EBA Shabutian core-and-blade components (Nishiaki 2010a), on the other hands, the assemblage can be dated to an intermediate stage between the two, probably the Late Neolithic. However, it does not always mirror the date of the structural complex, because many of the flint artifacts occurred in the surface and fill layers, not the floor deposits. Meanwhile, the distribution range of the two cultural components roughly overlaps with each other. Thus the truth may be that the flint assemblage scattered on the then ground surface was disturbed in the course of the construction of the complex. The question is the time lag between the two, but it seems negligible first because no clear stratigraphic gap intervened between the two, and second because as noted above, the structural complex itself is also suggestive of a Late Neolithic date. It is our tentative perspective that the flint assemblage and the complex do not always constitute a unified body but share a similar date.

Fakat Bidewy 2

Fakat Bidewy 2 is located on the ridgeline of a small limestone hill ca. 1 km NW of Fakat Bidewy 1. It contains several cairns and forms a small-scale cairn field commanding the surrounding intermountain plains. The site was found for the first time in the spring field season of 2010, and briefly examined in the spring of the next year together with the nearby site of Fakat Bidewy 1 (Fujii, Adachi and Yamafuji 2013).

Cairn 01

Due to time constraints, our operation was focused only on Cairn 01 relatively well-preserved. The trench-excavation revealed that the feature consisted of the following two components: a small cairn ca. 3.5-4.8 m in diameter and ca. 0.1-0.25 m in preserved height, and a two-rowed upright slab wall ca. 5.8 m long and ca. 0.6 m in maximum height (Fig. 10). Both components were constructed with undressed or partly dressed limestone slabs ca. 30-60 cm long, which were probably pulled up from an exposed limestone bedrock layer underlying the feature. Traces of such operations were left as remarkable irregularities on the base of the feature.

The cairn was essentially empty, and small finds scattered in and around it were limited to several dozen flint artifacts and a glazed pottery sherd only. The flint assemblage was even more *ad hoc* in nature than that of Fakat Bidewy 1, being marked by the predominance of change-of-orientation flake cores (Fig. 9: 15) and very small retouched flakes (Fig. 9: 18-20). Thus it seems that the flint assemblage is slightly later in date than that of Fakat Bidewy 1. Here again, tool class specimens account for ca. 20 % of the grand total.

Fig. 10: Fakat Bidewy 2: plan and section of Cairn 01 (left) and its general view (right; looking WNW).

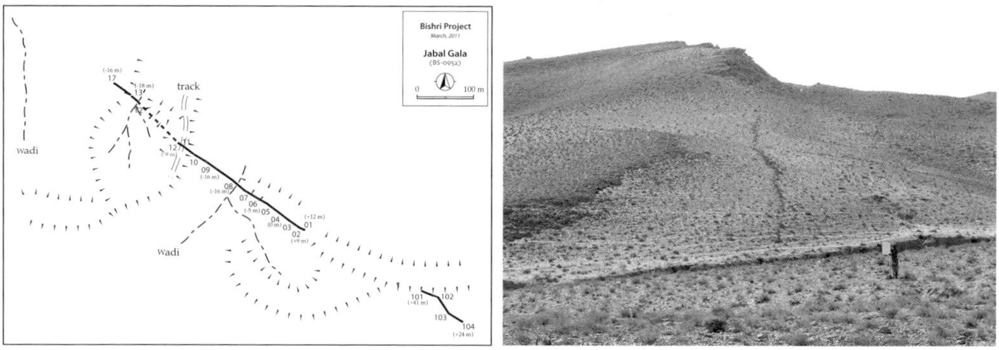

Fig. 11: Jabal Gala: site plan (left) and partial view (right; looking SE).

The small hill included another cairn field (i.e. Fakat Bidewy 3), where a dozen similar features were recorded. It follows that the hill contains at least twenty cairns associated with a freestanding wall. As discussed below, they probably fall into the pseudo-wall cairn characteristic of Chalcolithic open sanctuaries in the Jafr Basin.

Jabal Gala

The site of Jabal Gala stretches along the ridgeline of a small foothill ca. 0.5 km NW of Fakat Bidewy 2. Unlike the two nearby sites, it consists only of an elongated stone-built wall (Fig. 11). We briefly examined it again in the spring field season of 2011.

The surface observation and partial cleaning confirmed that: 1) the freestanding wall is constructed with undressed limestone cobbles up to ca. 50 cm long, measuring ca. 800 m in total length (including two interrupted parts), ca. 0.5-1 m in width, and ca. 0.2-0.5 m in preserved height; 2) as with the Fakat Bidewy sites, it adopts the two rowed upright slab wall technique with rubble core; and 3) it stretches roughly in a linear fashion regardless of the undulating topography and is associated with a small stone concentration roughly at regular interval. As discussed below, these traits are reminiscent of the K-line in the Final Chalcolithic or Early Bronze Age southern Levant. The surface collection was limited to a few simple ware sherds and several undiagnostic flint flakes.

Middle Bronze Age Cairn Fields

As mentioned at the beginning, the Bronze Age cairn was the primary target at the initial stage of our research project. The series of general surveys, conducted intermittently from the spring of 2007 until the summer of 2009, located a total of 35 cairn fields or some 400 individual cairns. The investigations were focused on 8 cairn fields of them, where we excavated a total of 53 cairns. The research outcome has been summarized elsewhere (Fujii and Adachi 2010), and no repetition is needed. We will only note that the cairn fields can be dated to the first half of the Middle Bronze Age (hereafter MBA) on the basis of several C-14 dates and the comparative studies of small finds, and that the Bishri MBA cairn started with a fully developed form associated with a large cist and a single or double ring wall and ended with a simple cobble mound without any internal feature (Figs. 12-13).

Cross-check with the Jafr Chronology

The series of investigations has proved that the northwestern piedmont zone of Mt. Bishri possesses a variety of archaeological footprints of prehistoric and protohistoric mobile groups. The problem is where to place them in the general scheme of the Levantine Badia archaeology, but the scarcity of comparative materials makes it difficult to approach the issue. The only exception is the Jafr Basin in southern Jordan, where similar features have been confirmed and synthesized into a local chronology (Fujii 2013). The following is a chronological review of the Bishri sites viewed from the Jafr chronology (Fig. 15).

PPNA

PPNA sites have yet to be located in the Jafr Basin. This makes sense, however, when we consider that PPNA settlements have a tendency to converge on water-rich lowlands such as the Jordan valley and the Mediterranean coast (Belfer-Cohen and Bar-Yosef 2000; Kuijt and Goring-Morris 2002). It is our present perspective that the Jafr chronology still includes more than one millennium hiatus between the Late Natufian settlement of Wadi Qusayr 139 (Fujii 2005a: 42-44) and the Middle to Late PPNB (hereafter M-LPPNB) outpost of Wadi Abu Tulayha.

This is not always the case of the Bishri, because the small flint workshop at Wadi al-Hajana 1 has exemplified that the early Holocene land use history in the area dates back to the Khiamian, probably its final phase (Figs. 2-5). Available evidence suggests that the Khiamian sporadic land use was focused on flint exploitation, and that the flint knappers derived from the Middle Euphrates River Basin. No doubt the Khiamian mobile group did not bring along livestock, but we can argue that such a pioneering task-specific expedition into the arid margin probably paved the way for the subsequent pastoral penetration.

PPNB

Although no clear evidence for the Early PPNB has been identified in the Jafr Basin, the M-LPPNB agro-pastoral outposts have been located at Wadi Abu Tulayha (Fujii 2006a, 2007a, 2008, 2009) and Wadi Ghuwayr 17 (Fujii, Quintero and Wilke 2011; Fujii, Adachi, Endo *et al.* forthcoming a). This period witnessed the beginning of short-range pastoral transhumance. The former site produced a small percentage of domestic sheep's and goats' bones (Hongo

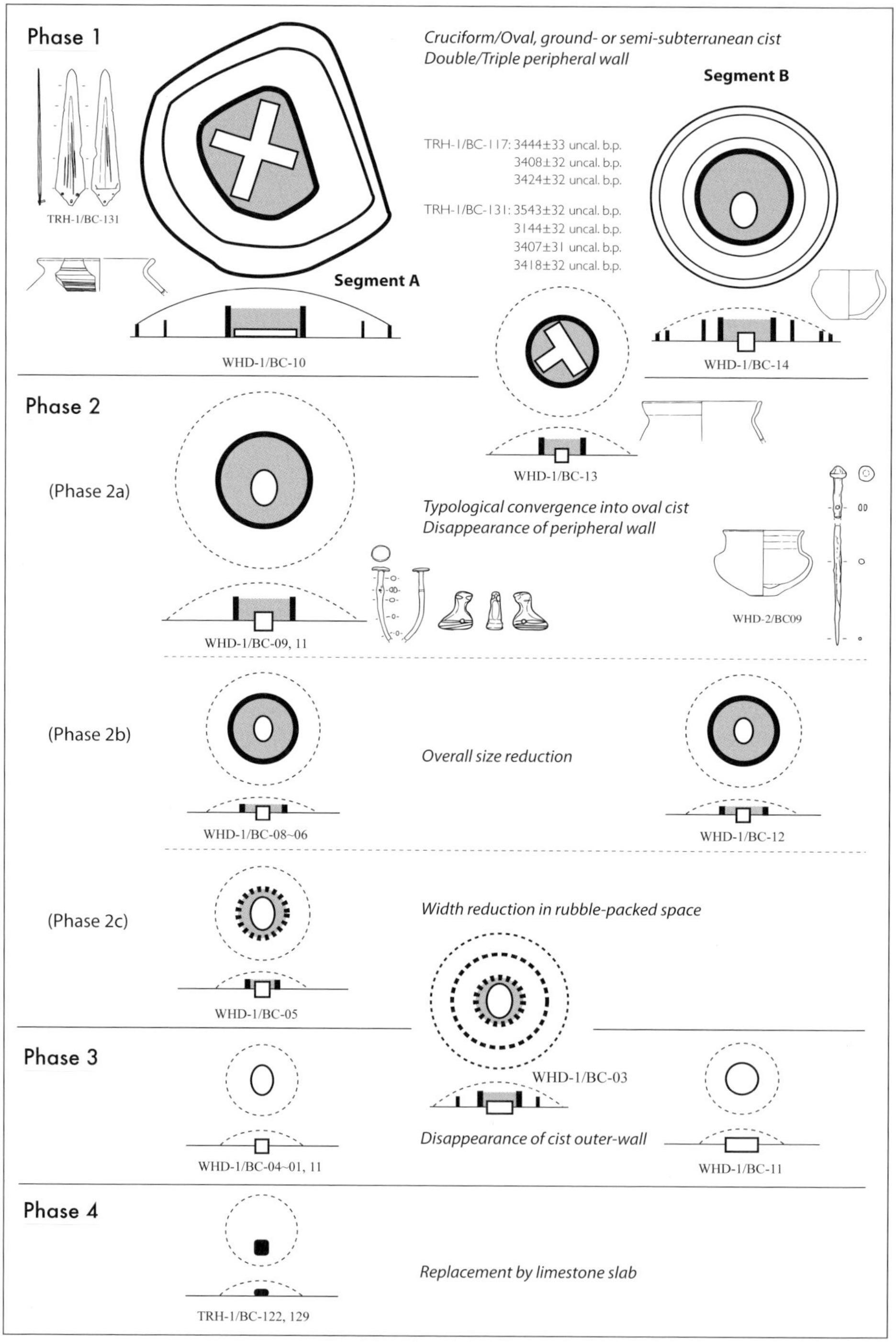

Fig. 12: Tentative chronology of the Middle Bronze Age cairns in the Bishri Hills (partly revised from Fujii and Adachi 2010: fig. 8).

Fig. 13: Wadi Hedaja 1: general view of Cairn 10 (looking NE).

et al. 2013), whereas the latter yielded a few petroglyphs depicting a probable scene of seasonal pasturing around the outpost (Fujii, Quintero and Wilke 2011, fig. 30: 2, 4). Available evidence suggests that the seasonal outposts was sustained by a mixed economy including the pastoral transhumance probably derived from farming communities to the west, hunting of wildlife, and basin-irrigation agriculture within the flooded area of a stone-built barrage (Fujii 2007b, 2010, 2013; Fujii, Adachi, Quintero *et al.* 2011; Fujii, Adachi, Endo *et al.* forthcoming b). Burial practice at the outposts is marked by the *façade-side burial cairn*, a small cairn constructed along a front wall of a disused structure (Fujii 2006a: 14-15; 2006b: 7-8) (Fig. 14: 1). This unique burial custom was inherited, though loosing its substance, down to the two post-PPNB open sanctuaries referred to below.

No contemporary sites have yet been found in the Bishri. (The Final PPNB Bishri is thought to have witnessed the sporadic flint exploitation, but it is roughly equivalent to the PPNC in the Jafr chronology and will be referred to below.) Given the absence of PPNB sites in a more limited sense, it would follow that the pastoral nomadization in the Bishri was delayed by nearly one millennium than the Jafr. However, this is a tentative perspective at the present stage of our knowledge, and further scrutiny is needed to ascertain the truth.

Later Half of the Neolithic

The PPNC and Late Neolithic Jafr Basin saw the sudden disappearance of the seasonal agro-pastoral outpost and the basin-irrigation barrage system. What appeared instead were large-scale open sanctuaries isolated in the desert. This drastic replacement is suggestive of a shift

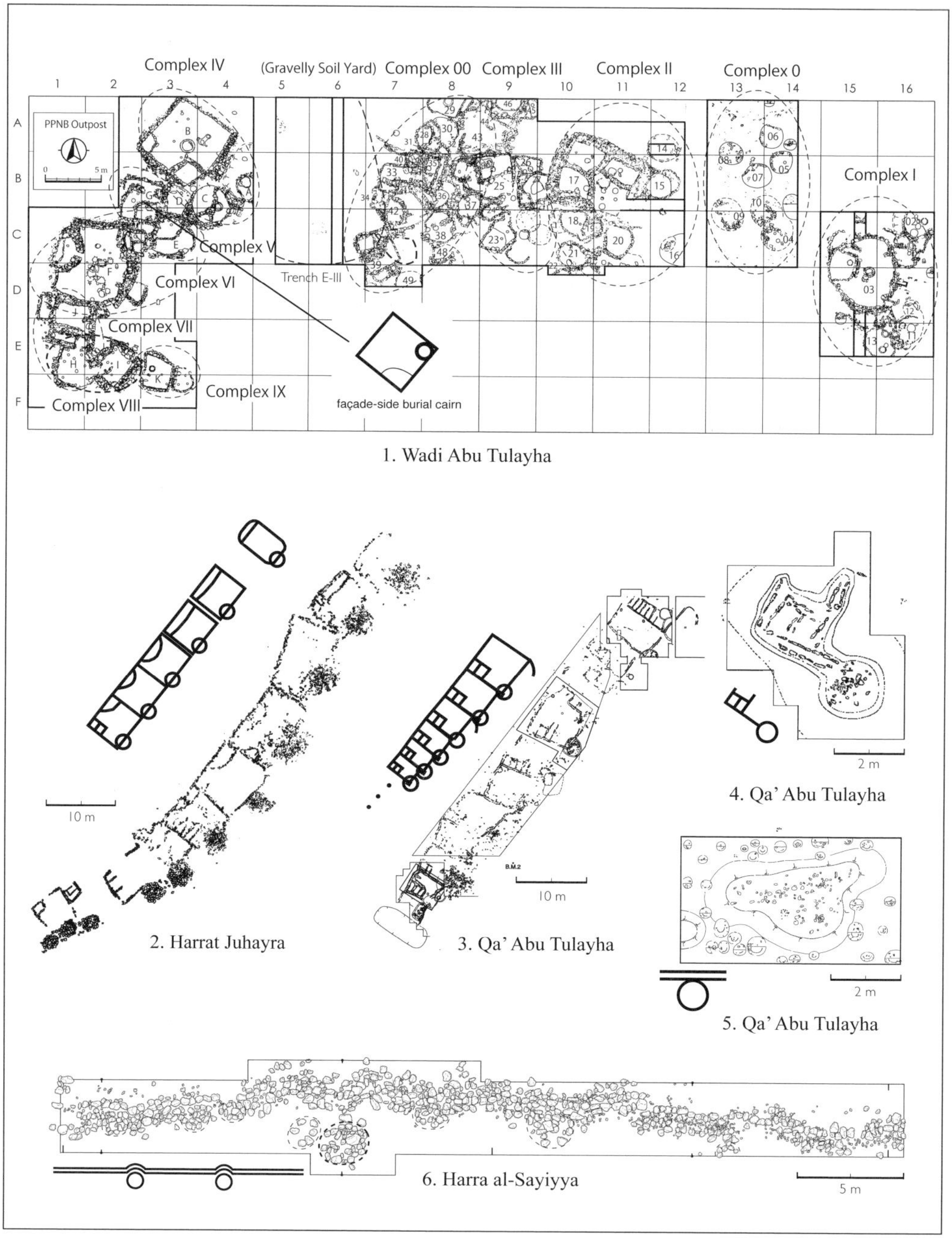

Fig. 14: Various burial features at the Jafr Sites.

in lifestyle from the pastoral transhumance (based on the fixed outposts and the well organized water-use facilities) to the pastoral nomadism (not associated with any large-scale social infrastructure other than the open sanctuaries). The post-PPNB open sanctuaries consist of empty houses associated with a burial cairn. This unique burial feature, tentatively called the *pseudo-house burial cairn*, still occupies the right corner of a front wall of a rectangular structure and, in this sense, can be regarded as a subsequent form of the PPNB façade-side burial cairn. The pseudo-house is literally devoid of any trace of actual life, serving as a simple receptacle for an abutting burial cairn.

The pseudo-houses are laterally connected to form an elongated *pseudo-settlement* again lacking in any trace of real life. The pseudo-settlement can be divided into the PPNC type confirmed at Harra al-Juhayra (Fujii 2005b) (Fig. 14: 2) and 'Awja 2, 5 (Fujii, Yamafuji and Nagaya forthcoming; Fujii, Adachi, Endo and Yamafuji in press), and the Late Neolithic type attested to at Qa' Abu Tulayha West (Fujii 2001, 2002b) (Fig. 14: 3) and 'Awja 1, 4 (Fujii, Yamafuji and Nagaya forthcoming; Fujii, Adachi, Endo and Yamafuji in press). What differentiate the two are overall composition and minor typology. While the former starts with a freestanding pseudo-house and gradually shifted into a connected form, the latter exhibits a fully developed from the beginning. While the former is equipped with an elongated cell along its rear wall, the latter incorporates a few rectangular cells into its rear left corner. However, both of them share the two-rowed upright slab wall technique, a traditional construction method in the Neolithic Badia. Available evidence indicates that the pastoral nomadization in the Jafr Basin is traceable through the process of the replacement of the outpost by the pseudo-settlement, at its initial stage, and the typological change of the pseudo-settlement itself, in the subsequent phase, respectively.

To date, no parallel example of the PPNC pseudo-settlement has been found in the Bishri. Possible evidence for the PPNC (i.e. the Final PPNB in the northern Levant) is limited to the open-air flint workshops at Wadi al-Hajana 1 and 2 (Fig. 6). The *ad hoc* flint exploitation activities hint at the initial pastoral infiltration into the Bishri, but the absence of C-14 dates and faunal evidence makes the assumption still tentative.

What followed the flint workshops is the unique structural complex found at Fakat Bidewy 1 (Figs. 7-9). The series of remarkable traits - the isolated location, the use of the two-rowed upright slab wall technique, the constant combination of a rectangular pseudo-house and a small empty cairn arranged along the front wall the former, the lateral connection of such homogeneous units and its consequent formation of an elongated complex, and the absence of any trace of actual life – allows us to regard the enigmatic complex as a pseudo-settlement or an open sanctuary of post-PPNB pastoral nomads. The Bishri pseudo-settlement takes on a fully developed form from the beginning and, therefore, probably falls into the Late Neolithic type. The sudden appearance of the unique feature might possibly suggest that the pastoral nomadization in the Bishri was triggered under the influence from the southern Levant.

One thing should be noted, however. While the Jafr Late Neolithic pseudo-settlement is essentially of a single room type and incorporates only a few small cells into its rear left corner, the Bishri contemporary pseudo-settlement adopts the tripartite layout and divides the whole indoor space into a total of fifteen homogeneous cells. This typological difference is significant in considering the genealogy of the initial pastoral nomadism in each area. While the former type seems to have its remote roots in local PPNB settlements such as Beidha (Kirkbride 1966; Byrd 2005), the latter appears to have derived directly from surrounding PPNB settlements such as Bouqras (Akkermans *et al.* 1983) and el-Kowm 2 (Stordeur 2000). In this light, we can argue that the pastoral nomadization at both edges of Badia ash-Sham proceeded roughly synchronically as a whole, yet within local contexts in its details – a reasonable perspective when we consider the existence of the preceding PPNB population in both areas.

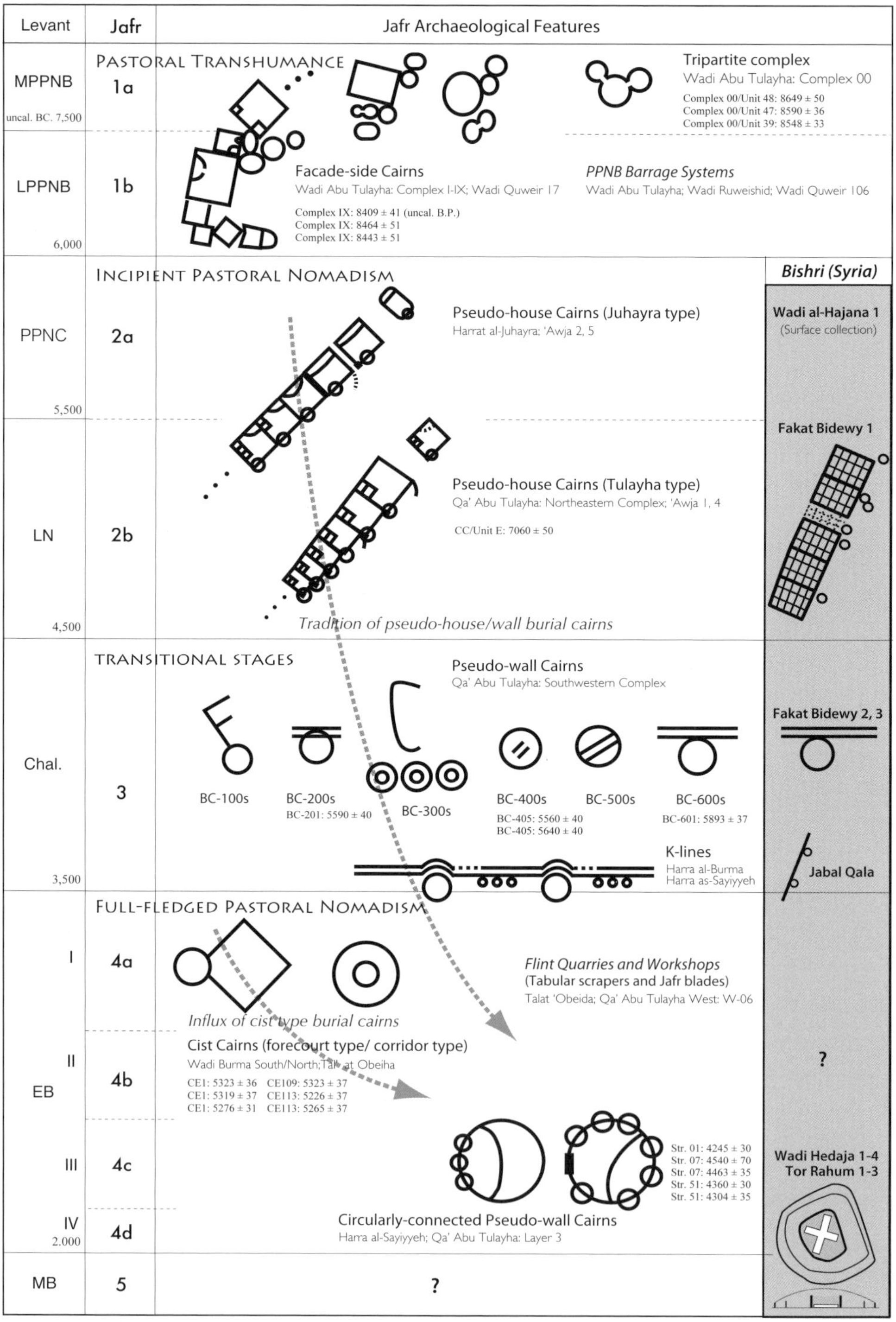

Fig. 15: Comparative chronology of burial features in the Jafr Basin and the Bishri Hills.

Chalcolithic

The Chalcolithic Jafr Basin witnessed the typological simplification from the pseudo-house burial cairn to the *pseudo-wall burial cairn*. This new type of burial feature is marked by the combination of a small cairn and a short freestanding wall, the latter of which can be regarded as a remnant of the front wall of a pseudo-house where a burial cairn was attached. The two-rowed upright slab wall technique was still in use, corroborating the techno-genealogical consistency between the two. The horizontal stratigraphy of Qa' Abu Tulayha West suggests that the pseudo-wall cairn began with the incipient form partly retaining the general profile of a pseudo-house cairn (i.e. BC-100s; Fig. 14: 4) and, through various intermediate types, ended eventually with a simple combination of the two components (i.e. BC-600s; Fig. 14: 5) (Fujii 2002a, 2003). Another line of evidence suggests that the final form of the pseudo-wall cairn was followed by their laterally-connected body, namely, the *K-line* (Evernari *et al.* 1958; Glueck 1958, 1959; Haiman 2000) in Cisjordan or *Hatt Shabib* (Abujaber 1992; Fujii 2004a: 40-46, 2004b: 286-295; Kirkbride 1948; MacDonald *et al.* 2000) in Transjordan (Fig. 14: 6). We have located several dozen pseudo-wall burial cairns and K-lines. The remarkable increase in the number of contemporary burial features probably indicates the substantial population growth of pastoral nomads in the Chalcolithic Jafr Basin.

There is little doubt that the unique feature excavated at Fakat Bidewy 2 falls into the BC-600s type of pseudo-wall burial cairn of the Jafr chronology (Fig. 10). Thus it can probably be dated to the Late Chalcolithic. Likewise, the elongated freestanding wall found at Jabal Gala has much in common with the K-line of the Jafr chronology and, therefore, can tentatively be dated to the end of the Chalcolihtic or the beginning of the EBA (Fig. 11). The frequency of these burial features suggests that as with the contemporary Jafr, the Chalcolithic Bishri saw the remarkable development of pastoral nomadism.

Early Bronze Age

The Early Bronze Age of the Jafr Basin is marked by the appearance of the cist-type burial cairn and the establishment of large-scale cairn fields such as Wadi Burma (Fujii 2004b: 27-38, 2005a: 17-30) and Tal'at 'Ubayda (Fujii 2005a: 30-41). Both of these episodes are suggestive of the advent of a full-fledged pastoral society backed with a tribal system. The cist type cairn (and the large-scale cairn field as its aggregation) can be regarded as an exotic cultural component, because it no longer incorporates a pseudo-wall and, instead, takes the form of a standard cairn tomb that consists of a clearly defined cist and a cobble mound covering it. Its proto-type can be found at Late Neolithic to Chalcolithic open sanctuaries in the Negev and the Jordan Valley, suggesting that the new wave originated in the west or the southwest (Belfer-Cohen and Bar -Yosef 2000; Kuijt and Goring-Morris 2002). As a matter of fact, the EBA cairn fields in the Jafr Basin is focused on its western edge and rarely found in its central and eastern parts. It is our present perspective that the cist cairn changed from the forecourt type common in Wadi Burma to the corridor type characteristic of Tal'at 'Ubayda. Meanwhile, the Jafr tradition still survived in the core of the basin, taking the form of a circularly connected pseudo-wall cairn (Fujii 1998, 1999, 2000). Incidentally, the EBA Jafr Basin is known for the large-scale flint exploitation as well (Quintero *et al.* 2002), which is also suggestive of the increase in contact with urbano-farming societies to the west (Fujii 2011).

In contrast, the Bishri EBA falls on a chronological hiatus. Although a few sporadic candidates have been reported from the surrounding areas (Falb *et al.* 2005; Lönnqvist *et al.* 2011; Nishaki

2010b; Numoto and Kume 2010), no clear evidence for EBA burial features, to say nothing of contemporary settlements, have been located in our research field. This is strange, all the more because the pastoral nomadization at both edges of Badia ash-Sham appears to have gone hand-in-hand until the end of the Chalcolithic. A possible interpretation is that the EBA Bishri became empty, but it should be noted that our research field covers only a small part of the Bishri Hills. Further scrutiny is needed to ascertain the authenticity of the inexplicable phenomenon.

Middle Bronze Age

The situation was entirely reversed in the subsequent period, and archaeological footprints of early pastoral nomads suddenly disappeared from the Jafr Basin. This possibly means that the Jafr was entirely deserted in the second millennium B.C., although we should keep in mind the possibility that the gradual downsizing of burial features causes the decline in their archaeological visibility.

The MBA Bishri, on the other hand, witnessed the sudden appearance of large-scale cairn fields (Fig. 12). The Bishri MBA cairn fields probably represent communal cemeteries of pastoral nomads who migrated around the Mari Kingdom (Fujii and Adachi 2010). Similar cairn fields have been found in the southeastern piedmont of Mt. Bishri (Lönnqvist 2010; Lönnqvist *et al.* 2011), at Bahouz in the lower Euphrates River Basin (Du Buisson 1948; Kepinski 2010), and in the northern fringe of the Palmyra Basin (Anfinset 2009; Anfinset and Meyer 2010; Morandi Bonaccossi and Iamoni 2012; al-Maqdissi *et al.* 2008; Ruset 2011), suggesting that the MBA inland Syria witnessed the first zenith of pastoral nomadism.

The question is the origin of the Bishri MBA cairn culture. Suggestive in this regard is the fact that the Bishri cairn fields appeared in a fully developed form from the beginning despite the possible chronological hiatus in the preceding EBA. This probably means that the culture was exotic in origin. Seeing that the corridor type cist cairn marked its initial phase, it is not impossible that the culture was established first in the southern Levant and then gradually spread, or possibly moved, northward. The chronological blank in the MBA Jafr and the EBA Bishri might possibly be understood within the framework of such a traditional migration hypothesis (De Vaux 1971; Kenyon 1966, 1979; Wright 1971). However, it is unwise to connect the two separate areas directly when little is known about the status in the intermediate areas. This is even more so, because the issue requires us to pay close attention to the northeastern quarter of the Arabian Peninsula as well (Nayeem 1990; Potts 1990; Steimer-Herbet 2004; Weeks 2010). We will only note that the large-scale cairn field appeared in the Bishri more than one millennium later than the Jafr.

Concluding remarks

The above discussion has proved that the general sequence of the Bishri pastoral preshitory and protohistory can be traced within the framework of the Jafr chronology. Of significance is the fact that the Late Neolithic pseudo-settlement and the Chalcolithic pseudo-wall burial cairn were shared between the two separate areas. This unexpected fact indicates that the pastoral nomadization proceeded roughly synchronically at the northern and southern edges of Badia ash-Sham, although available evidence suggests that the indigenous population played a proactive role in the episode. No less important is the remarkable contrasts between the two areas in the Bronze Age. This issue is closely related to the time lag of the establishment of a full-fledged pastoral society at both ends of Badia ash-Sham and requires further verification. Anyhow, these new perspectives have opened the way for incorporating the Bishri sites into the general scheme of the Levantine Badia archaeology as its northernmost aspects. We are eagerly looking forward to the day when we may resume the investigation in Syria.

Acknowledgements

This paper is a revised version of our oral presentation at the workshop 'Settlement Dynamics and Human Landscape Interaction in the Steppes and Deserts of Syria' that constituted a part of the ICAANE-8 held in Warsaw in 2012. We wish to express our deep appreciation to Dr. Daniele Morandi Bonacossi for his invitation to the exciting meeting and this challenging publication. Our thanks also go to all the participants and anonymous referees, whose comments and questions were very useful for making our points clearer.

Bibliography

Abujaber, R. S.
1992 Pioneer settlement over the Jordan, *SHAJ* 4: 391-395.

Akashi, C. and K. Tanno
2009 Botanical survey of Bishri hills: related with the excavation of the Rujum Hedaja 1. In M. al-Maqdissi and K. Ohnuma (eds.), Preliminary reports of the Syria-Japan archaeological joint research in the region of ar-Raqqa, Syria, 2008, *Al-Rafidan* 30: 188-190.

Akkermans, P. A., J. A. K. Boerma, A. T. Clason, S. G. Hill, E. Lohof, C. Meiklejohn, M. Le Mière, G. M. F. Molgat, J. J. Roodenberg, W. Waterbolk-van Rooyen and W. van Zeist
1983 Bouqras revisited: Preliminary report on a project in eastern Syria, *PPS* 49: 335-372.

Akkermans, P. M. M. G. and G. M. Schwartz
2003 *The Archaeology of Syria: From Complex Hunter-Gatherers to Early Urban Societies (ca. 16,000-300 BC)*, Cambridge.

Al-Maqdissi, M., F. Braemer and J.-M. Dentzer
2011 *Hauran V: La Syrie du Sud du Néolithique à l'Antiquité Tardive, Recherches Récentes, Actes du Colloque de Damas 2007*, Beyrouth.

Al- Maqdissi, M., Y. Kanhouch, M. Cremaschi and D. Morandi Bonacossi
2008 Présentation sommaire des travaux archaéologiques de la mission Syro-Italienne: Les fouilles archéologique à Mishrifeh - Qatna et la prospection de la Palmyrène occidentale, *Studia Orontica* 1: 5-20.

Anfinset, N.
2008 Towards a specialized food production - A look at Jebel Bishri. In M. Lönnqvist (ed.), *Jebel Bishri in Context: Introduction to the Archaeological Studies and the Neighbourhood of Jebel Bishri in Central Syria*, Oxford: 1-13.
2009 Palmyrena - Palmyra and the Surrounding Territory, Joint Syrian-Norwegian Project, Surface Survey North of Palmyra, April and May 2009, Preliminary report: Prehistoric Periods, Bergen. (http://www.org.uib.no/palmyrena/documents/2009_II.pdf.; last access: Dec. 10, 2012)

Anfinset, N., J. C. Meyer
2010 The hinterland of Palmyra, *Antiquity* 084/324, Project gallery.

Bar-Yosef, O.
1982 Pre-Pottery Neolithic sites in southern Sinai, *BA* 45/1: 9-12.
Belfer-Cohen, A. and Bar-Yosef, O.
2000 Early sedentism in the Near East: A bumpy ride to Village life. In I. Kuijt (ed.), *Life in Neolithic Farming Communities: Social Organization, Identity, and Differentiation,* New York: 19-37.
Borrel, F.
2010 Bi-directional Neolithic blade technology in the northern Levant during the 7th–8th millennia cal. BC: New insights from Mamarrul Nasr 2, Syria, *JFA* 36/2: 132-150.
Borrel, F., E. Boëda, M. Molist, H. Al-Shakel and O. Vicente
2011 New evidence and revised interpretations of Neolithization process of central Syrian Desert: The Neolithic complex of Mamarrul Nasr, *Paléorient* 37/2: 35-46.
Borrel, F., E. Boëda, M. Molist and H. Al-Shakel
2013 The first half of the 7th millennium cal. B.C. in the el Kowm Basin, central Syria: Umm el-Tlel revisited. In O. P. Nieuwenhuyse, A. Russell, R. Bernbeck and P. M. M. G. Akkermans (eds.), *Interpreting the Late Neolithic of Upper Mesopotamia,* Turnhout: 277-287.
Breamer, F., J-C. Échallier and A. Taraqji (eds.)
2004 *Khirbet al Umbashi: Villages et Campements de Pasteurs dans le «Désert Noir» (Syrie) à l'Âge du Bronze*, Bibliothèque Archéologique et Historique tom 171, Beirut.
Buccellati, G.
1966 *The Amorites of the Ur III period*, Napoli.
1992 Ebla and the Amorites. In C. H. Gordon and G. A. Rendsburg (eds.), *Eblatica: Essays on the Ebla Archives and Eblaite Language*, vol. 3, Winona Lake: 83-104.
1997 Amorites. In E. M. Meyers (ed.), *The Oxford Encyclopedia of Archaeology in the Near East* vol. I, New York: 107-111.
Buccellati, G. and M. Kelly-Buccellati
1967 Archaeological survey of the Palmyrene and the Jebel Bishri, *Archaeology* 20: 305-306
Byrd, B. F.
2005 *Early Village Life at Beidha, Jordan: Neolithic Spatial Organization and Vernacular Architecture,* Oxford.
Cauvin, J.
1977 Les fouilles de Mureybet (1971-1974) et leur signification pour les origins de la sédentarisation au Proche-Orient, *AASOR* 44: 19-48.
1981 L'occupation Néolithique de la region d'El-Kowm: résultats 1978-79. In J. Cauvin and P. Sanlaville (eds.), *Préhistoire du Levant*, Pairs: 471-483.
1990 Nomadisme néolithique en zone aride: l'oasis d'el Kowm (Syrie). In P. Matthiae, M. van Loon and H. Weiss (eds.), *Resurrecting the Past: A Joint Tribute to Adnan Bounni*, Istanbul: 41-47.
1994 *Naissance des Divinités, Naissance de l'Agriculture*, Paris.
Cauvin, J. and D. Stordueur
1977 Radiocarbon dating el-Kowm: Upper Palaeolithic through Chalcolithic. In O. Bar-Yosef and R. S. Kra (eds.), *Late Quaternary Chronology and Paleoclimates of the Eastern Mediterranean,* Cambridge: 201-204.

Charpin, D.
2004 Histoire politique du Proche-Orient Amorrite (2002-1595). In D. Charpin, D. O. Edzard and M. Stol (eds.), *Mesopotamien: Die Altbabylonische Zeit,* Fribourg: 23-480.
2010 The desert routes around the Djebel Bishri and the Sutean nomads according to the Mari archives. In K. Ohnuma, S. Fujii, A. Tsuneki, Y. Nishiaki, S. Miyashita and H. Sato (eds.), *Al-Rafidan special volume (Formation of Semitic Tribal Communities: Integrated Research in the Middle Euphrates, Syria)*: 239-245.

Charpin, D. and N. Ziegler
2003 Mari et le Proche Orient à époque amorrite: essai d'histoire politique, *Florilegium Marianum* V: 20-27.

De Vaux, R.
1966 Palestine in the Early Bronze Age. In I. E. S. Edwards, C. J. Gadd and N. G. L. Hammond (ed.), *Cambridge Ancient History* vol. I, 2 (3rd edition): 208-237.

Du Buisson, M.
1948 *Baghouz, l'Ancienne Corsôtê: Le Tell Archaïque et la Nécropole de l'Âge du Bronze,* Leiden.

Durand, J.-M.
1987 Documents pour l'histoire du royaume de Haute-Mésopotamie, I/C: les trois routes de l'Euphrate à Qatna à travers le désert, *Mari, Annales de Recherches Interdisciplinaires* 5: 159-167.
2004 Peuplement et société à l'époque amorrite. In: C. Nicolle (ed.), *Amurru 3: Nomades et Sédentaires dans le Proche Orient Ancient,* Paris: 111-197.

Evernari, M., Y. Aharoni, L. Shanan and N. H. Tadmor
1958 The ancient desert agriculture of the Negev: III, early beginning, *IEJ* 8: 231-268.

Falb, C., K. Krasnik, J.-W. Meyer and E. Vial
2005 *Bräber des 3. Jahrtausends v. Chr. im Syrischen Euphrattal, 4: Der Friedhof von Abu Hamad,* Saarbrücken.

Fujii, S.
1998 Qa' Abu Tulayha West: An interim report of the 1997 season, *ADAJ* 42: 123-140.
1999 Qa' Abu Tulayha West: An interim report of the 1998 season, *ADAJ* 43: 69-89.
2000 Qa' Abu Tulayha West: An interim report of the 1999 season, *ADAJ* 44: 149-171.
2001 Qa' Abu Tulayha West, 2000: An interim report of the fourth season, *ADAJ* 45: 15-39.
2002a Qa' Abu Tulayha West, 2001: An interim report of the fifth season, *ADAJ* 46: 19-37.
2002b Pseudo-settlement hypothesis: Evidence from Qa' Abu Tulayha West, southern Jordan, *Archaeozoology of the Near East* 5: 181-194.
2003 Qa' Abu Tulayha West, 2002: An interim report of the sixth and final season, *ADAJ* 47: 195-223.
2004a Harra al-Burma K-lines and Wadi Burma kite site: A preliminary report of the first operation of the al-Jafr Basin Prehistoric Project, Phase 2, *ADAJ* 48: 25-50.
2004b Harra al-Burma cairn line, Wadi Burma cist enclosures, and Harrat al-Sayyiyeh K-line: A preliminary report of the second operation of the al-Jafr Basin Prehistoric Project, Phase 2, *ADAJ* 48: 285-304.
2005a Wadi Burma North, Tal'at 'Ubayda, and Wadi al-Qusair: A preliminary report of the third operation of the al-Jafr Basin Prehistoric Project, 2004, *ADAJ* 49: 17-55.

2005b Harrat al-Juhayra pseudo-settlement: A preliminary report of the al-Jafr Basin Prehistoric Project, 2004, *ADAJ* 49: 57-70.

2006a Wadi Abu Tulayha: A preliminary report of the 2005 spring and summer excavation seasons of the Jafr Basin Prehistoric Project, Phase 2, *ADAJ* 50: 9-31.

2006b A PPNB agro-pastoral outpost at Wadi Abu Tulayha, southern Jordan, *Neo-Lithics* 2/06: 4-17.

2007a Wadi Abu Tulayha: A preliminary report of the 2006 summer field season of the Jafr Basin Prehistoric Project, Phase 2, *ADAJ* 51: 373-402.

2007b Wadi Abu Tulayha and Wadi Ruweishid ash-Sharqi: An investigation of PPNB barrage systems in the Jafr Basin, *Neo-Lithics* 2/07: 6-16.

2008 Wadi Abu Tulayha: A preliminary report of the 2007 summer final field season of the Jafr Basin Prehistoric Project, Phase 2, *ADAJ* 52: 445-478.

2009 Wadi Abu Tulayha: A preliminary report of the 2008 summer final field season of the Jafr Basin Prehistoric Project, Phase 2, *ADAJ* 53: 173-209.

2010 Domestication of runoff water: Current evidence and new perspectives from the Jafr Pastoral Neolithic, *Neo-Lithics* 2/10: 14-32.

2011 'Lost property' at Wadi Qusayr 173: Evidence for the transportation of tabular scrapers in the Jafr Basin, southern Jordan, *Levant* 43/1: 1-14.

2013 Chronology of the Jafr pastoral prehistory and protohistory: A key to the process of pastoral nomadization in the southern Levant. In W. Abu-Azizeh and M. Tarawneh (eds.), *Syria* 90 (Thematic file on the proceedings of the workshop *Current Research on Protohistoric Settlement in Desert Areas of Jordan*): 49-125.

in press A half-buried cistern at Wadi Abu Tulayha: A key to tracing the process of pastoral nomadization in the Jafr Basin, southern Jordan. In B. Finlayson and G. Rollefson (eds.), *Jordan's Prehistory: Past and Future Research*, Amman.

Fujii, S. and T. Adachi

2010 Archaeological investigations of Bronze Age cairn fields on the northwestern flank of Mt. Bishri. In K. Ohnuma *et al.*, *Al-Rafidan special volume (Formation of Semitic Tribal Communities: Integrated Research in the Middle Euphrates, Syria)*, Tokyo: 61-77.

2013 Wadi al-Hajana 1: A Khiamian outpost in the northwestern piedmont of Mt. Bishri, Central Syria. In F. Borrell, J. J. Ibáñez and M. Molist (eds.), *Stone Tools in Transition: From Hunter-Gathers to Farming Societies in the Near East (The 7th Conference on PPN Chipped and Ground Stone Industries of the Fertile Crescent),* Barcelona: 45-57.

Fujii, S., T. Adachi, C. Akashi and K. Suzuki

2011 Wadi al-Hajana 1: A preliminary report of the 2010 excavation season, *Al-Rafidan* 32: 134-145.

Fujii, S., T. Adachi, H. Endo and M.Yamafuji

in press 'Awja sites: Further investigation of Neolithic open sanctuaries in southernmost Jordan, *ADAJ* 57.

Fujii, S., T. Adachi, H. Endo, M.Yamafuji, Y. Arimatsu and K. Nagaya

Forthcoming a Wadi Nadiya 1: Another Neolithic barrage system in the al-Jafr Basin, southern Jordan, *ADAJ* 56.

Forthcoming b Excavations at Wadi Nadiya 2 and complementary investigations of the Jafr Neolithic barrage system, *ADAJ* 57.

Fujii, S., T. Adachi, L. A. Quintero and P. J. Wilke
2011 Wadi Ghuwayr 106: A Neolithic barrage system in the northeastern al-Jafr Basin, *ADAJ* 55: 189-212.
Fujii, S., T. Adachi and M. Yamafuji
2013 Fakat Bidewy 1 and 2: Archaeological investigations around Bir Rahum, 2011 (Spring), *Al-Rafidan* 34: 3-12.
Fujii, S., L. A. Quintero and P. J. Wilke
2011 Wadi Ghuwayr 17: A Neolithic outpost in the northeastern al-Jafr Basin, *ADAJ* 55: 159-188.
Fujii, S., M.Yamafuji and K. Nagaya
forthcoming 'Awja 1-3: Neolithic and Chalcolithic open sanctuaries in southernmost Jordan, *ADAJ* 56.
Garrard, A., D. Baird, S. Colledge, L. Martin and K.Wright
1994 Prehistoric environment and settlement in the Azraq Basin: An interim report on the 1987 and 1988 excavation seasons, *Levant* 26: 73-109.
Geyer, B. (ed.)
2002 *Conquête de la Steppe et Appropriation des Terres sur les Marges Arides du Croissant Fertile,* Paris.
Geyer, B. and J.-Y. Monchambert
2003 *La Basse Vallée de l'Euphrate Syrien: Du Néolithique à l'Avènement de l'Islam: Géographie, Archéologie et Histoire*, vol. 1 & 2, Beyrouth.
Glueck, N.
1958 The seventh season of archaeological exploration in the Negeb, *BASOR* 152: 18-38.
1959 An aerial reconnaissance of the Negeb, *BASOR* 155: 2-13.
Goring-Morris, A. N.
1993 From foraging to herding in the Negev and Sinai: The early to late Neolithic transition, *Paléorient* 19/1: 65-89.
Haiman, M.
1992 Cairn burials and cairn fields in the Negev, *BASOR* 287: 25-45.
2000 The 'K-line' at Har Romem in light of the survey of the map of Har Ramon (203), *'Atiqot* 39: 21-29.
Hongo, H., L. Omar, H. Nasu and S. Fujii
2013 Faunal remains from Wadi Abu Tulayha: A PPNB outpost in the steppe desert of southern Jordan. In B. de Cupere *et al.* (eds.), *Ancient Near Eastern Studies Supplement Series*, Leuven: 1-25.
Issar, A. S. and M. Zohar
2007 *Climate Change: Environment and History of the Near East*, Berlin.
Jaubert, R. and B. Geyer
2006 *Les Marges Arides du Croissant Fertile: Peuplements, Exploitation et Contrôle des Ressources en Syrie du Nord*, Lyon.
Joannès, F.
1997 Palmyre et les routes du désert au début du deuxième millénaire av. J.-C., *Mari, Annales de Recherches Interdisciplinaires* 8: 393-415.
Kenyon, K. M.
1966 *Amorites and Canaanites*, London.
1979 *Archaeology in the Holy Land* (4th edition), London.

Kepinski, C.

2010 The burial mounds of the Middle Euphrates (2100-1800 B.C.) and their links with Arabia: the subtle dialectic between tribal and state practices. In L. Weeks (ed.), *Death and Burial in Arabia and Beyond*, Oxford: 165-171.

Kirkbride, A.

1948 Shebib's wall in Transjordan, *Antiquity* 87: 151-154.

Kirkbride, D.

1966 Five seasons at the pre-pottery Neolithic village of Beidha in Jordan, *PEQ* 98: 8-72.

Kuijt, I. and N. Goring-Morris

2002 Foraging, farming, and social complexity in the Pre-Pottery Neolithic of the southern Levant: A review and synthesis, *JWP* 16/4: 361-440.

Kupper, J.-R.

1957 *Les Nomades en Mésoptamie au Temps des Rois de Mari*, Paris.

Lönnqvist, M.

2006 Archaeological surveys of the Jebel Bishri: The preliminary report of the Finnish Mission to Syria, *Kaskal* 3: 203-240.

2010 Tracing tribal implications among the Bronze Age tomb types in the region of Jebel Bishri in Syria. In K. Ohnuma *et al.*, *Al-Rafidan special volume* (*Formation of Semitic Tribal Communities: Integrated Research in the Middle Euphrates, Syria*), Tokyo: 165-173.

Lönnqvist, M. (ed.)

2008 *Jebel Bishri in Context: Introduction to the Archaeological Studies and the Neighbourhood of Jebel Bishri in Central Syria,* Oxford.

Lönnqvist, M. and K. Lönnqvist (eds.)

2011 *Jebel Bishri in Focus: Remote Sensing, Archaeological Surveying, Mapping and GIS Studies of Jebel Bishri in Central Syria by the Finnish project SYGIS,* Oxford.

Lönnqvis, M., K. Lönnqvist and M. Törmä, with contributions by Whiting, M. S. and M. Nunez

2011 Vistas on ancient pastoral nomadism in the central areas of the mountain. In M. Lönnqvist and K. Lönnqvist (eds.), *Jebel Bishri in Focus: Remote Sensing, Archaeological Surveying, Mapping and GIS Studies of Jebel Bishri in Central Syria by the Finnish project SYGIS,* Oxford: 135-166.

MacDonald, B., L. Herr, M. P. Neeley, S. Quaintance and A. Bradshaw

2000 The Tafila-Busayra archaeological survey: Phase 2 (2000), *ADAJ* 46: 395-411.

Matthews, W.

2003 *Pastoral Nomadism in the Mari Kingdom (ca. 1830-1760)*, Cambridge.

Molist, M.

2008 Foyers et fours du site de Mureybet. In J. J. Ibanez (ed.), *Le Site Néolithique de Tell Mureybet (Syrie du Nord)*, vol. I, Oxford: 95-101.

Molist, M. and M.-C. Cauvin

1990 Une nouvelle séquence stratifiée pour la préhistoire en Syrie semi-désertique, *Paléorient* 16/2: 55-63.

Morandi Bonacossi, D. and M. Iamoni

2012 The early history of the western Palmyra Desert region, The change in the settlement patterns and the adaptation of subsistence strategies to encroaching aridity: A first assessment of the desert-kite and tumulus cultural horizons, *Syria* 89: 31-58.

Nayeem, M.A.
1990 *Prehistory ad Protohistory of the Arabian Peninsula,* vol. 1: Saudi Arabia, Hyderabad.
Nilhamn, B.
2008 Nomadic life in central and eastern Syria: A perspective from the present life on Badiyah to the Amorite nomadism in the Bronze Age. In M. Lönnqvist (ed.), *Jebel Bishri in Context: Introduction to the Archaeological Studies and the Neighbourhood of Jebel Bishri in Central Syria*, Oxford: 15-30.
Nishiaki, Y.
2000 *Lithic Technology of Neolithic Syri*a, Oxford.
2010a Early Bronze Age flint technology and flake scatters in the north Syrian steppe along the Middle Euphrates, *Levant* 42/2: 170-184.
2010b Archaeological evidence of the Early Bronze Age communities in the Middle Euphrates steppe, north Syria. In K. Ohnuma *et al., Al-Rafidan special volume (Formation of Semitic Tribal Communities: Integrated Research in the Middle Euphrates, Syria)*, Tokyo: 37-48.
Numoto, H. and S. Kume
2010 Surveys and sondage at the cemeteries near the site of Tell Ghanem al-'Ali. In K. Ohnuma *et al., Al-Rafidan special volume (Formation of Semitic Tribal Communities: Integrated Research in the Middle Euphrates, Syria)*, Tokyo: 49-60.
Ohnuma, K., S. Fujii, A. Tsuneki, Y. Nishiaki, S. Miyashita and H. Sato (eds.)
2010 *Al-Rafidan special volume (Formation of Semitic Tribal Communities: Integrated Research in the Middle Euphrates, Syria),* Tokyo.
Poidebard, A.
1934 *La Trace de Rome dans le Désert de Syrie: Le Limes de Trajan à la Conquête Arabe, Recherches Aériennes (1925-1932)*, Paris.
Potts, D. T.
1990 *The Arabian Gulf in Antiquity, vol. I: From Prehistory to the Fall of the Achaemenid Empire,* Oxford.
Quintero, L. A., P. J. Wilke and G. O. Rollefson
2002 From flint mine to fan scraper: The late prehistoric Jafr industrial complex, *BASOR* 327: 17-48.
Ruset, H. M. H.
2011 *Pastoral Nomads on the Syrian Steppe in the Early Bronze Age: A Study of Burial Monuments in the Hinterlands of Palmyra,* MA thesis submitted to University of Bergen.
Staubwasser, M. and H. Weiss
2007 Holocene climate and cultural evolution in late prehistoric - early historic West Asia, *Quaternary Research* 66/3: 372-387.
Steimer-Herbet, T.
2004 *Classification des Sépultures à Superstructure Lithique dans le Levant et l'Arabie Occidentale (IVè et IIIè millénaires avant J.-C.)*, Oxford.
Stordeur, D.
1993 Sédentaires et nomads du PPNB final dans le désert de Palmyre (Syrie), *Paléorient* 19: 187-204.
2000 *El Kowm 2: Une Île dans le Désert,* Paris.

Van Loon, M.
1968 The oriental institute excavations at Mureybit, Syria, Preliminary report on the 1965 campaign, part I: architecture and general finds, *JNES* 27: 265-282.
Weeks, L. (ed.)
2010 *Death and Burial in Arabia and Beyond*, Oxford.
Weiss, H.
1986 *The Origins of Cities in Dry Farming Syria and Mesopotamia in the Third Millennium B.C.*, Guilford.
2013 The northern Levant during the Intermediate Bronze Age: Altered trajectories. In A. Killebrew and M. Steiner (eds.), *Oxford Handbook of the Archaeology of the Levant,* Oxford: 367-387.
Whiting, R. H.
1995 Amorite tribes and nations of second millennium western Asia. In J. Sasson (ed.), *Civilizations of the Ancient Near East*, vol. II, New York: 1231-1242.
Widell, M.
2007 Historical evidence for climate instability and environmental catastrophes in northern Syria and the Jazira: The chronicle of Michael the Syrian, *Environment and History* 13/1: 47-70.
Wossink, A.
2009 *Challenging Climatic Change: Competition and Cooperation among Pastoralists and Agriculturalists in Northern Mesopotamia (c. 3000-1600 BC),* Leiden.
Wright, H. T.
1971 The archaeology of Palestine from the Neolithic to the Middle Bronze Age, *JAOS* 91: 276-293.
Zarins, J.
1989a Jebel Bishri and the Amorite homeland: The PPNB phase. In O. M. C. Haex, H. H. Curvers and P. M. M. G. Akkermans (eds.), *To the Euphrates and Beyond,* Rotterdam: 29-51.
1989b Pastoralism in southwest Asia, the second millennium BC. In J. Clutton-Brock (ed.), *The Walking Larder,* Cambridge: 127-155.
1990 Early pastoral nomadism and the settlement of lower Mesopotamia, *BASOR* 280: 31-65.
1992a Archaeological and chronological problems within the greater southwest Asian arid zone, 8500-1850 BC. In R. W. Ehrich (ed.), *Chronology in Old World Archaeology* (3rd Edition), Chicago: 42-62.
1992b Pastoral nomadism in Arabia: Ethnoarchaeology and the archaeological record. In O. Bar-Yosef and A. Khazanov (eds.), *Pastoralism in the Levant*, Madison: 219-239.

MINNA LÖNNQVIST
University of Oulu, Finland

The Emergence of Pastoral Transhumance at Jebel Bishri in Central Syria

Tracing *Transhumance* at Jebel Bishri

The mountain of Jebel Bishri in Central Syria (Fig. 1.) has been traditionally associated with the habitat of the early Amorites and Arameans, who in their pristine existence have been seen as nomads (passim, e.g., Kupper 1957, Heltzer 1981). The plentiful evidence of Amorite tribes and pastoralists from the Mari archives dating from the 18th century B.C. has coloured our view of the pastoralism in the region as having exclusively been transhumance throughout the past (Lönnqvist *et al.* 2010a: 165). Nevertheless, transhumance is indeed a typical mode of pastoralism in the region today, but there is no conclusive evidence that it is the only mode of pastoralism or has been such.

Nowadays the increased desertification in the Jebel Bishri region has caused instability in the village-based pastoral transhumance of particular tribes living on the western piedmont area of the mountain. There is an open direction to a more nomadic way of life, and some pastoral tribes living in the region are more nomadic than the others (Lönnqvist *et al.* 2010b). We need to keep in mind that pastoralists roam longer ways to find better pastures according to the changes in climate and grazing grounds. Pastures are vulnerable to micro-climatic variations and over-stocking (Glatzer 1992: 298-300). Access to pastures is also often dependent on tribal territorialism. In semi-arid and arid areas states can seldom dictate pastoral movements (Casimir 1992: 165-166).

The mountain of Jebel Bishri belongs to the Palmyrides and is the continuation of the Arabian platform. Its peak reaches ca. 867 metres over the sea level (Wirth 1971: 55). It is the landmark of significant topographic prominence while coming along the right bank of the Euphrates from the Lower Mesopotamia. It can be experienced from a far. The Finnish archaeological survey and mapping project SYGIS started to investigate the area by remote sensing and ground surveying in 1999/2000. The purpose was to study the relationship of the pastoral people of the mountain with those of the surrounding plains and villages. Jebel Bishri belongs to the so-called arid zones, but there are contrasts of different kinds of environments and landscapes. The finds opened vistas to the cultural development of the region largely depending on the environment and its contrasts. (See Lönnqvist *et al.* 2011).

The mountain is today desert-steppe which receives less than 200 mm of rain annually (Lönnqvist *et al.* 2011: 1-2), and the area is limited with the green valley of the Euphrates river where the precipitation can fall under and over the limit depending whether the question is of the Upper, Middle or Lower Euphrates. It is a pastoral landscape, and the present-day Bedouin tribes living in the region have originally arrived from the Arabian peninsula and Iraq. Many have largely changed their pastoral subsistence economy from camel nomadism to small-livestock rearing in the neighbourhood of the river zone. Certain environmental parameters, such as proximity of the water sources, seem to have affected the development of pastoralism towards small-livestock rearing.

Studia Chaburensia 4 (2014), pp. 93–109.

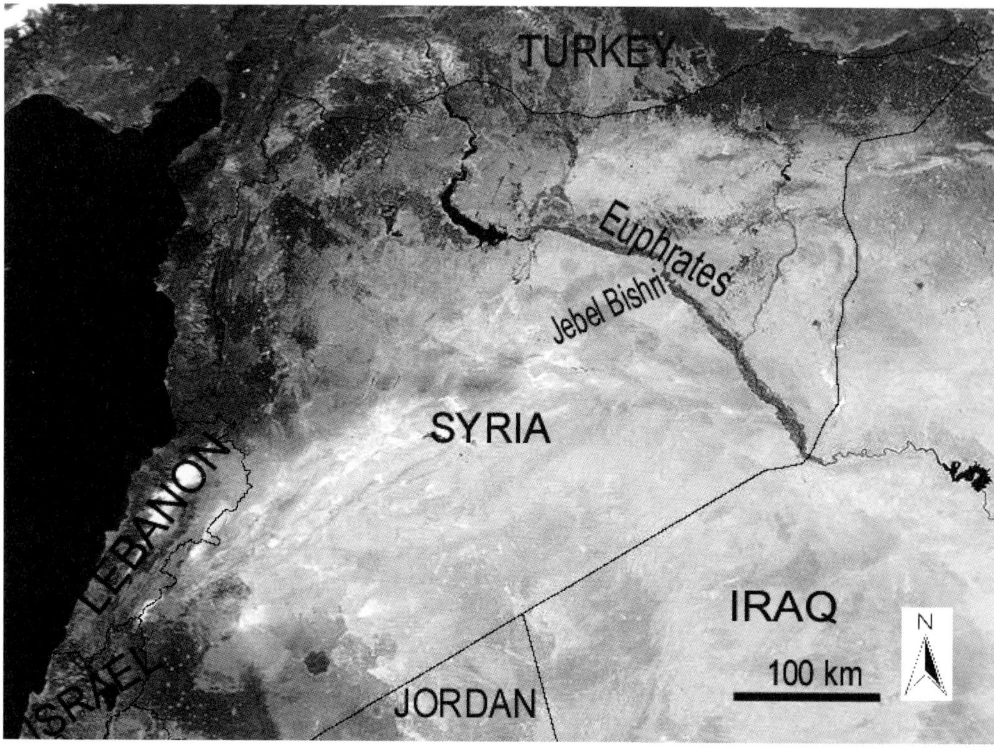

Fig.1: The mountainous desert-steppe region of Jebel Bishri in Central Syria.
Satellite mapping by Minna Lönnqvist.

The western oases in the foot of Jebel Bishri and the valley of the Middle Euphrates are the arena of early cultivation and agriculture, largely due to springs and river fans that gradually became supported by irrigation. But the region has also been an arena of early goat and sheep domestication and the emergence of nomadism in the Levant (see, e.g., passim Cauvin 2000). When we studied Jebel Bishri by remote sensing and archaeological field surveys, the long-term perspective actualized: the survey we carried out took into account all the periods equally in order to understand the development of the area in this marginal environment. The contrasts between surrounding valleys and mountain as well as valleys and hills in pastoral landscape were modelled in 3D (Figs. 2, 3).

Some curious elements were revealed in tracing the emergence and development of the pastoralism in the region and especially one of its modes known as transhumance (Lönnqvist *et al.* 2011: 3-5). The long-lived system of highland-lowland interaction, that is so typical also of transhumance, became evident in the landscape incorporating mobility and sedentism. According to the structuralist F. Braudel, in the Mediterranean landscape mountains come first. However, they and the people living in those regions have been largely neglected by the historians. The mountains seem to resist the pace of history, but the lowland and highland people have always held contact with each other (Braudel 1972: 25-41).

Fig: 2: 3D model of Jebel Bishri following the Euphrates river. Constructed by fusing Landsat-7 satellite image data with SRTM 2000 shuttle mission radar data. The Finnish project SYGIS, Markus Törmä 2003.

Fig. 3: 3D model of the pastoral landscape in the central area of Jebel Bishri. Constructed by fusing Quick-Bird satellite image data with ASTER image data. The Finnish project SYGIS, Markus Törmä 2012.

The Beginnings of the Highland–Lowland Interaction

In our studies by judging the archaeological data from Jebel Bishri it became clear that the highland–lowland interaction had already started during the Palaeolithic era for seasonal reasons. D.O. Henry, while dealing with the Jordanian Palaeolithic data, already denoted this type of interaction as transhumance (Henry 1989: 167-170), even if the original term of transhumance is strictly speaking to be associated with pastoral nomadism. Henry's approach does not reflect the purist association of transhumance with pastoral economy, and therefore does not apply to the emergence and development of pastoral transhumance. The word transhumance associated with pastoral nomadism means crossing of different kinds of environmental zones in seasonal grazing. Originally in the definition of transhumance there is question of winter grazing in base-villages in valleys followed by seasonal movement to pastures in higher altitudes in summer, as defined in the Swiss Alpine landscape context.

The reason for Henry's nomenclature is that comparable cultural features were found in the flint industry of the hunter-gatherer groups on different environmental zones reflecting seasonal exploitation of these different zones and altitudes by the same groups. In our studies at Jebel Bishri also intensive interaction between the hunter-gatherer groups of the western plain and the mountain was revealed. The interaction between the El Kowm basin with Tar al-Sbai, a semicircular escarpment, at the western edge of Jebel Bishri is especially discernable in the Upper Palaeolithic and Epipaplaeolithic periods. It is apparent that the people camping in the El Kowm basin used Tar al-Sbai with its abris at the mountain edge for hunting and carcass processing at that time (Lönnqvist *et al.* 2011: 83-108). The seasonal route of gazelles entered the plains beneath (Legge and Rowley-Conwy 1987: 79). The Upper Palaeolithic period was crucial in transforming the human mind for creating art, and people therefore began to understand kinetic properties of surrounding landscapes and thus visibility in 3D for the exploitation of natural resources in a new way. The mountains also appear as prominent landscapes and offer vantage points for hunters, who are able to better exploit the properties of visibility (Lönnqvist *et al.* 2012). The visibility is also important for pastoralists in looking for grazing grounds and defining pastures and tribal territories.

In our survey Tar al-Sbai was identified as the location of huts as well, the earliest date of which may already start from the Epipalaeolithic and Neolithic periods according to the associated finds. The Finnish survey revealed that Epipalaeolithic groups from the Geometric Kebaran and the Natufian phase had also reached in their movement from the Levantine corridor, like from the El Kowm basin, as far as to the area of a palaeolake at Nadra in the central eastern area of Jebel Bishri. Pre-Pottery Neolithic groups, also identified in the El Kowm basin and the Euphrates valley, became visible in the archaeological record of the palaeolake district at Nadra (Lönnqvist *et al.* 2011).

The Emergence of Cultivation and Agriculture in the Region

The role of cultivation and agriculture as the basis for the emergence of pastoral nomadism has been extensively discussed by anthropologists and archaeologists for decades. Alternative independent lines of development have, however, been suggested to have taken place in desert and steppe areas, too. M. Sahlins (Sahlins 1968: 33) thought that pastoralism is opposite to agriculture, even if not in its earliest and present forms. Furthermore, irrigation agriculture as a stimulus for the emergence of specialized pastoralism has been suggested (Lees and Bates 1974).

The dependence of pastoral nomads more or less on agriculture and its products is seen as an axiom in several studies: they tend to stress this central relationship of both subsistence economies. D. Eickelman (Eickelman 1981: 65), for example, thinks that nomadic pastoralism has apparently always been part of the economic system in which agriculture and trade are linked. According to the view, if pastoral nomads are not cultivating themselves, they acquire the grain from neighbouring agricultural societies.

A. Khazanov (Khazanov 1994: 19) takes a view that, although there are purely nomadic pastoralists, they in any case are somehow linked to agricultural societies and that especially the semi-nomadic pastoralists are more or less tied to agricultural groups.

The transition from hunter-gatherer economy to agriculture took place locally in the Near East. Landscapes and environmental conditions played as a background for changes in subsistence economy that occurred rather simultaneously in different localities in the Levant (Goring-Morris and Belfer-Cohen 2011). Environmental change,[1] population pressure (Boserup 1965) or food crisis (Cohen 1977), have been seen as major stimuli for taking up cultivation and agriculture in some studies. The environmental context of the Levantine corridor, into which the El-Kowm basin in the piedmont of Jebel Bishri and the Middle Euphrates fan with Abu Hureyra belong, was favourable habitat both for the emergence of agriculture and animal husbandry in the transition from the Epipalaeolithic period to the Neolithic one in archaeological terms (Lönnqvist *et al.* 2011: 3-5). The Epipalaeolithic phase had become the periodic threshold, especially the Natufian phase, to semi-sedentary hamlets and cultivation. The finds of Abu Hureyra some 100 km northwest from Jebel Bishri in Central Syria provide some of the earliest signs of agriculture with domesticated rye dating from 9000 BC (Moore *et al.* 2000). However, some scholars (see Weiss and Zohary 2011; Goring-Morris and Belfer-Cohen 2011) have recently given doubts to the early datings acquired from the Abu Hureyra seed samples because of the taphonomy of the site for the origin of domestication.

But it is interesting to study that, whether there were areas in which instead of the development from initial cultivation to agriculture as a food producing economy, the development after the hunter-gatherer phase took directly footsteps to another mode of mobility, namely pastoral nomadism. Both were subsistence economies based on human-animal interrelationship as a food producing economy. The possible direct transition in some marginal zones is an interesting question to which we shall return shortly.

The Domestication of Caprovines and the Beginnings of Pastoralism

Recent reports support the domestication of goats and the emergence of pastoralism as early as 8000 cal BC (Makarewicz and Tuross 2012) which after all is close to the 1960s datings for the domestication of sheep from Zawi Chemi Shanidar cave in Iraqi Kurdistan. In any event, earlier the validity of the Shanidar datings have been disputed in various studies (Zeider 1999: 12). The osteological finds from the Shanidar cave indicated that sheep were domesticated at the beginning of the 9th millennium BC, some 1000 years earlier than sheep domestication had been thought to have taken place. Still, however, the major component of meat diet consisted of big game (Perkins 1964: 1565-1566). The researchers of Abu Hureyra

1 The Oasis theory by Pumpelly, followed by V.G. Childe, see, e.g., Trigger 1980, 61-63 and more recently Bar-Yosef 2011.

also believed that sheep and goat husbandry already existed at the beginning of the Pre-Pottery Neolithic period at the site; the goats outnumbered sheep ca. five to one (Legge and Rowley-Conwy 2000: 462-463).

The seasonal route of gazelles heading towards north leads through the El Kowm basin in the area, where they were captured (Legge and Rowley-Conwy 1987). The kites, large V-shaped gazelle traps built of stone rows, were located in this Levantine corridor. Similar constructions have been identified in numbers in the basaltic areas of the Black Desert of Jordan (Rees 1929: 395-40; Betts 1987: 214-215), in the Palmyrene (Poidebard 1934 and Legge and Rowley-Conwy 2000: 446), in Abu Hureyra district (Moore et al. 2000) and in the Hemma plateau of Syria (Van Berg *et al.* 2004). In the Hemma Plateau rock art depicting kite formations have also been identified. Abu Hureyra had initially based its subsistence economy on gazelle killing in the Epipalaeolithic and partly during the Pre-Pottery Neolithic A (PPNA) period, while food producing economies were emerging (Moore *et al.* 2000).

Gazelle was the first main source of protein of the ancient people at Abu Hureyra, and A.J. Legge and P.A. Rowley-Conwy (Legge and Rowley-Conwy 1987) have suggested that the shortage of the gazelle meat and consequently protein stimulated people towards animal husbandry. This especially meant the switch from gazelle to caprines and their husbandry from the period 2A to 2B of Abu Hureyra. During the PPNA period the importance of gazelle killing had continued beside the initiative ovicaprine domestication (Moore *et al.* 2000: 478, 497). Mixed farming was the agricultural system which prevailed at Abu Hureyra during the clear caprine husbandry in the period 2B (Moore *et al.* 2000: 497). Legge and Rowley-Conwy (Legge and Rowley-Conwy 2000) also had become to the assumption that on the desert margins the transition from hunting to pastoral nomadism may have been much more intimate than earlier thought, compared to the afore mentioned assumptions in which the necessity for the agricultural basis of pastoralism has been exclusively emphasised.

N. Goring-Morris (Goring-Morris 1993) has done similar observation in the Negev Highlands and Sinai among the Pre-Pottery Neolithic foragers and pastoralists. L. K. Horwitz (Horwitz 2003) has equally identified three strategies in animal exploitation in the Sinai Peninsula, the Judean Desert and Eastern Jordan desert regions during the Pre-Pottery Neolithic B (PPNB) period. What was interesting is that there seemed to have been transition from hunter-gatherer economy of the Epipalaeolithic era to the pastoralism of the PPN groups in these marginal zones, too.

As far as the studies of Goring-Morris, Legge, Rowley-Conwy (Legge and Rowley-Conwy 2000) and Horwitz are concerned, C. Makarewicz and N. Tuross (Makarewicz and Tuross 2012) have most recently expressed somewhat similar ideas and see in the transition a close relationship with the replacement of meat protein from hunting to pastoralism, as the role of hunting was replaced by pastoralism. O. Bar-Yosef has identified seasonality in sites of southern Sinai and their connection to the littoral sites, but the bones recovered at the sites were mainly those of ibex and possibly goats. The pastoral nomadism had surely arrived in Sinai by the PPNB period (Cauvin 2000).

From Qdeir to Pastoral Remains on the Mountain

The site of Qdeir in the El Kowm basin (Stordeur and Taha 1996; Cauvin 2000) on the western piedmont of Jebel Bishri, less than 100 km southeast from Abu Hureyra, provides some of the earliest pieces of evidence concerning the transition to actual pastoral nomadism during the PPNB period. The steppe of Palmyra and the Damascus basin have provided similar evidence. The temporary huts excavated at the site of Qdeir are thought to represent nomadism. Desert burin sites have also been seen as indication of pastoral activities at these kinds of sites (Cauvin 2000: 190-198). The emergence of specialized pastoral nomadism characterized by the increasing use of secondary products such as milk and wool, has also been dated to the PPNB period (Grigson 1995: 257), although more pronounced specialization took place during the Chalcolithic and Early Bronze Age. During the latter periods the relationship between crop and livestock was more systematically associated (Levy 1983: 32; Rosen 1988: 503).

The emergence of pastoral nomadism and its form known as transhumant pastoralism in the region of Jebel Bishri deals with different kinds of ecozones: the plain, either on the western plateau or the Euphrates river valley, and the relationship of these areas with the mountain. In the landscapes transhumance deals with different altitudes. For archaeologist it is a challenge to trace interrelationships in material finds. Even the patterns of seasonal routes of transhumants have been successfully reconstructed in some archaeological studies: e.g. identifying the existence of lowland pottery in the higher, seasonally inhabited caves (Gilbert 1983: 106). The movement between different ecozones also applies to vertical nomadism in its seasonal habitat tracking in contrast to horizontal nomadism that is pastoral ranging in the similar ecological areas. Vertical nomadism is seasonal movement between complementary environments, exemplified in movement between lowland and highland regions. Typically vertical nomadism involves small livestock herding like goats and sheep (Eickelman 1981: 64).

As already indicated, the remains at Tar al-Sbai at the western edge of Jebel Bishri included Upper Palaeolithic and Epipalaeolithic flint industry that was related to the El Kowm basin. Apart from a few Neolithic flints, obsidian and pottery pieces which find some parallels in the El Kowm material the finds did not bring any conclusive evidence of Neolithic pastoralism at the edge, although from the environmental point of view pastoralism should have been exercised (Lönnqvist *et al.* 2011: 185-187).

The plain of Nadra in the area of a palaeolake identified by our project in the central northeastern district of the mountain provided small settlements with Epipalaeolithic and PPN flints. Also Chalcolithic pottery, such as a piece of Al-Ubaid pottery, as well as Bronze Age finds were recovered in the district. In the vicinity there was a large ring-tumulus overlooking a site. The recovered tabular flints and burins appeared aside PPN perforators. Finds in the neighbourhood of Nadra showed that small-scale settlements dating already from the Epipalaeolithic and the Pre-Pottery Neolithic period existed in the high country areas of the mountain. However, beside some Geometric Kebaran and Natufian types of flint implements, the Epipalaeolithic microliths had types typical of dry areas comparable to the Harifian industry typical for steppes and deserts (Lönnqvist *et al.* 2011: 142-146).

More pronounced structures characteristic for pastoral cultures such as cairns/tumuli and large enclosures resembling the megalithic structures found at Tar al-Sbai seem to date from the Chalcolithic and Bronze Age (Lönnqvist *et al.* 2011: 135-166, 179-195). During the Chalcolithic and Bronze Age the mountain had become the realm of ancestors and animals as the most common structures are cairns/tumuli for the deceased and corrals, animal pens, and curvilinear stone circles of various kinds. Burial sites and their internal structures seem to reflect tribalism and the chiefdom system (Lönnqvist *et al.* 2010a). The majority of cairns/tumuli were clearly allocated on the mountain, not in the river plain. It is evident that some village-pastoralists of the Euphrates river plains wanted also to become buried on the mountain.

According to M. Sahlins (Sahlins 1968), pastoral nomadic societies tend to form chiefdoms, and we have suggested that ring-tumuli, imposing cairn structures on Jebel Bishri, may represent tombs built for individuals of higher status such as chiefs. The shaft-graves encountered on the riverine side slopes of Jebel Bishri (Falb *et al.* 2005) may reflect different tribal groups or confederates compared to those buried in cairns/tumuli. As I have earlier shown, the distribution of Bronze Age shaft-graves in the Levant tend to follow the line of agropastoralism, while cairns and tumuli stretch beyond the border of agropastoral coastal and river plains to the semi-arid and arid areas and are included in the pastoral nomadic remains identified as their technocomplex (Lönnqvist *et al.* 2010a).

In our project some clusters of stone enclosures that mostly were pastoral pens, corrals, or hut-bases that were identified on the mountain were studied by site catchment analysis (Lönnqvist *et al.* 2011: 167-178). The structures were clearly found in the satellite imagery offering high spatial resolution, such as the QuickBird imagery, concentrating on the highest areas of the mountain. This region is still the greenest one on the mountain and most favourable for seasonal grazing, according to our environmental cluster analysis carried out with satellite imagery. As previously mentioned, the enclosures probably mostly date from the Chalcolithic and the Bronze Age. They were apparently used for long period of time; some may already have been in use during the Neolithic period. The grazing area of each pen was calculated in the site catchment by using GIS to study, to analyse and illustrate the radius of the pastoral movement during grazing around the pens. The distance to the nearest water sources were calculated, and it seems that the grazing in the radius of 1 km was favoured between the water sources and pens (Lönnqvist *et al.* 2011: 167-178). The area was located so in the higher altitudes of the northeastern sector of the mountain that the winter waters ran down to the palaeolake district which forms the plain of Nadra.

We have no conclusive evidence that the stone circles were exclusively used by transhumant pastoralists of the river plain or the western oases, some of the pastoralists may have originated from the valley of Nadra.

Contacts to the River Zone and the Emergence of Villages

By the beginning of the Holocene the Euphrates valley faced largely climatologically stimulated geomorphological changes so that more accumulated land became open for human settlement and initial cultivation. Neolithic settlements were followed by Chalcolithic villages (Geyer and Besançon 1996: 5-15). Pastoral transhumance needed base-villages. On the Euphrates, beside Abu Hureyra Bouqras (Akkermans *et al.* 1983) is also a site that was concentrating on animal breeding during the PPNB period. Surveys along the Euphrates carried out K. Kohlmeyer (Kohlmeyer 1984: 104-105), however, provide otherwise scanty PPNB evidence on the right bank. Our Finnish project also brought to light only some signs of PPN activities on the river terraces (Lönnqvist *et al.* 2011: 127).

The transition from the Pre-Pottery Neolithic period to the Chalcolithic one, especially to the Ubaid period ca. 5200 BC (Carter and Philip 2010), has attracted some attention in recent studies. Both the Ubaid and following Uruk phases, or as sometimes dabbed as "cultures", have been seen as Mesopotamian in their origin; the periods also received names after the core sites in Mesopotamia (Carter and Philip 2010; Algaze 1993). "The Uruk world system" (cf. Algaze 1993) had reached the Euphrates banks of the river. Uruk activities had also been earlier recognized in the French studies of the El Kowm basin (Cauvin and Stordeur 1985: 191-195). The dawn of the Sumerian civilization has been attributed to these phases in various studies.

In our survey on the Euphrates side of Jebel Bishri clear Ubaid and Uruk impacts were recognized at early village sites (Lönnqvist *et al.* 2011: 218-220, 233-236). On the Euphrates side of the mountain we discovered a new tell site that had not been earlier found or mapped, and we named it Tell Kharita according to the site, namely the village of Kharita, ca. 22 km northwest of Deir ez-Zor (Lönnqvist *et al.* 2011). Apart from Al-Ubaid pottery Tell Kharita provided Uruk pottery. The Uruk pottery, but earlier handmade variety, was also found at Tell Tibne in close proximity to Halabiya. The Ubaid and Uruk impact is associated with sites that provide new proof of sedentarization closely beneath Jebel Bishri on the river side. As previously mentioned, a piece of Al-Ubaid pottery was already found in the central areas of the mountain (Lönnqvist *et al.* 2011: 143-144).On the river bank beneath the mountain the sedentarization becomes more visible in the Chalcolithic Age than during the Neolithic period. The Ubaid and Uruk impact in the finds attest clear cultural contacts or commercial links with the Lower Mesopotamia.

G. Buccellati (Buccellati 1990b) has suggested that the Uruk pots of the certain types were used for salt production, and we know that the later Amorites of the neighbourhood were especially involved in salt trade. The cuneiform texts indicate that Amorites lived as vassals of the Sumerians in the region. From Sumerians we also receive early information about the trade with these pastoral nomads of the region. The acme of the sedentarization process, however, is reflected on the Euphrates side of the mountain in the Early Bronze. Interestingly, compared to the other tell sites beneath Jebel Bishri, the site of Tell Kharita, which is attached to the mountain proper, offered a lot of pottery typical of pastoral activities: such as Early Bronze Age IV period (EBIV) material including hole mouth jars and caliciform pottery. The site may reflect sedentarization process of the pastoralists and gradual shift to transhumance between the river zone and the mountain (Lönnqvist *et al.* 2011: 243-245).

Y. Nishiaki has studied flint material in the river plain beneath Jebel Bishri at the sedentary site of Tell Ghanem Ali and in the steppe of Jebel Bishri. The site of Tell Ghanem Ali had emerged on the virgin soil in the Early Bronze Age. Despite of the technological similarities of the highland flints with the lowlands, the tell finds from Ghanem Ali are typologically different and functionally more varied than those found in the steppe. The evidence indicates that the domestic system similar to the lowland was only partly used in the highlands. The steppe did not provide tools for agricultural use (see Nishiaki 2010: 182-183).

In the case of Ghanem Ali the nomadic component of the population is not actually so visible in the finds as at Tell Kharita. The proximity of Tell Kharita to Jebel Bishri is more intimate, while Tell Ghanem Ali is clearly situated on the river plain and agricultural landscape. Also the large shaft-grave cemetery of Abu Hamed on the slopes of Jebel Bishri associated with Ghanem Ali (Falb *et al.* 2005) reflects more the agropastoralist lifestyle comparable to the Upper Euphrates and Middle Euphrates settled sites than the pastoral nomads of the steppe of Jebel Bishri (Lönnqvist *et al.* 2010a).

Reflections from the Cuneiform Sources

The god Martu, the eponymous god of the Amorites, was a Sumerian god. The text known as the Marriage of Martu (CBS 14061) (Chiera 1924) can been seen as a mythical reflection of a more nomadic stage of the Amorites than the picture transmitted from the later Mari documents. However, recently A. Porter (Porter 2012) seems to be critical of the more nomadic and pastoral state of the Amorites in the past, and she sees them as one original component of the common Mesopotamian society in which the roles of pastoralists and sedentarists were constantly changing. Porter's views are welcome to build new discussion in the field. However, she reads the Jebel Bishri evidence to the opposite direction for which the Jebel Bishri archaeological reports are allowing. The area where she has worked is more agropastoralist arena than the Palmyrene and Jebel Bishri region of the Finnish project that is desert-steppe, and the habitat of more nomadic pastoralists.The evidence of agropastoralist component in the Euphrates valley and most of the Mari archive evidence may allow her avenue of interpretation. But the same evidence does not necessarily apply to the Early Bronze Age data from the vast desert-steppe areas.

The gazelle killing and pastoralism go hand in hand in the story of Martu, and the possibility of the long-living mobile life-style of the steppes and deserts needs to be left open. The Tidnu tribes associated with the Amorites and referred to the Sumerian sources "come down" from the mountains; they are not indigenous people of the river plain. But already in the Early Bronze Age agropastoralists of the same or related cultural background were apparently populating the Middle Euphrates valley as also attested at the foot of Jebel Bishri. There were different degrees of mobility in different times and in various groups.

We need to recall that the Mari documents date from the period after the large-scale movement of the Amorites in the end of the third millennium and their general sedentarization by ca. 2000 BC (Lönnqvist 2008, 2009, 2010). The sedentarization had most likely affected to the degrees of nomadism that is reflected in the Mari documents. In Akkadian texts "high country" is thought to denote steppe areas east and west of the Euphrates contrasted with the valley floor (Buccellati 1990a: 96-97). The term nawū means pastureland, also identified as situating on river terraces called as plateaus (Buccellati 1990a: 96, Lafont 2000: 144). Our Finnish project identified some tabular flints from the plateaus on the Euphrates side beneath the mountain, like the material later found by the Japanese project and analysed by Nishiaki, indicating pastoral nomadism and that those areas had been used by pastoralists like in the Mari texts.

Evidence of different tribal and pastoral groups in the Mari cuneiform sources testify that the Sutean tribe (Kupper 1957: 98; Hetzer 1981: 57) apparently was more nomadic than the Benjaminite tribe, which carried out more horizontal grazing along the right river bank on the side of Jebel Bishri. The pastures of the Benjaminites stretched from Mari to Harran (Kupper 1957: 47-48). In my view the distribution of Benjaminites indicates that they were agropastoralists of the river zone. The Suteans were more tied with the mountain proper of Jebel Bishri and the Palmyrene desert, and therefore in my view were probably more nomadic. According to some scholars, the Suteans did not belong to the Amorites, like the Benjaminites, and according to others, they are seen as a related tribal group.

There was a special clan living in the region of Jebel Bishri which particularly is characterized as transhumant in its pastoral mode of living and which is called the hibrum (Anbar 1991: 79; Streck 2000: 97; Durand 2004: 160-161). The hibrum have been seen as a subgroup of the Benjamite tribe like the Rabbeans. The tribe of the Benjaminites who had its pastures on the right bank of the river on the side of Jebel Bishri and had several subtribes and vassal (see, e.g., Durand 2004: 157-173). The tribe had chiefs called merhûm that might have been equivalent in their power to kings."Hibrum ša nawem" refers to the nomadic entity and pasture of the hibrum clan in contrast to its more sedentary members (Durand and Streck

2000: 97). The story of goat-boy Bazi from the Old Babylonian period also illustrates transhumant type of pastoralism as do the Mari archives. Bazi left his father's house for grazing in the region of Jebel Bishri, a deified mountain (George 2008: 3).

Ethnoarchaeology also provides us with a window through transhumant pastoral life in the region today, and can transmit an echo of the past pastoral life styles as well. The Finnish project studied villages of transhumant pastoralists, the movement of desertification and different types of site abandoning processes. We studied desertification by satellite imagery and the effects of desertification on pastoral life in the past decades. Informants were valuable in understanding different reasons for site abandonments whether being seasonal or final. (Lönnqvist *et al.* 2011: 371-398).

Nowadays the Sba'a Bedouin tribe, after which the escarpment of Tar al-Sbai has received its name, frequents this edge of the mountain from the El Kowm plain, where there are small hamlets carrying the name of the tribe. Other transhumant Bedouins also live in the neighbourhood. The territory of the mountain belongs to the Fida'an Bedouin tribe which seems to be more mobile than the Sba'a. Year-around tent habitation was also observed by our project among local Bedouins beside the transhumants, who used base-villages for their seasonal grazing.

Concluding Remarks

The human–animal relationship is emphasized in the archaeological remains of Jebel Bishri. The highland–lowland interaction started in the region already during the Palaleolithic period. For hunter-gatherers and foragers the area was used for acquirying prey, also for processing the meat, bone, antler and skin. For pastoralists the grazing was the focus of the mountain environment. Grazing produced meat, milk, leather and wool.

From the Palaeolithic period onwards we have evidence of the concomitant relationship between the El Kowm basin on the western piedmont area and Tar al-Sbai at the western edge of the mountain. The flint assemblages in the central areas of the mountain were more of the desert-steppe types than those at the edges. From the Neolithic period onwards the mountain apparently began to serve as grazing grounds for pastoralists. It is possible that the mobile subsistence economies of hunter-gatherers and pastoralists overlapped some time on the mountain. Many sites, for example at Nadra, are multi-period sites of mobile groups. The change from one major type of mobile subsistence economy to another did not bring any prominent impact of agricultural endeavours to the mountain, however, although some cultivation, like today, may have existed.

Whether the sites of the El Kowm basin, where early indications of pastoral nomadism have been traced, already were involved in the systematic pastoral transhumance with the mountain edge during the Neolithic era is not clearly visible in our material evidence, although finds from Tar al-Sbai dating from the Neolithic period were recovered. The actual evidence of pastoral transhumance can only be followed to the Chalcolithic and Early Bronze Age in the archaeological record of Jebel Bishri.

As far as the finds from the river zone are concerned, the clear interrelationship with the plains more dependent on agricultural production intensified during the Chalcolithic and Early Bronze Age, in the period when also cairns/tumuli and stone circles of pastoral culture are prominent on the mountain. The mountain with its pastoral enclosures served seasonal grazing in higher altitudes probably for the sites in the plain of Nadra and the river zone.

The pastoralists developed a closer relationship with the river zone and probably became early involved in trade and gradual sedentarization in a larger scale. The archaeological data, especially at Tell Tibne and Tell Kharita, indicates that this closer relationship to the river zone beneath the mountain emerged sometime in the Al-Ubaid and Uruk period in specific, and that the pastoral element in the pottery starts to show in the Early Bronze Age material, when more pronounced number of sedentary sites emerged on the river side of Jebel Bishri. This was the time of the historical movement of the Amorites and their gradual sedentarization. From the Middle Bronze Age we, however, know two tribal groups that were especially involved in grazing on Jebel Bishri, namely the Suteans and the Benjaminites. The Suteans probably were more nomadic of the two groups.

Transhumance seems to have continued thousands of years in the region of Jebel Bishri and is still common mode of pastoralism among different Bedouin tribes living in the region. However, like today, there were other modes of pastoralism in the region. Some have continued to live mobile life all around the year. The interest of both hunter-gatherers and pastoralists was in acquiring meat. Actually in the ancient Akkadian name *Bašar*, i.e., Jebel Bishri some may also find the echo of the word "meat" (Hebrew) or "bringer of glad tidings" (Arabic) in some Semitic languages. The mountain could symbolically be seen as a "meat warehouse" or "a meat bazaar for mobile people" - the availability of meat has always meant good tidings for hunter-gatheres and pastoralists. The mountain has been the realm of animals, and also of ancestors.

Bibliography

Akkermans, P.A., J.A.K. Boerma, A.T. Clason, S.G. Hill, E. Lohof, C. Meiklehohn, M. Le Mière, G.M.F. Molgat, J.J. Roodenberg, W. Waterbolk-Van Rooyen and W. van Zeist

1983 Bouqras Revisited: Preliminary Report on a Project in eastern Syria, *Proceedings of the Prehistoric Society* 49: 335-372.

Anbar, M.

1991 *Les tribus amurrites de Mari.* Orbis Biblicus et Orientalis 108. Freiburg: Univeristätsverlag.

Algaze, G.

1993 *The Uruk World System: The Dynamics of Expansion of Early Mesopotamian Civilization.* Chicago: University of Chicago Press.

Bar-Yosef, O.

1981 Paleolithic and Neolithic Desertic Adaptations in The Sinai Peninsula, *Union Internacional de Ciencas prehistoricas y Protohistoricas, X Congreso.* Comison IX Cultra y Medio ambiente de Hombre fosil en Asia Mexico City: 35-61.

Bar-Yosef, O.

2011 Climatic Fluctations and Early Farming, *Current Anthropology* 52: S175- S193.

Betts, A.

1987 The Hunter's Perspective: 7th millennium BC Rock Carvings from eastern Jordan, *World Archaeology* 19: 214-225.

Boserup, E.

1965 *Conditions of Agricultural Growth: the Economics of Agrarian Change under Population Pressure*, London: George Allen & Unwin.

Braudel, F.
1972 *The Mediterranean and the Mediterranean World in the Age of Philip II*, Vol. I, London: Collins.

Buccellati, G.
1990a "River Bank", "High Country", and "Pasture Land": the Growth of Nomadism on the Middle Euphrates and the Khabur. In S. Eichler, M. Wäfler and D.Warburton (eds.), *Tall al-Hamīdīya 2. Symposion. Recent Excavations in the Upper Khabur Region 1986*, Göttingen: Vandenhoek & Ruprecht: 87-117.
1990b Salt at the Dawn of History: the Case of the Bevelled-Rim Bowl. In P. Matthiae, M. Van Loon and H. Weiss (ed.), *Resurrecting the Past: A Joint Tribute to Adnan Bounni,* Publications of Insitut historique-archéologique néerlandais de Staboul, Vol. 67, Leiden, Historich Archäologisch Instituut te Istanbul: 17-40.

Carter, R.A. and G. Philip
2010 Deconstructing the Ubaid. In R.A. Carter and G. Philip (eds.), *Beyond the Ubaid: Transformation and Integration in the Late Prehistoric Societies of the Middle East,* Studies in Ancient Civilizations 63, Chicago, Illinois: The Oriental Institute of the University of Chicago: 1-22.

Casimir, M.J.
1992 The Determinants of Rights to pasture: Territorial Organsiation and Ecological Constraints. In M. J. Casimir and A. Rao (eds.) *Mobility and Territoriality: Social and Spatial Boundaries among Roragers, Fishers, Pastoralists and Peripatetics*, New York/Oxford: Berg: 153-202.

Cauvin, J.
2000 *The Birth of the Gods and the Origin of Agriculture.* Cambridge: Cambridge University Press.

Cauvin, J and D. Stordeur
1985 Une occupation d' époque Uruk en Palmyrène: le niveaux supérieur d'el Kowm 2-Caracol. In J. Cauvin and O. Aurenche (eds.), *Cahiers de Euphrate* 4: 191-205.

Chiera, E.
1924 *Sumerian Religious Texts*, Crozer Theological Seminary Babylonian Publications. Vol. I. Upland, PA: Crozer Theological Seminary.

Cohen, M.N.
1977 *The Food Crisis in Prehistory: Overpopulation and the Origins of Agriculture.* New Haven: Yale University Press.

Durand, J.-M.
2004 Peuplement et Sociétés à l'époque Amorrite (I) Les clans bensim'alites. In C. Nicolle (ed.), *Amurru 3, Nomades et sédentaries dans le Proche-Orient ancien*, Compte rendu de la XLVIe Recontre Assyriologique Internationale (Paris, 10-13 juillet 2000), Éditions Recherche sur les Civilisations, Paris: 111-197.

Eickelman, D. F.
1981 *The Middle East: An Anthropological Approach.* Englewood Cliffs, New Jersey: Prentice-Hall.

Falb, C., K. Krasnik, J.-W. Meyer and E. Vila
2005 *Der Fiedhof von Abu Hamed*, Gräber des 3. Jahrtausends v. Chr. im syrischen Euphrattal. Saarwellingen: Saarländische Druckerei & Verlag.

George, A.
2008 Something old, something new, *The Middle East in London*, Nov., No. 3.
Geyer, B. and J. Besançon
1996 Environement et occupation du sol dans la vallée de l' Euphrate Syrien durant le Néolithique et le Chalcolithique, *Paléorient* 22: 5-15.
Gilbert, A.
1983 On the Origins of Specialized Nomadic Pastoralism in Western Iran, *World Archaeology* 15: 105-119.
Glatzer, B.
1992 Pastoral Territoriality in West Afghanistan: An Organisation of Flexibility. In M. J. Casimir and A. Rao (eds.), *Mobility and Territoriality: Social and Spatial Boundaries among Roragers, Fishers, Pastoralists and Peripatetics*, New York/ Oxford: Berg: 293-306.
Goring-Morris, N.
1993 From Foraging to Herding in the Negev and Sinai: the Early to Late Neolithic Transition, *Paléorient* 19: 65-89.
Goring-Morris, N. and A. Belfer-Cohen
2011 The Origin of Agriculture: New Data, New Ideas, *Current Anthropology* 52, No.S4 S195-S208.
Grigson, C.
1995 Plough and Pasture in the Early Economy of the Southern Levant. In T. E. Levy (ed.), *The Archaeology of Society in the Holy Land*, Hong Kong: 245-265.
Heltzer, M.
1981 *The Suteans*. Istituto Universitario Orientale. Seminario di Studi Asiatici. Series Minor XIII. Naples: Istituto Universitario Orientale.
Henry, D.O.
1989 *From Foraging to Agriculture: the Levant at the Age of the Ice Age*. Philadelphia: The University of Pennsylvania Press.
Horwitz, L.K. with contribution by O. Lernan
2003 Temporal and Spatial Variation in Neolithic Caprine Exploitation Strategies: A Case Study of Fauna from the Site of Yitah'el (Israel), *Paléorient* 29: 19-58.
Khazanov, A.
1994 *Nomads and the Outside World*. Second edition. Wisconsin: The University of Wisconsin Press.
Kohlmeyer, K.
1984 Euphrat-Survey, Sie mit Mitteln der Gerda Henkel Stiftung durchgeführte archäologische Geländebegehung im Syrischen Euphrattal, *Mitteilungen der Deutschen Orient-Gesellschaft* 116: 95-111.
Kupper, J.-R.
1957 *Les nomades en Mésopotamie au temps des rois de Mari*. Bibliothèque de la Faculté de Philoslophie et Lettres de l'Université de Liège. Fascicule CXLII. Paris: Les Belles Lettres.
Lafont, B.
2000 Irrigation Agriculture in Mari, In R.M Jas (ed.), *Rainfall and Agriculture in Northern Mesopotamia* (MOS Studies 3), Proceedings of the Third MOS Symposium (Leiden 1999), Publications de l'Institut historique-archéologique néerlandais de

Stamboul, Vol. LXXXVIII, Leiden: Nederlands Historisch-Archaeologisch Instituut te Istanbul: 129-145 .

Lees, S.H. and D.G. Bates

1974 The Origins of Specialized Pastoral Nomadism. A Systemic Model, *American Antiquity* 39: 187-193.

Legge, A.J. and A. Rowly-Conwy,

1987 Gazelle Killing in Stone Age Syria, *Scientific American* 257 (2): 76-83.

2000 The Eploitation of Animals. In A.M.T. Moore, G.C. Hillman and A.J. Legge, *Village on the Euphrates: from Foraging to Farming at Abu Hureyra*: 423-471.

Levy, T.E.

1983 The Emergence of Specialized Pastoralism in the Southern Levant, *World Archaeology* 15: 15-36.

Lönnqvist, M.

2008 Were Nomadic Amorites on the Move? Migration, Invasion and Gradual Infiltration as Mechanisms for Cultural Transition. In H. Kühne, R. Czichon and F. J. Kreppner (eds.), *Proceedings of the 4th International Congress of the Archaeology of the Near East, 29 March – 3 April 2004*, Freie Universität Berlin, Vol. 2, Social and Cultural Transformation: The Archaeology of Transitional Periods and Dark Ages, Excavation reports, Wiesbaden, Harrassowitz Verlag: 195-215.

2009 Jebel Bishri and the Role of Nomadism in the End of the Early Bronze Age. In P. Parr (ed.), *The Levant in Transition, Proceedings of a Conference held at the British Museum on 20-21 April, 2004*, Palestine Exploration Fund Annual IX, Leeds: Maney Publishing: 49-55.

2010 How to Control Nomads? A Case Study Associated with Jebel Bishri in Central Syria, West Semitic Nomads in Relation to the Urban World. In L. Kogan *et al.*, *City Administration in the Ancient Near East, Proceedings of the53e Reconte Assyriologique Internationale*, Vol. 2, Orientale et Classica XXXI, Babel und Bibel 5, Winona Lake, Indiana: Eisenbrauns: 115–139.

Lönnqvist, M. with contributions by K. Lönnqvist, M. Törmä, M. Nuñez and M. Stout Whiting

2010a Tracing Tribal Implications among the Bronze Age Tomb Types in the Region of Jebel Bishri in Syria. In K. Ohnuma (ed.), *Formation of Tribal Communities in the Bishri Mountains, Middle Euphrates, Al-Rafidan, Special Issue*: 165-173.

Lönnqvist, M., M. Törmä, M. Nuñez, J. Okkonen, M. Stout Whiting, H. Riihiaho and M. Nissinen

2010b Desertification and Ethnoarchaeology: Studying Hazards in the Nomadic Environment of Jebel Bishri in Syria. In P. Matthiae, F. Pinnock, L. Nigro and N. Marchetti with the collaboration of L. Romano, *Proceedings of the 6th InternationalCongress on the Archaeology of the Ancient Near East, May 5th-10th 2008*,"Sapienza"- Università di Roma, Vol. 1, Near Eastern Archaeology in the Past, Present and Future: Heritage and Identity, Ethnoarchaeological and Interdisciplinary Approach: Results and Perspectives, Visual Expression and Craft Production in the Definition of Social Relations and Status, Wiesbaden: Harrassowitz Verlag: 369-390.

Lönnqvist, M., M. Törmä, K. Lönnqvist and M. Nuñez

2011 *Jebel Bishri in Focus: Remote sensing, archaeological surveying, mapping and GIS studies of Jebel Bishri in central Syria by the Finnish project SYGIS.* British Archaeological Reports International Series 2230. Oxford: Archaeopress.

2012 Satellite Perspective on Highland–Lowland Interaction in Ancient Syria. In M. Shortis and M. Madden (ed.), *International Archives of Photogrammetry, Remote Sensing and Spatial Information Sciences* Vol. XXXIX-B4, XXII ISPRS Congress 25 August – 01 September 2012, Melbourne, Australia: 455–460.

Makarewicz, C. and N. Tuross,
2012 Finding Fodder and Tracking Transhumance: Isotopic Detection of Goat Domestication Processes in the Near East, *Current Anthropology* 53: 495-505.

Moore, A.M.T., G.C. Hillman and A.J. Legge
2000 *The Village on the Euphrates: From Foraging to Farming at Abu Hureyra*. Oxford: Oxford University Press.

Nishiaki, Y.
2010 Early Bronze Age Flint Technology and Flake Scatters in the North Syrian Steppe Along the Middle Euphrates, *Levant* 42: 170-184.

Perkins, D. Jr.
1964 Prehistoric Fauna From Shanidar, Iraq, *Science* 144: 1565-1566.

Poidebard, A.
1934 *Trace de Rome dans le Désert de Syrie: Le Limes de Trajan a la conquète arabe, recherches aériennes (1925-1932)*, Texte, Atlas, Biblithèque Acheoligique et Historique. Tome XVIII.

Porter, A.
2012 *Mobile Pastoralism and the Formation of the Near Eastern Civilization: Weaving Together Society*. Cambridge: Cambridge University Press.

Rees, L.W.B
1929 The Transjordan Desert, *Antiquity* 3: 389-407.

Rosen, Steven A.
1988 Notes on the Origins of Pastoral Nomadism: A Case Study from the Negev and Sinai, *Current Anthropology* 29: 498-506.

Sahlins, M.D.
1968 *Tribesmen*. Foundation of Modern Athropological Series. Englewood Cliffs, New Jersey: Prentice Hall Inc.

Stordeur, D. and A. Taha
1996 Ressemblances et dissemblances entre les sites nomades et sédentarires de la steppe syrienne au Vie milléniaire, *Les Annales Archéologiques Arabes Syriennes* (Numéro spécial: *Palmyra and the Silk Road*) 42: 85-98.

Streck, M.P.
2000 *Das amurritische Onomastikon der altbabylonischen Zeit*, Band 1, Die Amurriter, Die onomatische Forschung, Orthographie und Phonologie, Nominalmorphologie. Alter Orient und Altes Testament, Verffentlichungen zur Kultur und Geschichte des Orients und des Alten Testaments, Herausgeber: Manfried Dietrich, Oswald Lorentz Band 271/1. Münster: Ugarit-Verlag.

Trigger, B.
1980 *Gordon Childe: Revolutions in Archaeology*. London: Thames and Hudson Ltd.

Van Berg, P.-L., M. Vander Linden, S. Lemaitre, N. Cauwe and V. Picalause
2004 Desert-Kites of the Hemma Palteau (Hassake, Syria), *Paléorient* 30 (1): 89-100.

Weiss, E. and D. Zohary

2011 Neolithic Southwest Asian Founder Crops: Their Biology and Archaeobotany, *Current Anthropology* 52 (suppl. 4): S237-S252.

Wirth, E.
1971 *Syrien, eine geographische Landeskunde.* Wissenschäftlichen Länderkunden. Darmstadt: Wissenschaftliche Buchgesellschaft.

Zarins, J.
1992 Archaeological and Chronological Problems within the Great SouthwesteAsian Arid Zone, 8500-1850 BC. In R.W. Ehrich (ed.), Chronologies in Old World Archaeology, Third edition, Vol.1, Chicago: The University of Chicago Press: 42-62.

Zeider, M.A.
1999 Animal Domestication in the Zagros: A review of Past and Current Research, Paléorient 25: 11-25.

YOSHIHIRO NISHIAKI
The University of Tokyo

Steppe exploitation by Bronze Age communities in the Middle Euphrates Valley, Syria

The cultural changes that took place widely in Mesopotamia and neighboring regions at the end of the 3rd millennium BC have been well recognized in the archaeological literature, marking the boundary between the Early and the Middle Bronze Age (e.g., Weiss *et al.* 1993; Postgate 1994; Cooper 2006). In our study region, which encompasses the Middle Euphrates valley east of Raqqa, northern Syria (Fig. 1), those changes are manifested by a remarkable shift in the settlement pattern: the abandonment of tell-based settlements in the river valley and the establishment of large-scale cairn fields in the Bishri Mountains in the inland steppe (Ohnuma *et al.* 2010). Given that cairn fields belonged to pastoral nomad societies, these changes can be interpreted as reflecting the intensified use of the steppe by nomads (Fujii and Adachi 2010; Nishiaki 2010a). Humans had already been exploiting the steppe's resources since the Paleolithic and this was particularly so since the Neolithic, when pastoralism began

Fig. 1: Location of Tell Ghanem al-Ali and the Bishri Mountains.

Studia Chaburensia 4 (2014), pp. 111 – 124.

to develop. However, the expansion of steppe use characteristic of this period of the Bronze Age can be regarded as a special case, as it ultimately determined the shape of subsequent social systems and history in Mesopotamia (Buccellati 1966; Postgate 1994). In order to better understand the beginning of intensive steppe exploitation in the Early-Middle Bronze Age, site reconnaissance surveys were conducted in the Middle Euphrates steppe for five seasons starting in 2008 as part of the "Bishri Project" (Ohnuma *et al.* 2010).

The Bishri Project consisted of a wide range of fieldwork and laboratory studies, including excavations of Tell Ghanem al-Ali (Hasegawa 2010), a Bronze Age settlement of the 3rd millennium BC in the riverside lowlands (Fig. 1), cemeteries on the adjacent plateau (Numoto and Kume 2010) and cairn fields in the Bishri Mountains dating to the 2nd millennium BC (Fujii and Adachi 2010). In addition to these excavations, our site reconnaissance survey is valuable in itself. The survey area, on the right bank of the Euphrates River within a 10 km

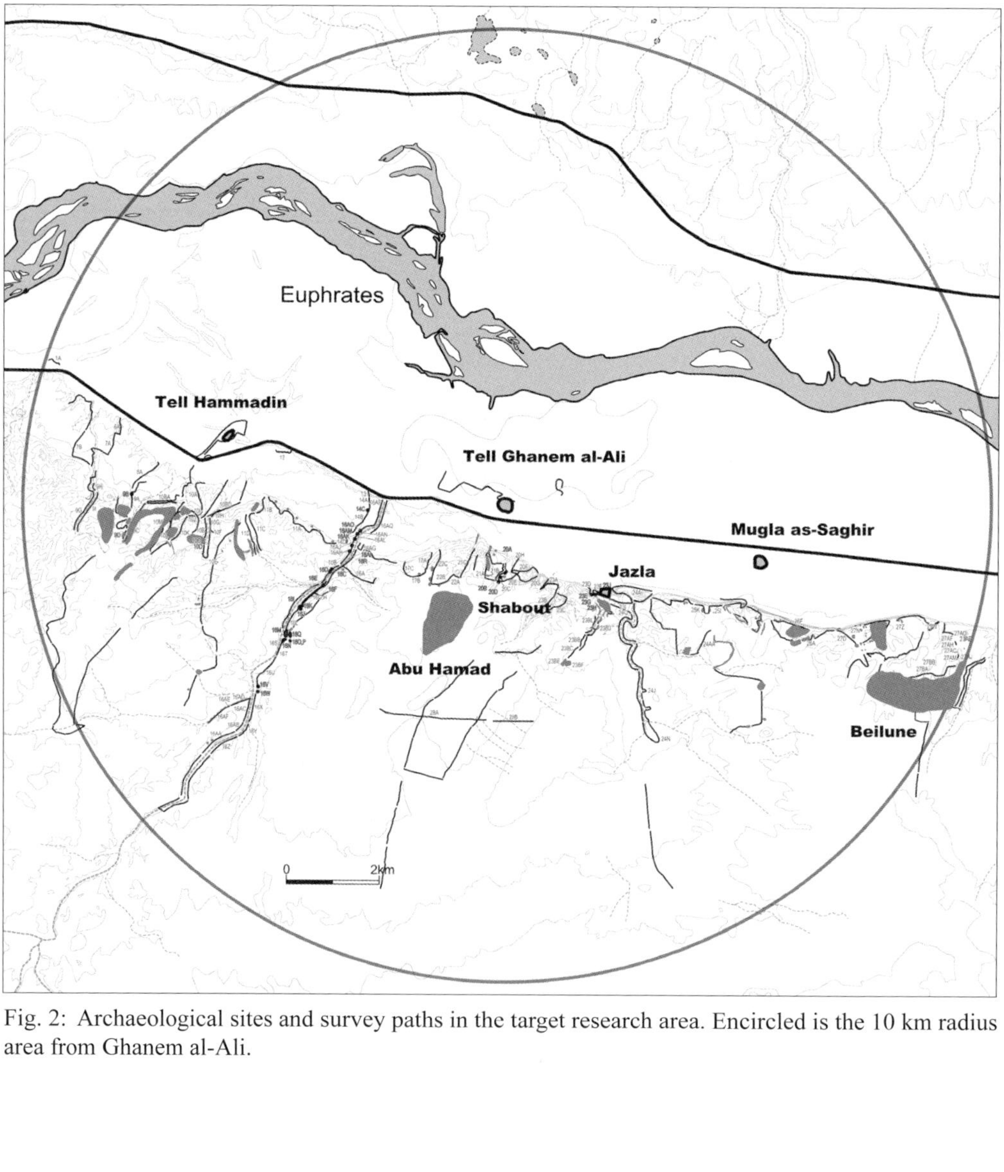

Fig. 2: Archaeological sites and survey paths in the target research area. Encircled is the 10 km radius area from Ghanem al-Ali.

radius from Ghanem al-Ali (Fig. 2), encompasses not only riverside lowlands but also inland steppes. The identified archaeological sites also varied in type, from tell settlements and cemeteries to open-air sites comprising only scatters of flint flakes; dates for these sites ranged between the 3rd and 2nd millennia BC. Thus, this particular survey covers diverse sites over an extensive area and across a wide range of time (Nishiaki 2010a). Therefore, the findings of this survey provide a platform on which to evaluate processes in the changing settlement pattern of Bronze Age society on the Middle Euphrates. This paper presents a synthetic view of the survey's results.

Bronze Age communities in the Middle Euphrates Valley

The Bishri Project examined processes in the formation of Bronze Age tribal society based in the Bishri Mountains, namely on the arid plateaus on the right bank of the Middle Euphrates (Fig. 1) (Ohnuma *et al.* 2010). They were thought to be a nomadic community tied by tribal bonds, known as the Amorites in cuneiform text, and that they directly contributed to the establishment of ancient dynasties such as Assyria and Babylon in the Mesopotamian plain at the beginning of the 2nd millennium BC (Buccellati 1966; Postgate 1994: 84-87). While some became city dwellers in Mesopotamia afterwards, others remained nomads in their homeland, the Bishri Mountains. Using archaeological evidence collected during fieldwork, the Bishri Project set out to test or verify this story proposed by philological studies.

The excavations conducted thus far indeed yielded archaeological evidence to partly verify this process (Ohnuma *et al.* 2010). They revealed that settlements of a certain scale (16 ha or more), such as Tell Ghanem al-Ali, had existed in the Euphrates lowlands in the 3rd millennium BC, and that contemporaneous cemeteries had been concentrated on the rims of the steppe plateau. These settlements and cemeteries disappeared in the 2nd millennium BC, when cairn fields emerged far inland in the Bishri Mountains. These observations suggest drastic changes in the society and its subsistence during the interface period of the 3rd and 2nd millennia BC, or the transitional period of the Early to Middle Bronze Age. Although these results are already quite significant, further documentation is needed on the detailed processes of change.

The following two issues deserve particular attention. The first involves the nature of society in the Middle Euphrates valley during the 3rd millennium BC. Archaeological and cuneiform evidence strongly suggest that the populations occupying the steppe of the Bishri Mountains in the 2nd millennium BC were tribal communities of nomads (Fujii and Adachi 2010; Lönnqvist 2010; Buccellati 1966; Charpin 2010; Nakata 2010). There is little doubt that they were descendants of lowland populations of the 3rd millennium BC. The question is therefore whether they had already formed the same type of tribal communities in the 3rd millennium BC, or whether they transformed their societal organization in the 2nd millennium BC. The second issue involves the timing of these events. It is almost certain that full-scale expansion to inland began during the 3rd and 2nd millennia BC, but the exact dates remain unclear. Precise dating is vital in order to explain the circumstances of expansion. For example, it would be useful to correspond our dates to significant changes in global climate that supposedly occurred at the end of the 3rd millennium BC, as well as notable historical events, such as the collapse of the Akkadian Empire (Costanza *et al.* 2007; Weiss *et al.* 1993; Walker *et al.* 2012). Although both issues are difficult to resolve conclusively with survey data alone, this paper attempts to discuss them by combining our results with those of other research teams.

Insights from general surveys in the Middle Euphrates Valley

Our fieldwork began in August 2007 with a preliminary inspection of the target area (Nishiaki 2008). The first intensive survey was then conducted between March and April 2008 (Nishiaki *et al.* 2009), followed by subsequent surveys in February–March 2009 (Nishiaki *et al.* 2011a), May 2009 (Nishiaki and Abe 2010), February–March 2010 (Nishiaki *et al.* 2011b), and March 2011 (Nishiaki *et al.* 2012). As mentioned above, the target area was the 10 km radius from Tell Ghanem al-Ali, where many *wadis* run from south to north (Fig. 2). The survey team walked along these *wadis* and areas in between, checking for the presence of archaeological sites and their content, when found. The uniqueness of these surveys lies in the detailed recording of surveyed areas regardless of the presence or absence of archaeological remains. Therefore, it is unnecessary to discuss whether the areas lacking archaeological records signify an absence of sites or an absence of research. The number of archaeological sites recorded during the five seasons of surveys was more than 200.

The paper now attempts to describe the characteristics of human subsistence and societal structure during the 3rd millennium BC, as suggested by settlement patterns. We then discuss how that society commenced expansion into the steppes at the end of the 3rd millennium BC.

Subsistence of the 3rd millennium BC society

The excavation of Tell Ghanem al-Ali revealed continuous occupation throughout almost the entire 3rd millennium BC (Hasegawa 2010; Nakamura 2010). Furthermore, it yielded remains of barley and other crops as well as domesticated sheep and goats. These finds suggest that village communities likely existed in the Euphrates lowlands whose subsistence was based on agriculture and stockbreeding over this millennium (Hongo *et al.* 2010).

On the other hand, of the archaeological sites found in the steppes, no tells such as Ghanem al-Ali were dated any earlier than the end of the 3rd millennium BC (see below). Only locations scattered with flint artifacts and burials could be dated back to the late 3rd millennium BC. Artifacts collected at the flint scatters consisted of simple flakes (Fig. 3). A comparative analysis of flakes recovered at Ghanem al-Ali and other radiocarbon dated archaeological sites in the region allowed us to date most of these flint scatters to the late 3rd millennium BC afterwards (Nishiaki 2010b). The scarcity of artifacts at each locality implies only extremely short occupations by small communities. Judging from the absence of farming tools, such as sickles and ground stones, and hunting tools such as arrowheads, the most likely interpretation is that the flint scatters represent places used for short-term visits by shepherds or pastoralists. Many of the simple flakes excavated from Bronze Age sites in the Levant are assumed to have been used for animal butchery (Greenfield 2006). Although no use-wear analysis has been conducted on the flakes from these surface scatters, they could also have been discarded after animal butchery in short-term occupation sites.

If this interpretation is correct, the question of who was responsible for these archaeological sites remains: shepherds dispatched from lowland tells or nomads based in the steppes on the plateau? The answer may be found by determining who left the graves at the boundary between the valley and steppes. A German archaeological team has already reported the presence of a large number of burials from the 3rd millennium BC on the steppes of the Middle Euphrates valley plateau (Falb *et al.* 2005). Some of these burials were also excavated as part of the Bishri Project (Numoto and Kume 2010). If these burials form the cemetery for

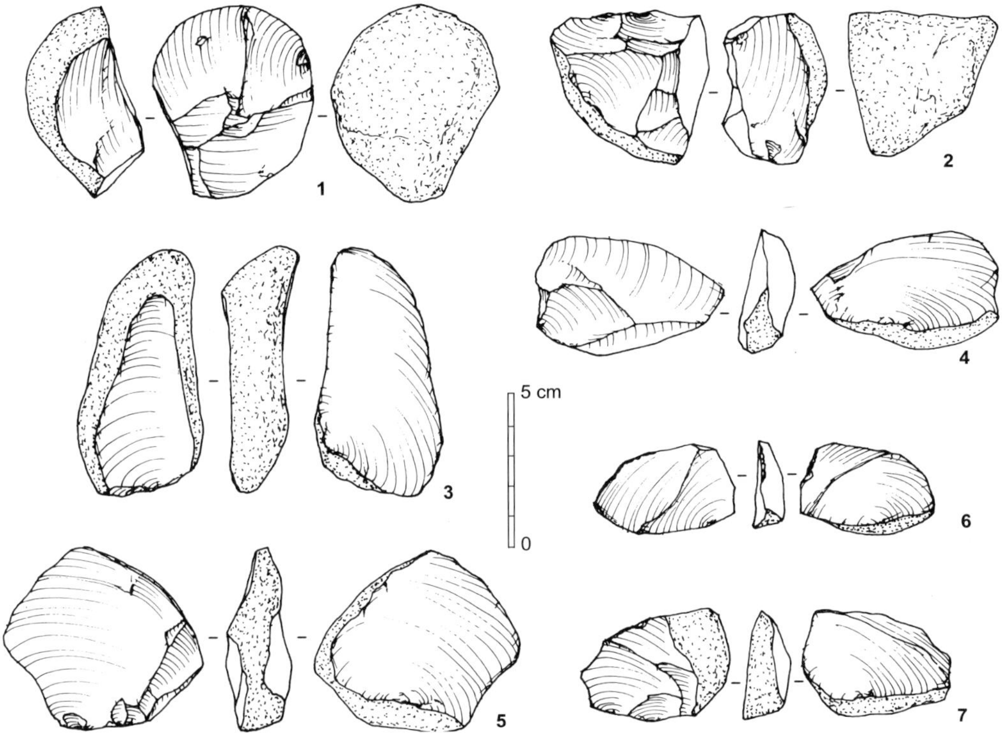

Fig. 3: Stone artifacts collected at flint scatters on the steppes (Nishiaki 2010b).

lowland communities, such as Ghanem al-Ali, it is likely that only lowlanders engaged in farming were present at the time and their regular or seasonal visits for grazing resulted in the flint scatters on the steppes. However, if the burials were left by communities other than lowlanders, this would suggest that nomads existed inland in the 3rd as well as the 2nd millennium BC.

Attributing the burials to a specific group remains a controversial topic. Meyer (Meyer 2005, 2010), who excavated burials at the site of Abu Hamad on the steppe near Ghanem al-Ali, attributes them to nomads, namely communities other than lowlanders. He noted that the burials differ in structure from those in lowland settlements known in the Upper Euphrates valley, and regarding quantity, the number of burials seems to exceed the population of lowlanders as estimated from the scale of tells. Meyer adds that the burials of lowlanders, like the inhabitants of Ghanem al-Ali, may be as yet undiscovered, located somewhere in the lowlands.

This paper presents a different view. First, burials were not found in the lowlands around Tell Ghanem al-Ali in spite of our rigorous, purposeful surveys. Second, as shown by excavations by Numoto and Kume, burial areas "on the plateau" are actually not limited to the plateau but rather spread out in succession, reaching close to lowland settlements (Numoto and Kume 2010: fig. 1). There is no clear spatial boundary between lowland settlements and plateau cemeteries. Then, the presumption made here is that burial areas on the plateau could include burials of both nomads and lowlanders. There may have been loose federations of farmers and nomads of the same tribe in the Bronze Age (Cooper 2006). In keeping with this, the argument here is that the communities on the steppe plateau might have been virtually the same as the lowlanders. Indeed, burial areas are widespread on the plateau and some burials are located at a distance of over 2 km from settlements, suggesting that not all burials are attributable to lowlanders. It is possible to assume that the individuals buried in these burials include those who were tribally identical but had not settled in the lowlands, namely nomads. The reality of Bronze Age society in the Middle Euphrates valley in the 3rd millennium BC may have been inseparable communities of farmers and nomads.

Tribalism of the 3rd millennium BC society

The next issue to be addressed involves social structure, especially whether or not tribal communities already existed in the study region during the 3rd millennium BC. The cultural anthropology literature defines modern tribal communities of the Middle East as those meeting the following four conditions (Evans-Pritchard 1940; Akahori 2009): a) having their own territory; b) being structurally homogeneous in each segment; c) being composed only of patrilineal families; and d) defining one's origin by paternal kinship. The study of cuneiform texts has revealed the existence of such a society, linked to modern Arab society, by at least the 2nd millennium BC and after, but written evidence for the earlier period is very limited (Postgate 1994: 84-87).

Archaeological evidence cannot tell us about a community's patrilineality (c) or about traits in origin (d) without testing ancient DNA from unearthed human bones. However, archaeology can provide more or less convincing evidence for the other two conditions (a and b). For example, cairn fields excavated in the Bishri Mountains dating to the 2nd millennium BC are distributed in clusters that form bigger groups (Fujii and Adachi 2010; Lönnqvist 2010). As long as the spatial structure of burial grounds is considered representative of the communities themselves, then this pattern represents exactly the segmental characteristic (b) of tribal communities. Our surveys have also shown similar distribution patterns for burials of the 3rd millennium BC on the steppe plateau. As recorded by the German research team (Falb *et al.* 2005) at the cemetery of Abu Hamad, the burials on the plateau also display a nested distribution pattern. It is indeed a segmental structure (b), identical to the Bishri cairns from the 2nd millennium BC.

Furthermore, the correspondence of lowland tells to plateau cemeteries fits with the first condition (a) in that the settlement area comprises the tribe's territory. Three examples have been found in the survey area of a tell and a plateau cemetery forming a pair (Fig. 4). In order from west to east, these include the Hamadin, Ghanem al-Ali, and Mugla as-Sagir settlements and their corresponding burial sites. Importantly, these pairs are separated from one another by areas lacking burials and these areas comprise deep wadis (Fig. 5). This pattern suggests that the groups occupying these settlements and to which the burial grounds belonged mutually respected geographical boundaries and maintained their own territories.

Although this concludes the full extent of potential conclusions based on the survey data, the assumption is that tribal communities already existed in the 3rd millennium BC. This view corresponds with previous studies discussing tribalism in societies of the 3rd millennium BC that were based on excavations of archaeological sites in riverside lowlands of the neighboring regions. For example, magnificent remains relating to ritual practice dating to the 3rd millennium BC have been discovered in the area of Tell Banat on the Upper Euphrates. The excavators believe these were a venue for ancestral worship repeatedly used by

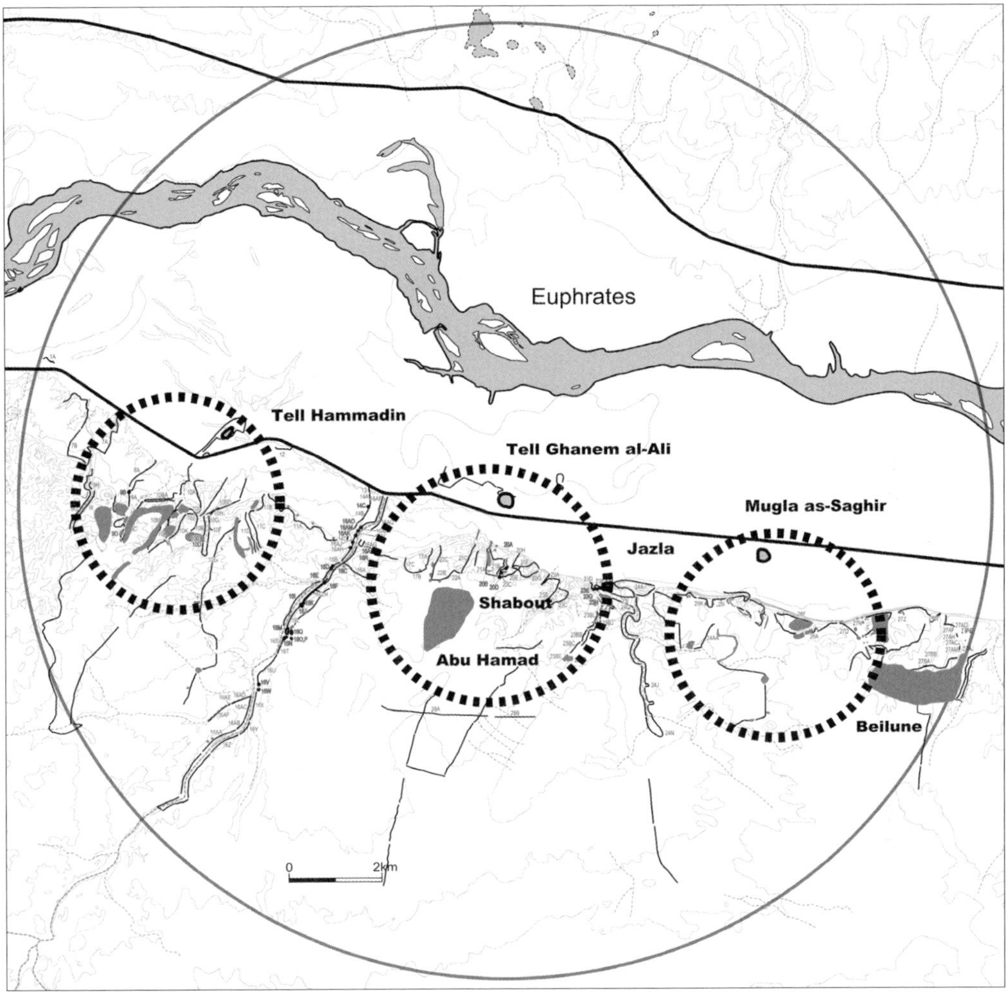

Fig. 4: Three pairs of lowland settlements and plateau burial grounds.

nomadic tribes (Porter 2002, 2008). Similarly, Peltenburg (Peltenburg 2008), who examined archaeological sites in regions further upstream, and Lyonnet (Lyonnet 2009), who reviewed available evidence on settlement structure in Mari and other settlements in the lower region, both argue for the existence of tribal communities in the Euphrates valley in the 3rd millennium BC. Our survey records support this viewpoint, namely that tribal communities were already established by the 3rd millennium BC.

An intriguing question remaining to be answered involves how these communities of the Middle Euphrates were connected, since they were segmented in a nested pattern. Cairn fields, which suddenly appear in large numbers in the Bishri Mountains in the 2nd millennium BC, are presumed to have functioned as sacred places, e.g., points of ancestral worship among tribal communities. Our current supposition is that communities of the Middle Euphrates valley in the 3rd millennium BC might also have had such places. They would not have been inland such as in the Bishri Mountains but in locations near lowlands with tells,

Fig. 5: The Wadi Kharrar, separating Hamadin from Ghanem al-Ali.

as in the case of the monumental construction at Tell Banat on the Upper Euphrates (Porter 2002, 2008). The most likely candidate is in valleys known as Beilune (Fig. 4), where a number of cairn clusters from the 3rd millennium BC were discovered over the course of our surveys (Nishiaki *et al.* 2011a, b). Although cairns from this time period have been occasionally discovered in other burial areas on the rims of the plateau, these valley areas are completely different, as they consist almost exclusively of cairn burials rather than shaft burials on the plateau (Fig. 6). Moreover, they also differ in that there is no accompanying lowland settlement. As already mentioned, all of the other burial areas proximate to the plateau are paired with corresponding lowland settlements (Fig. 4).

Based on pottery sherds collected from the cairns and radiocarbon dates for some of the looted ones (Fig. 7), there is no doubt that the cairn field at Beilune included cemeteries of the 3rd millennium BC. In comparison with other contemporary burial fields on the plateau, the differences point to the cairn field's special role for the regional society. Like the cairn field of the Bishri Mountains in the 2nd millennium BC, did it serve as a sacred place for communities in the 3rd millennium BC? However, at the current stage of research, this represents nothing more than a speculation that needs to be verified by excavations. At this time, none have been conducted.

Steppe exploitation after the end of 3rd millennium BC

The sections above refer to the nature of subsistence and society in 3rd millennium BC human communities in the Middle Euphrates valley. Those communities are presumed to have been semi-nomadic tribes with territories encompassing the Euphrates lowlands and nearby

Fig. 6: Valleys of Beilune and cairn fields.

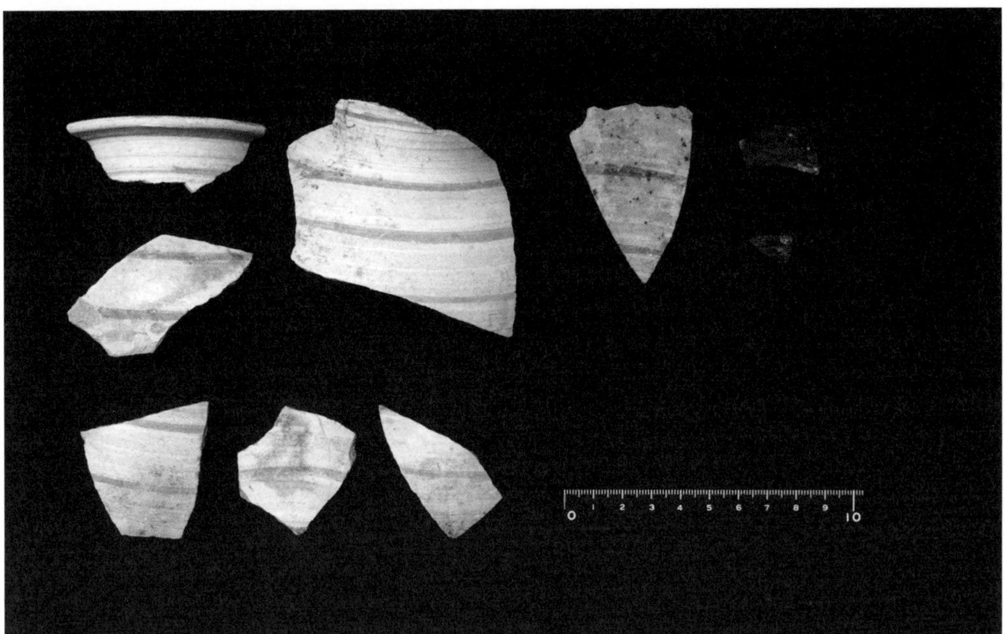

Fig. 7: Pottery collected at the cairns of Beilune.

Fig. 8: A Bronze Age mound at Jazla.

rims of the plateau. However, in the 2nd millennium BC, most of the settlements in the Euphrates lowlands, cemeteries at the plateau edge, and the cairn fields of Beilune were all deserted. Instead, cairn fields on a far larger scale emerged in the Bishri Mountains. The significant decrease of archaeological traces in the riverside areas suggests a shift in emphasis to a more nomadic life.

This section refers to the process of this change. Although the change may seem to have occurred suddenly, the surveys undertaken for this study indicate that it happened in stages. Two important data sets are presented to support this argument.

The first piece of evidence is the discovery of a group of sites near Jazla on the plateau (Nishiaki *et al.* 2009), which bridge the society of riverside lowlands in the 3rd millennium BC with that of inland steppes in the 2nd millennium BC, both in chronological and spatial terms (Fig. 2).

These sites probably include one discovered by Kohlmeyer (Kohlmeyer 1984), who identified Bronze Age artifact scatters near the Roman fortress of Jazla (Napoli 2000). However, our surveys revealed that the area comprised a large site complex consisting of at least three Bronze Age sites (Fig. 8).

The first site lies at the height of the plateau edge on which the Roman fortress itself was built. The second site is a small tell (Fig. 8), approximately 60 m in diameter, situated on a low terrace to the west across a wadi. The third site comprises the cemetery or burial ground that spreads out from the fortress and the mound far inland from the plateau. This site complex has extremely interesting aspects, including the first known settlement built on the plateau, which was abandoned not long after its construction. The recovery of ground stones and sickle blades indicates that the inhabitants practiced farming in the lowlands, while main-

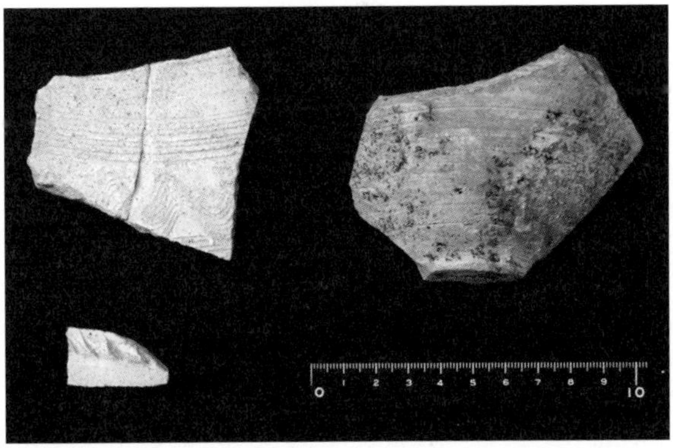

Fig. 9: Pottery collected at the mound near Jazla.

taining their settlement on the plateau. The latter aspect suggests an emphasis on exploitation of the steppe as well. The pottery collected at all three sites suggest a date between the end of the 3rd and the beginning of the 2nd millennium BC (Fig. 9) (Nishiaki *et al.* 2009).This typological dating was confirmed by three radiocarbon dates obtained in looters' pits at the mound and the cemetery, all falling in a period of ca. 2200-1875 BC (Nishiaki in prep). This period overlapped with one when mudbrick buildings disappeared at lowland settlements such as Ghanem al-Ali. Furthermore, it corresponds to a period after which the oldest cairn fields were established in the Bishri Mountains. In other words, the site complex at Jazla represents a highly critical period of transition in the local settlement history.

The second body of evidence revealing the process of steppe exploitation involves lithic artifacts. It is clear that areas with surface scatters of flint flakes rapidly increased in number on the steppes during the Early Bronze Age. In order to identify the exact period of this increase, chronological changes in lithic technology were reexamined using the stratigraphic sequence of lithic samples unearthed from Ghanem al-Ali (Nishiaki 2014). Rather than the conventional division into three phases of the Ghanem al-Ali sequence proposed by the excavators (Hasegawa 2010), this paper proposes that it be divided into four phases, according to radiocarbon dating (Nakamura 2010): 3100–2900 cal. BC, 2900–2650 cal. BC, 2650–2350 cal. BC, and 2350–2050 cal. BC. The fourth phase coincides with the period when mudbrick structures disappeared at this mound. Lithic analysis based on this division determined that flint flakes from the fourth phase most resembled those distributed on the plateau (Nishiaki 2014). This suggests that the increase of flake scatters on the steppes occurred close to the end of the 3rd millennium BC, after 2350 BC, a period shortly before the above-mentioned emergence of settlements in Jazla. In brief, it is assumed that immediately prior to the construction of settlements on the plateau, there was a period of intensified steppe use by communities inhabiting plateau edge settlements.

Conclusion

This paper has examined the communities of the 3rd millennium BC in the Middle Euphrates valley. Specifically, it considered subsistence, social structure and the changes in their settlement patterns, based on the results of our site reconnaissance surveys.

The reconstructed picture is as follows. In the 3rd millennium BC, semi-nomadic groups, with characteristics of tribal communities, already existed in the Middle Euphrates valley. Based in lowland settlements, they began to develop exploitation of the steppes at the end of the 3rd millennium BC. Soon thereafter, they moved to the plateau close to the lowlands and stayed for a couple of centuries. At the beginning of the 2nd millennium BC, these groups again deserted their settlements and established their bases in the inland Bishri Mountains.

The survey data show that the settlement changes from lowlands to the steppe did not occur all at once, but instead, began well into the 3rd millennium. Although it is a general consensus that the transition from the Early to Middle Bronze Age and the sudden increase of cairn fields in the Bishri Mountains took place after 2000 BC (Anastasio *et al.* 2004; Nakamura 2010), this apparent turning point in the local settlement history was, in fact, the conclusion of a process already underway in the previous centuries. This paper has not discussed the reasons behind this change. For example, it is unlikely that the global aridification that took place at the end of the 3rd millennium BC (known as the 4.2 k event)(Costanza *et al.* 2007; Weiss *et al.* 1993; Walker *et al.* 2012) was irrelevant. However, further study is required to determine why these communities chose to intensify the use of inland over all the other possible choices for coping with aridification. It is quite possible that chronologically relevant archaeological sites discovered during this study will serve as physical evidence for informing us about such a significant event in Mesopotamian history. The survey data presented here certainly deserves further analysis.

Bibliography

Akahori, M.
2009 Narrating tales of saints is making saints: Three different hagiographic traditions of a Muslim saint in the Egyptian Desert, *Orient* 42: 27-39.

Anastasio, S., M. Lebeau and M. Sauvage
2004 *Atlas of Preclassical Upper Mesopotamia*, Turnhout.

Buccellati, G.
1966 *The Amorites of the Ur III Period*, Naples.

Charpin, D.
2010 The desert routes around the Jebel Bishri and the Sutean nomads according to the Mari archives, *Al-Rafidan Special Issue*: 239-245.

Cooper, L.
2006 *Early Urbanism on the Syrian Euphrates*, London.

Costanza, R., L.J. Graumlich and W.L. Steffen
2007 *Sustainability or Collapse? An Integrated History and Future of People on Earth,* Cambridge.

Evans-Pritchard, E. E.
1940 *The Nuer: A Description of the Modes of Livelihood and Political Institutions of a Nilotic People,* Oxford.

Falb, C., K. Krasnik, J.-W Meyer and E. Vila
2005 *Gräber des 3. Jahrtausends v. Chr. im syrischen Euphrattal: 4. Der Friedhof von Abu Hamed,* Saarwellingen.

Fujii, S. and T. Adachi
2010 Archaeological investigations of Bronze Age cairn fields on the northwestern flank of Mt. Bishri, *Al-Rafidan Special Issue*: 61-77.

Greenfield, H.J.
2006 Slicing cut marks on animal bones: Diagnostics for identifying stone tool type and raw material, *Journal of Field Archaeology* 31/2: 147-163.

Hasegawa, A.
2010 Sondage at the site of Tell Ghanem al-Ali, *Al-Rafidan Special Issue*: 25-35.

Hongo, H., C. Akashi, L. Omar, K. Tanno and H. Nasu
2010 Zooarchaeology and ethnoarchaeobotany at Tell Ghanem al-Ali, *Al-Rafidan Special Issue*: 97-104.
Kohlmeyer, K.
1984 Euphrat-Survey: Die mit Mitteln der Gerda Henkel Stiftung Durchgeführte Archäologische Geländebegehung im Syrischen Euphrattal, *Mitteilungen der Deutschen Orient-Gesellschaft zu Berlin* 116: 95-118.
Lönnqvist, M.
2010 Tracing tribal implications among the Bronze Age tomb types in the region of Jebel Bishri in Syria, *Al-Rafidan Special Issue*: 165-173.
Lyonnet, B.
2009 Who lived in the third millennium "round cities" of northern Syria. In J. Szuchman (ed.), *Nomads, Tribes, and the State in the Ancient Near East,* Chicago: 179-200.
Meyer, J.-W.
2005 Auswertung und Zusammenfassung der Ergebnisse. In C. Falb, K. Krasnik, J.-W. Meyer and E. Vila (eds.), *Gräber des 3. Jahrtausends v. Chr. im syrischen Euphrattal: 4. Der Friedhof von Abu Hamed*, Saarwellingen: 359-365.
2010 The cemetery of Abu Hamad: A burial place for pastoral groups? *Al-Rafidan Special Issue*: 155-163.
Nakamura, T.
2010 The Early Bronze Age chronology based on 14C ages of charcoal remains from Tell Ghanem al-Ali, *Al-Rafidan Special Issue*: 119-129.
Nakata, I.
2010 Nomads and farmers in the orbit of the Mari kingdom in the 18th century (B.C.E.) Syria: A few observations on Merhum-officials and their roles, *Al-Rafidan Special Issue*: 231-238.
Napoli, J.
2000 Les emparts de la forteresse de Djazla sur le Moyen-Euphrate, *Syria* 77: 117-136.
Nishiaki, Y.
2008 Prehistoric survey at the northern edge of Jebel Bishri, Raqqa, *Al-Rafidan* 29: 151-152.
2010a Archaeological evidence of the Early Bronze Age communities in the Middle Euphrates steppe, North Syria, *Al-Rafidan Special Issue*: 37-48.
2010b Early Bronze Age flint technology and flake scatters in the North Syrian steppe along the Middle Euphrates, *Levant* 42/2: 170-184.
2014 Dating simple flakes: Early Bronze Age flake production technology at Tell Ghanem al-Ali, the Middle Euphrates, Syria. *Journal of Lithic Studies* 1/1 (in press).
Nishiaki, Y., S. Kadowaki and S. Kume
2009 Archaeological survey around Tell Gahnem Al-'Ali, *Al-Rafidan* 30: 145-153, 160-163.
Nishiaki, Y. and M. Abe
2010 Archaeological survey around Tell Ghanem Al-Ali (III), *Al-Rafidan* 31: 125-128.
Nishiaki, Y., M. Abe, S. Kadowaki, S. Kume and H. Nakata
2011a Archaeological survey around Tell Ghanem Al-Ali (II), *Al-Rafidan* 32: 189-215.
Nishiaki, Y., S. Kadowaki, H. Nakata, K. Shimogama and Y. Hayakawa
2011b Archaeological survey around Tell Gahnem Al-'Ali (IV), *Al-Rafidan* 32: 125-133.

Nishiaki, Y., S. Kadowaki, S. Kume and K. Shimogama

2012 Archaeological survey around Tell Gahnem Al-'Ali (V), *Al-Rafidan* 33: 1-6.

Numoto, H. and S. Kume

2010 Surveys and sondage at the cemeteries near the site of Tell Ghanem al-Ali. *Al-Rafidan Special Issue*: 49-60.

Ohnuma, K., S. Fujii, Y. Nishiaki, S. Miyashita, A. Tsuneki and H. Sato (eds.)

2010 *Al-Rafidan Special Issue: Formation of Tribal Communities –Integrated Research in the Middle Euphrates, Syria*, Tokyo.

Peltenburg, E. J.

2008 Enclosing the ancestors and the growth of socio-political complexity in Early Bronze Age in Syris, *Scienze Dell'antichita, Storia Archeologia Antropologia* 14/1: 215-247.

Porter, A.

2002 The dynamics of death: Ancestors, pastoralism, and the origins of a third-millennium city in Syria, *Bulletin of the American Schools of Oriental Research* 325: 1–36.

2008 Evocative topography: Experience, time, and politics in a landscape of death, *Scienze Dell'antichita, Storia Archeologia Antropologia* 14/1: 195-214.

Postgate, J.N.

1994 *Early Mesopotamia: Society and Economy at the Dawn of History*, London and New York.

Walker, M.J.C., M. Berkelhammer, S. Bjorch, L.C. Cwynar, D.A. Fisher, A.J. Long, J.J. Lowe, R.M. Newnham, S.O. Rasmussen and H. Weiss

2012 Formal subdivision of the Holocene series/epoch: a discussion paper by a Working Group of INTIMATE (Integration of ice-core, marine and terrestrial records) and the Subcommission on Quaternary Stratigraphy (International Commission on Stratigraphy), *Journal of Quaternary Science* 27/7: 649-659.

Weiss, H., M.-A. Courty, W. Wetterstrom, F. Guichard, L. Senior, R. Meadow and A. Curnow

1993 The genesis and collapse of third millennium North Mesopotamian Civilization, *Science* 261: 995-1004.

ATSUNORI HASEGAWA and KATSUHIKO OHNUMA
University of Tsukuba, Kokushikan University

New Evidence from Tell Ghanem Al-Ali

A Third Millennium Site in the Middle Euphrates, Syria

In 2007, a Syrian-Japanese Archaeological Joint Research in the Bishri region began fieldwork. To date, we have carried out sounding excavations five times at Tell Ghanem al-Ali. The aim of this article is to show the latest results of this work.

Tell Ghanem al-Ali is located 50 km east of the city of Raqqa, and 2.5 km south of the Euphrates. It is located on the river terrace of the Euphrates and measures about 290m (west to east), about 250 m (north to south), and 10m in height. However, the west and south slopes of the tell have been artificially flattened in modern times. So, the size of the tell today is smaller than that of the original (Fig.1). According to a contour map that was made in 1962 by an Italian company, the tell used to have an irregular oval plan with maximum width of about 400 m (west to east) and about 300 m (north to south), reaching about 12 ha. According to these measurements, Tell Ghanem al-Ali was of a middle size among Bronze Age sites located along the middle Euphrates.

We have carried out soundings of eight squares. The main objective of the sounding was two-fold. The first was to confirm the chronological sequence of Tell Ghanem al-Ali. The second was to investigate the town plan of the site.

For the first objective, we set a step trench on the northern slope of the mound. We call it Square 2. For the second objective, we chose to investigate the east and west slopes of the mound, where many architectural ruins were identified. We named them Squares 1, 7 and 8. In addition to these main four squares, we carried out soundings of four smaller squares. They were named Squares 3 to 6. Squares 3 and 6 are located on the northwest part of the mound. Squares 4 and 5 are on the south part of the mound. This paper will discuss the results from the main four squares and Square 6.

Chronology of Tell Ghanem al-Ali

In Square 2, a step trench along the northern slope of the mound (Fig. 2), we identified eight building levels before we reached virgin soil (Al-Maqdissi, M. and K. Ohnuma 2008, 2009). Based on these structures, it is possible to classify them into three phases (Hasegawa 2010).

The Third Phase

The third phase, which is the latest phase, includes building levels 1 to 4. They consist of dwelling architecture constructed of stones. They are rectangular, and their longitudinal orientation is mainly in the north-south axes. In some rooms, we found evidence of plastered floors. Some ovens were unearthed. It is possible that there was a town wall in this phase, as well.

Studia Chaburensia 4 (2014), pp. 125–136.

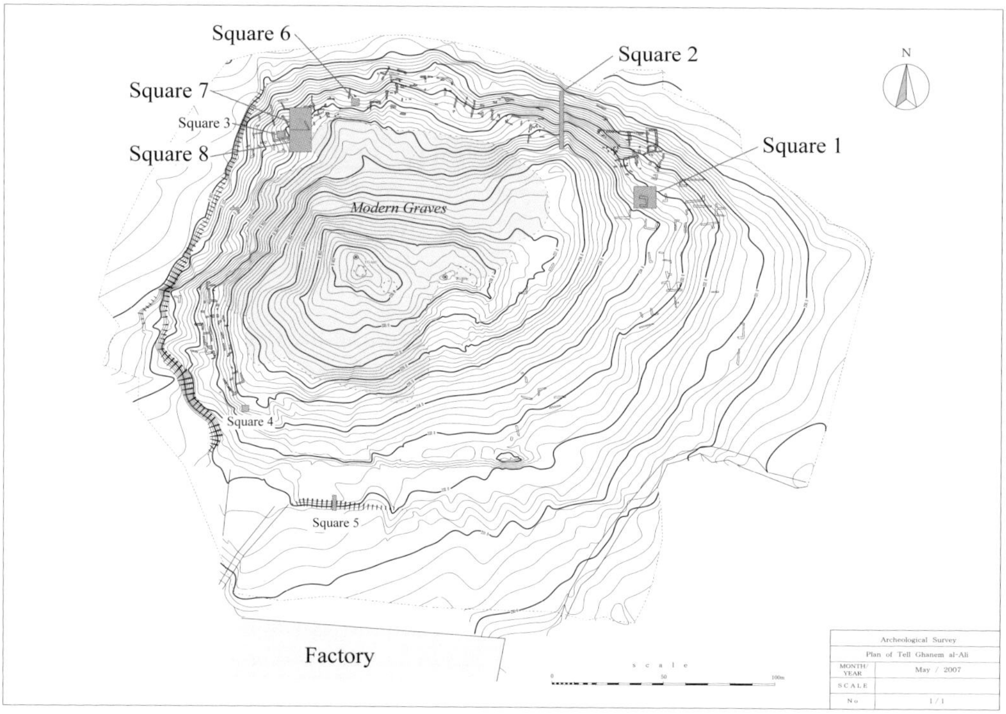

Fig. 1: Contour map of Tell Ghanem al-Ali.

For example, in the building level 2, three adjoining rooms were identified (Fig. 3). Their arrangement is in the north-south direction. All of them have an entrance through their west walls. Between a southernmost room and a central one, there is a doorway. To the south of these three rooms, another room was identified. One of its walls, running along the east–west axis, is shared by the three adjoining rooms. This room has a gypsum-plastered floor. In some rooms, we found fire installations. In building levels 3 and 4, massive stone walls were found. A massive wall in building level 3 is hidden in the east section baulk of Square 2. It continues towards the north and remains more than 1m in height. Large stones are used for parts of this wall.

In building level 4, a massive wall is formed by a pair of stone walls (Fig. 4). They extend from north–west to south–east and run in parallel. The north wall is thin but very high. It remains about 1.8 m in height. On the other hand, the south wall is very thick. It is about 2m in width. This massive wall is constructed with stones larger than those used for dwellings in the upper level. This apparent double wall structure is laid out in a different direction compared to the walls in the upper level. It is possible that they were a part of a town wall.

The Second Phase

The second phase includes building levels 5 and 6. The buildings in level 5 are poorly preserved, but one thick wall built of large stones was identified in building level 6. Its orientation is mainly in the northwest to southeast. Both the size and the orientation of this wall are quite different from those in the third phase.

Fig. 2: Square 2, from the north.

Fig. 3: Building level 2 of Square 2, from the south.

Fig. 4: Building level 4 of Square 2, from the west.

In contrast to the southern part of Square 2, which mostly belongs to the third phase, the central part of Square 2 had very thick topsoil, which was disturbed up to approximately 1.3 m below the surface of the highest point of the tell. Under these top layers, we found a foundation built of large stones. Upon this stone foundation, some mud-bricks were identified.

The First Phase (Fig. 5)

The first phase includes building levels 7 and 8. In each level, three adjacent rooms were identified. The walls are constructed of mud-bricks measuring about 30 × 60 cm. They run to the northwest and southeast. It is noteworthy that unlike the walls in the upper phases, the walls in this phase do not have stone foundations.

The size of the mud-bricks and the axes of the walls are similar between levels 7 and 8. However, the arrangement of mud-bricks is different. In level 7, mud-bricks are placed side by side along a long axis, making the width of the walls 60cm. In contrast, in the level 8, they are placed along a short axis, and the walls are 30cm wide.

Below Level 8

Below building level 8 was an 40 cm thick layer of ash, which contained many charcoal inclusions. Below the ash layer, we encountered a layer of brown soil. This layer contained a few fragments of potsherds and charcoal, but no structures. The next layer consisted of darker soil that was wet and homogeneous, like silt. Neither potsherds nor other remains were present. It is certain that this is the virgin soil at Tell Ghanem al-Ali. It is 3.4m deep from the surface at the north end of Square 2.

For the third millennium B.C., there are a lot of chronological terms, for example Early Bronze I to IV, Early Jazira, and so on (Pruß 2004; Pfälzner 1997; Lebau 2000) To analyze the pottery found from Tell Ghanem al-Ali, I would first like to use the chronological scheme suggested by Anne Porter, based on the pottery from the middle Euphrates.

The pottery found inside the structures of the third phase, which is the latest phase, includes Plain Simple Ware. There are a few fragments of Euphrates Fine Ware, such as Black Euphrates ware. It seems that similar pottery was found in Phases 3 and 4 of Porter's Chronology (Porter 2007). Almost all the pottery recovered from the second phase is typical Plain Simple Ware. It seems that they belong to Phase 4 of the Porter's Chronology. In addition, many sherds of Cooking Pot Ware with everted rims were discovered.

The most distinctive pottery found from the first phase has circular or crescent motifs incised on their bodies (Fig. 6). But most examples of Plain Simple Ware lack decoration, and it is difficult to find a similar type in the Porter's chronology. It is also difficult to find the comparative examples in the middle Euphrates area. However, circular designs can be found in the assemblage of Amuq G (Braidwod and Braidwod 1960). By considering the radiocarbon dates, it may be possible to suggest that the first phase of Tell Ghanem al-Ali can be placed between the end of the 4th millennium B.C. and the beginning of the 3rd millennium B.C.

According to the result of radiocarbon dating, some charcoal samples from upper ash layers indicate similar dates. They are dated to the end of 3rd millennium B.C. We may be able to assign these layers to the Fourth Phase, in spite of the fact that no structures were found there.

Fig. 5: Building level 8 of Square 2, from the south.

Fig. 6: Pottery found from First Phase.

Square 6

Square 6 was established to investigate the occupational sequence of Tell Ghanem Al-Ali, and its intra-site functions. During two seasons of excavations, one pit-grave (Fig.7) and two building levels were unearthed.

The stratigraphy of this square is as follows: (1) Layer 1 (Surface deposit), 10 to 15 cm in thickness; (2) a pit-grave, dug into the underlying Building Level 1; (3) a pit found in the corner of the square; (4) Building Level 1, with walls run in the north-south and east-west directions; and (5) Level 2a/b, with walls that run in the northwest-southeast and northeast-southwest directions.

The pit-grave that was discovered in Square 6 had an irregular outline that suggests a non-systematic construction. The pottery recovered from the pit-grave in association with a skeleton suggests that it belongs to the Middle Bronze Age, particularly the final phase of the Middle Bronze Age I.[1]

On the other hand, the pottery recovered from the other pit and Building Levels 1 and 2a/b is dated to the Early Bronze Age III/IVa. This strongly suggests that the occupation of Tell Ghanem Al-Ali was at its height in the Early Bronze Age, and in the Middle Bronze Age it was only sporadically used for non-systematic structures such as pit-graves.

Town layout of Tell Ghanem al-Ali

The top of the mound was covered with modern graves of the local villagers (green-colored area). Nevertheless, we are able to identify a considerable number of stone lines elsewhere on the tell. Some lines have one or two corners, and others form rectangular shapes. Some lines in the contour map show the plans of stone lines and white lines seen on the surface.

To investigate the architectural plans visible on the surface, and to identify the period from which they originate, we set Square 1 on the east slope, and Squares 7 and 8 on the west slope of the mound. To understand the town plan, we surveyed the surface of the mound in its north and east parts.

Square 1

In Square 1, three main structures were unearthed. The first structure was found in the southwestern part of Square 1. The second structure was lying at the north of the first one. The third structure was identified to the east of the second structure.

From the results of the investigation in Square 1, we can point out three features. The first is that some walls show signs of reuse and reconstruction, although their foundations were unchanged, and always in the north-south and east-west axes. Inside its room, we found another stone wall which has a shape of "T" like this. It seems that this room was reconstructed in a short time. The second feature that we identified is the presence of three styles of wall construction. The first form of wall construction utilizes only stone. Most of the walls in Square 1 are of this type. The second type of wall construction uses mud-bricks laid upon a stone foundation. The last type of wall is made of mud-brick, but does not incorporate a stone foundation. The third feature is that Square 1 did not contain any gypsum-plastered floors, which were uncovered in Square 2.

1 Personal communication with Michel Al-Maqdissi in August of 2009.

Fig. 7: Pit grave of Square 6, from the north, and pottery found from it.

Squares 7, 8

In the northwest part of the mound, we set a small trench, Square 3, and found a thick plastered floor. The fact that a plaster-floor was found within the small test area of Square 3 but not in Square 1 is interesting, and suggests that the structures within Square 1 and Square 3 may have served different functions. Near Square 3, we set two 10 m x 10 m trenches, which we call Squares 7 and 8.

In Square 7, after removing the surface layer, we encountered a structure in the southern half. It consisted of three walls of stone. The main wall is 70 cm in width, and runs east-northwest. At least four rooms were identified, and it is supposed that they were to have been a part of a multi-roomed building. The room located in the southern end of Square 7 was well preserved, and expanded towards south, which meant that almost all of the main rooms were in Square 8.

The other rooms were found in a fragmentary condition. Unfortunately, we could not identify any plastered floors such as that found in Square 3.

The most characteristic features were five fire installations. Three of them were in the main room, one was in the corner, and the others were located in the eastern edge of Square 7 (Fig. 8). Their plan is circular, and about 70 cm in diameter. They had thin walls constructed of clay. Only the lower parts remained. Pebbles were found at their bases, and, they may have served as foundation layers for the fire pits. There was white ashy soil inside them, and burnt soil and gray ash with charcoal inclusions around them. They were similar to baker's ovens, or *tannur*, that are still in used in local villages today. The fire installation located on the corner of the main wall was built into the wall. It should be noted that stones used for the floor of this fire installation include re-used saddle querns. Five fire installations found in Square 7 have almost the same structure. And, their bottom levels are also almost the same. It seems, therefore, that they were used nearly at the same time.

Fig. 8: Fire installations of Square 7, from the southeast.

Square 8 is located in the immediate south of Square 7. In the northern part of Square 8, we discovered a rectangular building that was the continuation of the main room identified in Square 7. The long axis of the building is two rooms divided by a mud-brick wall. The walls are usually constructed of stones.

Inside the northern room, there was a shallow pit about 3m in diameter, which contained ash and many fragments of charcoal. After removing the ash layer, we unearthed a cluster of stones at the center of the pit. In addition, we unearthed at least three plaster basins made of gypsum. They were attached to the southern wall of the northern room. They have an irregular square shape. One of these gypsum basins is located between the stone cluster and the mud-brick wall. Given the presence of these features, in addition to the fire installations of the main room, which seem to have been *tannurs*, it is possible that this room was a kind of a workshop, such as a bake-house, although we cannot clarify its function at the moment. Almost all of the pottery unearthed from three squares is Plain Simple Ware.

A few fragments of Black Euphrates Ware were also unearthed. Therefore, these squares belong to the third phase of Tell Ghanem al-Ali, which was established in Square 2. It is likely that the buildings unearthed were used during the Early Bronze Age III to the early part of the Early Bronze Age IVa periods.

Surface survey in the eastern and northern half of the mound

A considerable number of buildings were detected on the surface, and thus their outlines were planned. In the eastern half of the mound, we recorded them intensively. One of the largest buildings was detected on the surface. It measures 18 m x 14 m and consists of five or six rooms with a large courtyard. Another large structure was a multi-roomed building that con-

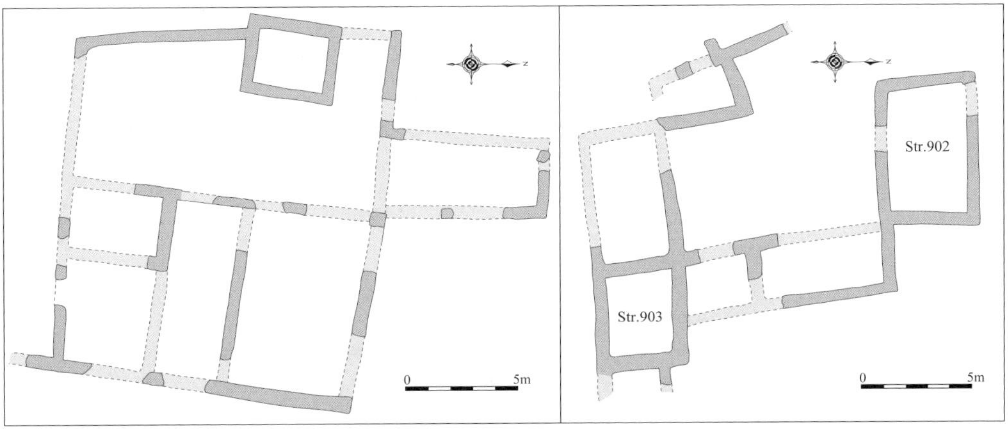

Fig. 9: Some typical building plans (right: A multi-roomed building, left: large room with courtyard).

sisted of five or more rectangular rooms in a row. It measured 20 m long from east to west, and 5.5 m wide from north to south (Fig. 9). They are fairly typical building plans for this site.

Almost all of the buildings detected by the surface survey are multi-roomed houses consisting of a series of rectangular small rooms, with axes orientated east to west. Therefore, we can conclude that this type of house represents the main dwelling structures at that time. We also identified many *tannurs* attached to the outer walls of these houses, and they indicate domestic nature of these buildings. No public buildings were identified.

Although the building outlines were not always clearly detectable, those that were visible provide us with a general settlement layout, at least for the last phases of EB occupation. We could not identify any straight streets or well-defined blocks of buildings. Many vacant plots came to a dead end bounded by building walls. Although there are some rough similarities in building layout and their direction, it seems likely that the late EB people at Ghanem al-Ali built their houses one after another without having any specific town plan.

The relationship between the tell and a cemetery

Before finishing my article I will consider the relationship between Tell Ghanem al-Ali and a cemetery located nearby, by briefly discussing about a clay figurine found on the surface of Tell Ghanem al-Ali. It has a characteristic decoration, a lot of small holes on its head (Fig.10). When compared with other clay figurines from the Early Bronze Age, it is an unusual type.

However, similar figurines have been reported from Hammam et-Turkman (van Loon 1988: 566, Pl. 176: figs. 34, 45), Tell Chuera (Orthmann 1990: 30, Abb.26: 4) and Abu Hamad (Falb *et al.* 2005: 79, Abb. 29: 4; Abb. 41: 8).

Abu Hamad is a cemetery site which is dated to the Early Bronze Age and figurines have been unearthed from burials (Falb *et al.* 2005). Similar burial sites are known on the edge of Euphrates Plateau (Kume 2010) and detailed discussion of the relationship between tells and cemeteries in this area (Nishiaki 2010; Kume 2010; Meyer 2010).

The fact that this figurine was found from Tell Ghanem al-Ali suggests strong relation between people who were buried in a cemetery such as Abu Hamad and those who lived at Tell Ghanem al-Ali.

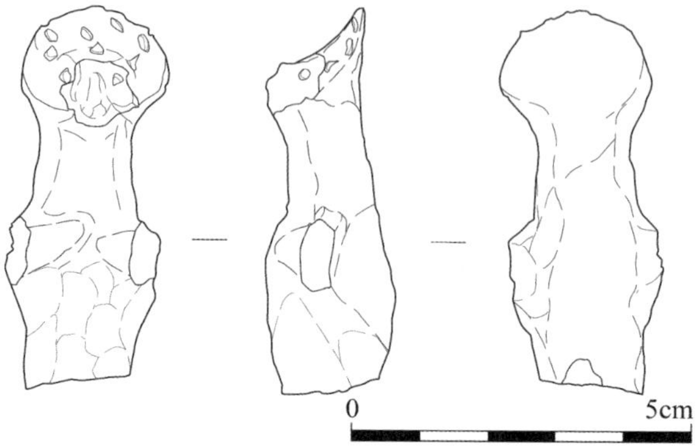

Fig. 10: Clay figurine found from Tell Ghanem al-Ali.

Summary

The results of our suggest that Tell Ghanem al-Ali began to be occupied in the end of the 4th millennium B.C. and the beginning of the 3rd millennium B.C. It was mainly occupied from the Early Bronze Age III to the early phase of the Early Bronze Age IVa periods. The main house type of Tell Ghanem al-Ali was a multi-roomed building consisting of several rectangular rooms. It was apparently not uncommon for houses to have small fire installations attached to their walls.

The townscape of Tell Ghanem al-Ali in the EB period was not deliberately planned, and was constructed in a relatively random manner. After the abandonment of Bronze Age settlements, human activities at Tell Ghanem al-Ali appear to have declined at the site, and seem to have been limited to the construction of small, isolated burials.

Bibliography

Al-Maqdissi, M. and K. Ohnuma (eds.)

2008 Preliminary Reports of the Syria-Japan Archaeological joint research in the region of ar-Raqqa, Syria, 2007, *Al-Rāfidān* 29: 117-193.

2009 Preliminary Reports of the Syria-Japan Archaeological joint research in the region of ar-Raqqa, Syria, 2008, *Al-Rāfidān* 30: 135-225.

Al-Khalaf, M. and J.-W Meyer,

1993/1994 Abu Hamad, *Archiv Für Orientforschung* 40/41: 196-200.

Braidwood, R. J. and L. Braidwood

1960 *Excavations in the Plain of Antioch I, The Earlier Assemblages Phases A-J.* Chicago - Illinois.

Cooper, L.

2006 *Early Urbanism on the Syrian Euphrates*, New York - London.

Falb, C., K. Krasnik, J.-W. Meyer and E. Vila

2005 *Gräber des 3. Jahrtausends v. Chr. im syrischen Euphrattal: 4. Der Friedhof von Abu Hamed,* Saarwellingen.

Hasegawa, A.

2010 Sondage at the Site of Tell Ghanem al-Ali. In K. Ohnuma *et al.* (eds.), *Al-Rāfidān Special Issue*: 25–35.

2011 Archaeological Research in the Bishri Region-Report of the fifteenth Working Season-: II. Sondage at Tell Ghanem al-Ali, Square 7 and 8, *Al-Rāfidān* 32: 158-163.

Kohlmeyer, K.

1984 Euphrat-Survey: Die mit Mitteln der Gerda Henkel Stiftung durchgeführte archäologische Geländebegehung im syrischen Euphrattal, *Mitteilungen der Deutschen Orient-Gesellschaft zu Berlin* 116: 95-118.

Lebeau, M., A. Pruβ, M. Roaf and E. Rova

2000 Stratified Archaeological Evidence and Compared Periodizations in the Syrian Jezirah during the Third Millennium BC. In C. Marro and H. Hauptmann (eds.), *Chronologies des pays du Caucase et de l'Euphrate aux IV-IIIe millénaires*, Paris: 167-192.

Meyer, J.-W.

2010 The cemetery of Abu Hamad: a burial place of pastoral group?, in: K. Ohnuma *et al.* (eds.), *Al-Rāfidān Special Issue*: 155-163.

Nishiaki, Y.

2010 Archaeological evidence of the Early Bronze Age communities in the Middle Euphrates steppe, North Syria. In K. Ohnuma *et al.* (eds.), *Al-Rāfidān Special Issue:* 37-48.

Numoto, H. and S. Kume

2010 Survey and Sondage at the cemeteries near the site of Tell Ghanem al-Ali, In K. Ohnuma *et al.* (eds.), *Al-Rāfidān Special Issue*: 49-60.

van Loon, M. (ed.)

1988 *Hammam et-Turkman I: Report on the University of Amsterdam's 1981-1984 Excavations in Syria,* Vol.1, Istanbul - Leiden.

Ohnuma, K., S. Fujii, Y. Nishiaki, S. Miyashita, A. Tsuneki and H. Sato (eds.)

2010 *Al-Rāfidān Special Issue: Formation of Tribal Communities -Integrated Research in the Middle Euphrates, Syria*, Tokyo.

Orthmann, W.

1990 *Tell Chuera*, Damascus - Tartous.

Pfälzner, P.

1997 Wandel und Kontinuität im Urbanisierungsprozeß des 3. Jtsds. V.Chr. in Nordmesopotamien. In G. Wilhelm (ed.), *Die orientalische Stadt : Kontinuität, Wandel, Bruch,* Saarbrücken: 239-265.

Porter, A.

2007 The Ceramic Assemblages of the Third Millennium in the Euphrates Region. In M. Al-Maqdissi, V. Matoïan and C. Christophe (eds.), *Céramique de l'âge du bronze en Syrie II, L'Euphrate et la region de Jézireh*, Beyrouth: 3-21.

2011 *Mobile pastoralism and the formation of Near Eastern civilizations: weaving together society,* Cambridge.

Pruß, A.

2004 Remarks on the Chronological Periods. In S. Anastasio, M. Lebeau and M. Sauvage (eds.), *Atlas of Preclassical upper Mesopotamia,* Subartu XIII, Turnhout: 7-21.

Strommenger, E. and K. Kohlmeyer

2000 *Tall Bi'a-Tuttul-III: Die Schichten des 3. Jahrtausends v. chr. im Zentralhügel E*, Saarbrücken.

Tsuneki, A., A. Hasegawa and A. Sultan
2011 Archaeological Research in the Bishri Region-Report of the Eighth Working Season-: 5. Sondage and Surface Research at Tell Ghanem al-Ali, *Al-Rāfidān* 32: 181-189.
Wilkinson, T., J.
2004 *On the Margin of the Euphrates, Settlement and Land Use at Tell es-Sweyhat and in the Upper Lake Assad Area, Syria*, Chicago - Illinois.

SHOGO KUME and AHMAD SULTAN
National Research Institute for Cultural Properties, Tokyo,
Directorate-General of Antiquities and Museums, Syria

Burials, nomads, and cities:

A perspective to changing nomad-sedentary relations on the Syrian Middle Euphrates during the third and second millennium BC

In the late 1960s and 1970s, a series of papers on nomad-sedentary relations and tribe state interactions in the ancient Near East were published by M. Rowton (e.g. Rowton 1973, 1974). His famous model of 'dimorphic society' essentially attempted to characterize the nature of the second millennium BC Mari society as the dichotomy of tribe/state institutions. Although there is increasing debate about the nature of the 'dimorphism' of nomadic and sedentary populations proposed by Rowton and interpreted by others, it still remains influential on the issues of nomad-sedentary relations in the archaeology of the Near East (e.g. Liverani 1997; collected papers in Szuchman 2009; Porter 2012 and references therein).

Conspicuous archaeological evidence of massive extramural cemeteries in the Early Bronze Age on the Syrian Euphrates has potential to make a contribution to shed light on the issues, since the possible deceased of the cemetery has been frequently attributed to pastoral nomads on the steppe (e.g. Akkermans and Schwartz 2003: 253). This hypothesis is apparently supposed to exist nomadic pastoralists isolated from sedentary way of life in the valley (Roobaert and Bunnens 1999: 165; Meyer 2005, 2010), reflecting more or less Rowton's concept of 'dimorphism.' On the other hand, recent archaeological survey in the region reveals that such extramural cemeteries in the upland steppe have obviously been paired with settlement mounds in the valley, suggesting that they represent the inhumations of inhabitants of such settlements (Nishiaki 2010a). The conflicting views are seemingly a simple question whether the deceased is pastoral nomads or sedentary. However, it is also a crucial question in practice, since the answer can be closely associated with not only the nature of the 'dimorphism' of nomadic and sedentary populations in the Syrian Euphrates Valley during the Early to Middle Bronze Age, but also discussions on stability and change of economic and political strategies adopted by the populations, which is including social responses to the so-called 4.2ka BP abrupt climate change, the appearance and spread of influential ethnic groups or the Amorites, and developments of tribally organized polities (as recent discussions, see Buccellati 2008; Cooper 2006; Laneri, Pfälzner and Valentini 2012; Ohnuma *et al.* 2010; Porter 2012; Ristvet and Weiss 2013; Schwartz and Nichols 2006; Szuchman 2009; Weiss 2012, 2014; Wossink 2009).

In this paper, we will argue that agro-pastoral strategies instead of specialized agricultural and pastoral economy have been adopted in the Early Bronze Age on the Syrian Middle Euphrates, suggesting that there were no 'real nomads' as a candidate for the deceased of massive contemporary cemeteries in the region. As a case study, an Early Bronze Age site of Tell Ghanem al-'Ali and a nearby cemetery that was recently investigated by us are selected.

Studia Chaburensia 4 (2014), pp. 137–150.

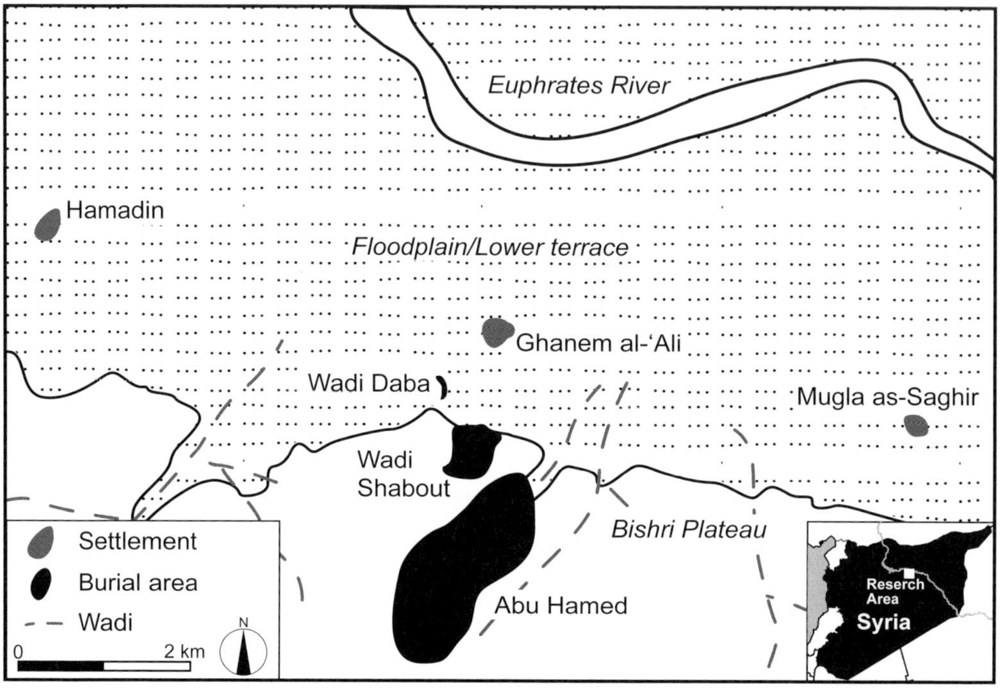

Fig. 1: Tell Ghanem al-'Ali and its surroundings, showing settlements and burial areas mentioned in the text.

Through a reconstruction of modeled mortality dynamics at the site, we will suggest that such massive cemeteries in the region represent the inhumations of inhabitants of nearby settlement mounds. We will further argue that such agro-pastoral strategies have drastically shifted to seemingly 'dimorphic' urban-/steppe-oriented strategies in the subsequent Middle Bronze Age through a comparison of settlements patterns and burial practices during the Early to Middle Bronze Age transition. We emphasizes that the appearance of seemingly dimorphic communities in the Middle Bronze Age in the region was a phenomenon of a flexible transformation of the Early Bronze Age agro-pastoral communities as social responses to environmental and socio-political changes in the transition.

Early Bronze Age extramural cemetery near Tell Ghanem al-'Ali

The Early Bronze Age site of Tell Ghanem al-'Ali is situated on the right bank of the Syrian Middle Euphrates, approximately 40 km southeast of ar-Raqqah, Syria (Fig. 1). The excavations conducted by the Syrian-Japanese mission since 2007 attested a continuous occupational sequence of Early Bronze Age at the site (Fig. 2) (Hasegawa 2010). Extensive and systematic radiocarbon dating of collected charcoal samples from the stratigraphic sequence has also confirmed the continuous Early Bronze Age occupation during the end of the fourth and third millennium BC, ranging from c. 3100 cal. BC to c. 2050 cal. BC, although the latest phase of the occupation without any architectural remains between c. 2400/2350 cal. BC to 2050 cal. BC was apparently sparse (Nakamura *et al.* 2010; Nishiaki this volume).

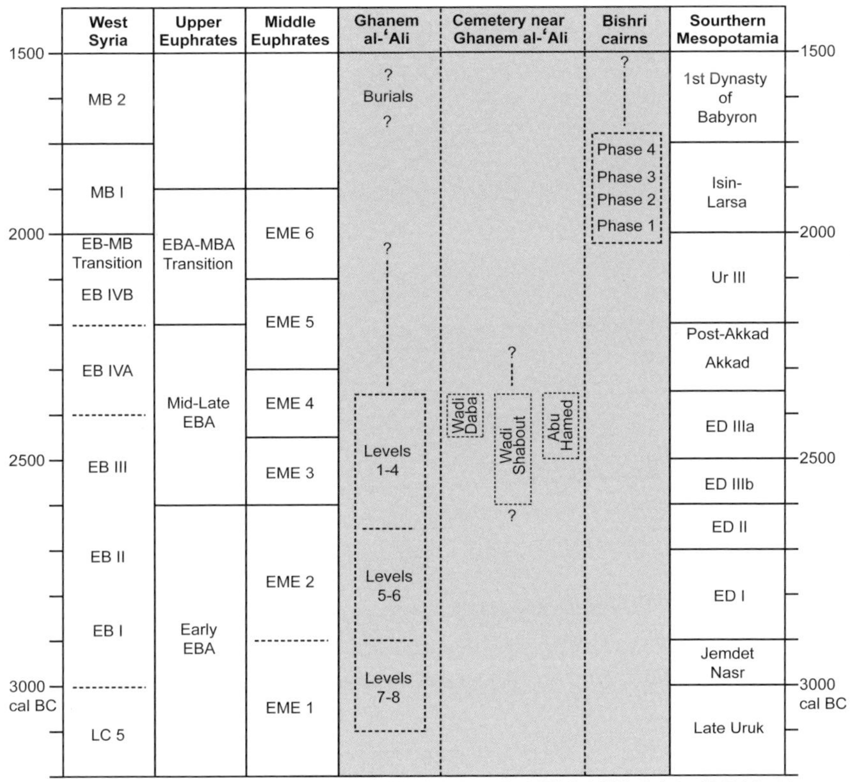

Note 1: LC=Late Chalcolithic; EB(A)=Early Bronze (Age); MB(A)=Middle Bronze (Age); EME=Early Middle Euphrates; ED=Early Dynasty
Note 2: This table is based on various sources, including Akkermans and Schwartz 2003; Cooper 2006; Falb et al. 2005; Fujii and Adachi 2010; Hasegawa 2010; Nakamura et al. 2010; Porter 2007; Pruß 2004; Wilkinson 2004

Fig. 2: Chronological table of Tell Ghanem al-ʿAli, nearby burial areas, and cairns in the Jebel Bishri.

To the south of the settlement, a massive Early Bronze Age extramural cemetery is situated on the upper terrace of the Euphrates or the Bishri Plateau. Unfortunately, graves in the cemetery have seriously been damaged by grave robbers, but the cemetery has contained estimated thousands of clustered graves in an area of 3.5 x 1.5 km (Fig. 3). Several grave-clusters have also been located on the middle terrace of the Euphrates in the modern village of Ghanem al-ʿAli. In 1990, a German mission directed by J.-W. Meyer investigated part of the cemetery named Abu Hamed (Falb *et al.* 2005). The mission has documented 26 grave-clusters in the burial area, containing possibly over 1,000 graves. Subsequently, since 2008, the Syrian-Japanese mission to Tell Ghanem al-ʿAli has investigated different areas of the cemetery, which was provisionally named Wadi Shabout and Wadi Daba burial areas after neighboring wadis for the convenience of our researches (Fig. 1). Because of the wide distribution and disturbed condition of the graves in the massive extramural cemetery, our researches principally focused on documentation of spatial variability of particular burial types in Wadi Shabout and Wadi Daba burial areas through survey and cleaning of the plundered graves (Numoto and Kume 2010a), while soundings of relatively well-preserved graves were also conducted. In particular, uncovered earth-cut shaft graves from Wadi Daba Area 1C, containing two undisturbed infant burials, have provided new evidence for diverse aspects

Fig. 3: Spread of plundered graves on the steppe of the Bishiri Plateau (photo: courtesy of S. Kadowaki).

of mortuary practices on the Early Bronze Age Syrian Euphrates, including social context of grave construction, social differentiation by age and gender, and mortuary rituals like secondary burial practice and funerary feasts (Numoto and Kume 2010b; Kume *et al.* 2011; Kume, Miyata and Kadowaki 2012).

Burials for nomads? Mortality dynamics of Early Bronze Age Tell Ghanem al-'Ali

As we mentioned at the beginning of this paper, there are two candidates for the deceased of the massive extramural cemetery near Tell Ghanem al-'Ali. The excavators of Abu Hamed, for example, have proposed that nomadic pastoralists in the steppe of the Bishri Plateau, rather than inhabitants of the nearest settlement, i.e., Tell Ghanem al-'Ali, were buried in the cemetery (Meyer 2005, 2010: 162). Reasoning behind this hypothesis essentially focuses on massive numbers of graves in the cemetery. More specifically, the excavators assume that the estimated thousands of graves with a limited time-range of Early Bronze III to an early phase of Early Bronze IVA or approximately 150 years (Fig. 2) are disproportionate to the number of the dead from the settlement in the valley during the contemporary period. Because such massive extramural cemeteries have well-documented in the Early Bronze Age on the Syrian Euphrates in particular (Orthmann 1980; Carter and Parker 1995; Cooper 2006), the 'nomads hypothesis' for the deceased of the cemeteries have frequently been applied to the phenomenon (Roobaert and Bunnens 1999: 165; see also Akkermans and Schwartz 2003: 253). On the other hand, our survey results around Tell Ghanem al-'Ali revealed that settlements in the valley including Tell Ghanem al-'Ali, Tell Hamadin, and Tell Mugla as-Saghir were obviously associated with massive extramural cemeteries on the plateau, and that the individual pairings of a settlement and a cemetery appear to represent territorial cohesions of agro-pastoral communities that explored resources from both the valley as well as the steppe instead of the dichotomy of sedentary and nomadic way of life (Nishiaki 2010a: 44-45). The results suggest that the deceased of the cemeteries were inhabitants of the nearest settlements in the Euphrates Valley rather than apparent pastoral nomads on the steppe of the Bishri Plateau. T. J. Wilkinson also suggests the possibility of such close settlement-cemetery linkage through the position of extramural cemeteries on the upper terrace of the Euphrates as an indicator of loss of the floodplain and associated settlements by riverine erosion (Wilkinson 2004: 176 [note 88]).

The above outline of two different hypotheses for the deceased of the extramural cemetery has indicated that the arguments, in nature, tend to reflect two different views on nomad-sedentary relations in the Syrian Middle Euphrates Valley during the Early Bronze Age, i.e., the dichotomy of nomad and sedentary populations or variations of subsistence strategies and habitation patterns by the same agro-pastoral communities (Liverani 1997). Accordingly, pursuing the deceased of the cemetery can be closely associated with broad-scale discussions on political, economic and cultural strategies in the period. A complete discussion on the issues may be challenges of the future, but as a start, this paper attempts to answer the simple question whether the number of estimated thousands of graves in the massive extramural near Tell Ghanem al-'Ali is too large for deaths of the settlement. Toward the goal, estimation of mortality was conducted through three demographic parameters of settlement size, population growth rate and survivorship rate.

First, assuming on-site population densities of 100 persons or 200 persons per hectare (Wilkinson 2004: 140) on the basis of ethnographic analogy (cf. Aurenche 1981; Zorn 1994), calculated population of Tell Ghanem al-'Ali was estimated. In practice, one problem of this method of estimating populations is that measuring of a settlement area is not necessarily easy task because of modern land development and riverine deposition or erosion activities. For example, survey maps of Tell Ghanem al-'Ali prepared by the Syrian-Japanese mission in 2007 and an Italian surveying company in 1961-62 characterize the size of the settlement as c. 7 ha and c. 12 ha respectively. Loss of the 5 ha area of the settlement before 2007 is due to artificial erosions by expansion of agricultural land and construction of a factory at the western and southern foot of the tell (Hasegawa 2008, 2010: 25). In this article, we used the value of 9.5 ha as the settlement size that was acquired from Google Earth through its measuring function. It is obvious that the value is approximate, but the size of 9.5 ha is more or less intermediate between c. 7 ha and c. 12 ha collected from surveying in the field. As a result, using population densities of 100 persons or 200 persons per hectare described above, 950-1900 persons can be estimated as a maximum range of population of Tell Ghanem al-'Ali at a certain point.

Second, reconstruction of population dynamics of Tell Ghanem al-'Ali was challenged, using a general population growth rate. Archaeological evidence and radiocarbon dates indicated that intensive habitation at Tell Ghanem al-'Ali occurred over around 750 years between c. 3100 cal. BC and c. 2350 cal. BC as described above (Fig. 2). In practice, an attempt to estimate dynamics of past human population during such period is required 'the broad spectrum of evidence' including archaeological, anthropological and palaeoenvironmental data that provide proxies for population size and growth, mortality patterns, and fertility (Chamberlain 2009: 275-276). Such interdisciplinary approach is obviously beyond our data at hand. Rather, we adopted a simple method for estimating population dynamics of the settlement in accordance with a case study in Bronze Age Cyprus (Webb and Frankel 2009). Under conditions that there are no catastrophic events (i.e. natural disasters, pandemic disease and armed conflict), stable or quasi-stable populations demonstrate an upward trend in general (Webb and Frankel 2009: 62; cf. Chamberlain 2006: 26-27). According to F. Hassan, equivalent societies generally demonstrate a population growth rate of 0.52 % per year, in which the population doubling time is 133 years or seven generations (Hassan 1981: 140 cited in Webb and Frankel 2009: 62). If we apply the general population growth rate of 0.52 % to the intensive habitation time range of 750 years at Tell Ghanem al-'Ali by assuming that the peak of the number of the population was 2450 cal. BC, just 100 years before the timing of apparent abandonment of the settlement, Table 1 demonstrates two modeled population changes by population densities of 100 persons and 200 persons per hectare during the 133 years or seven generations chronological stages.

Last, cross-cultural studies of survivorship in different ages suggest that only one third of the population can grow into adult levels (Hassan 1981: 103-123 cited in Webb and Frankel 2009: 62; Chamberlain 2006: 43). Based upon the survivorship rate and modeled population growth curves proposed above, Tables 2 and 3 show estimations of the number of adults and the total number of deaths of adults by the 133 years chronological stages of Tell Ghanem al-'Ali. The number of adults was calculated as: numbers of modeled populations in individual stages shown in Table 1 times the survivorship rate of one third, while the total number of deaths of adults was as: the number of adults shown in Table 2 times seven by assuming that 133 years equal seven generations as described above.

Table 1. Modeled populations by 133 years chronological stages at Tell Ghanem al-'Ali

Population densities	Stage 1 (3100-2967 BC)	Stage 2 (2966-2833 BC)	Stage 3 (2832-2699 BC)	Stage 4 (2698-2565 BC)	Stage 5 (2564-2431 BC)	Stage 6 (2430-2297 BC)
100/ha (n=950)	59	119	238	475	950	475
200/ha (n=1900)	119	238	475	950	1900	950

Table 2. Estimations of adult populations by 133 years chronological stages at Tell Ghanem al-'Ali

Population densities	Stage 1 (3100-2967 BC)	Stage 2 (2966-2833 BC)	Stage 3 (2832-2699 BC)	Stage 4 (2698-2565 BC)	Stage 5 (2564-2431 BC)	Stage 6 (2430-2297 BC)
100/ha (n=950)	20	40	79	158	317	158
200/ha (n=1900)	40	79	158	317	633	317

The Early Bronze Age cemetery near Tell Ghanem al-'Ali was intensively used over some 250 years between c. 2600 cal. BC and c. 2350 cal. BC in a longer chronological view (Fig. 2), while the intensive use might be some 150 years between c. 2500 cal. BC and c. 2350 cal. BC in a shorter view (Falb et al. 2005). On the basis of the general estimation of the number of deaths of adults shown in Table 3, we can suggest how many adults of Tell Ghanem al-'Ali died during the both longer and shorter periods, i.e., between c. 3000 persons (100/ha) and c. 6000 persons (200/ha) for the former, and between c. 2000 persons (100/ha) and c. 4000 persons (200/ha) for the latter (Fig. 4). Obviously, we should be careful in the evaluation of the results, since the above general estimates of deaths are naturally based upon many assumptions, and even they may be some type of fictional story. However, the estimated deaths between 2000 and 6000 persons of Tell Ghanem al-'Ali allow us to suppose existence of

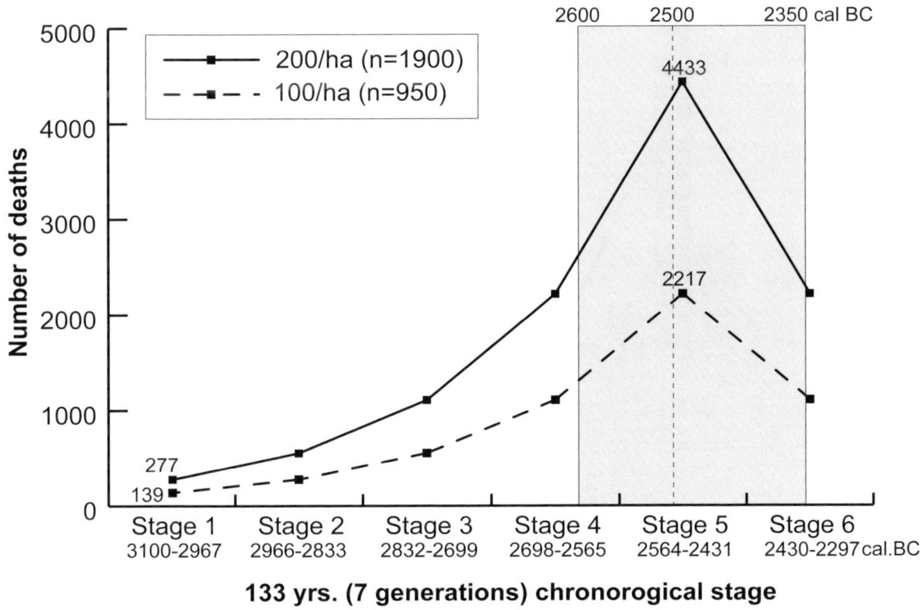

Fig. 4: Modeled curve of deaths of adults during the individual 133 years chronological stages at Tell Ghanem al-'Ali (shaded area shows the time range of intensive use of the cemetery).

thousands of graves for the deceased like the extramural cemetery near the settlement, even if multiple inhumations frequently occurred as has been attested from earth-cut shaft graves in the cemetery.

Table 3: Estimations of the total number of deaths of adults during the individual 133 years chronological stages at Tell Ghanem al-'Ali

Population densities	Stage 1 (3100-2967 BC)	Stage 2 (2966-2833 BC)	Stage 3 (2832-2699 BC)	Stage 4 (2698-2565 BC)	Stage 5 (2564-2431 BC)	Stage 6 (2430-2297 BC)
100/ha (n=950)	139	277	554	1108	2217	1108
200/ha (n=1900)	277	554	1108	2217	4433	2217

Nomads and cities: settlement dynamics on the Syrian Middle Euprates during the Early-Middle Bronze Age transition

If the deceased of the Early Bronze Age extramural cemeteries in the Syrian Middle Euphrates Valley were inhabitants of the neighboring settlements like Tell Ghanem al-'Ali, Tell Hamadin, and Tell Mugla as-Saghir instead of possible mobile pastoralists on the steppe of the Bishri Plateau, then why were such massive extramural cemeteries constructed during the Early Bronze Age in particular? Observations of settlement trends in the region may answer the question. Fig. 5 illustrates diachronic changes of settlement numbers in the Syrian Middle Euphrates Valley between the Pre-pottery Neolithic B Period and the Hellenistic/Roman/Byzantine Periods on the basis of survey results by K. Kohlmeyer in the 1980s (Kohlmeyer 1984).

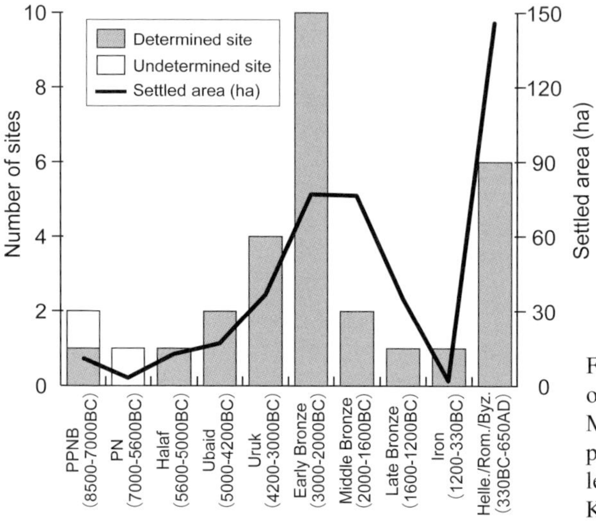

Fig. 5: Diachronic changes of the number of sites and settled area (ha) in the Syrian Middle Euphrates Valley between the Prepottery Neolithc B (PPNB) and the Hellenistic/Roman/Byzantine Periods (after Kohlmeyer 1984).

Obviously, there are two peaks of site numbers in the Early Bronze Age and the Hellenistic/Roman/Byzantine Periods. On the other hand, only a few sites were established in other periods, suggesting sparse occupations of sedentary people in the valley during the periods. It seems likely that the growth and decline of settlement numbers were closely associated with the appearance of massive extramural cemeteries in the Early Bronze Age Syrian Middle Euphrates Valley.[1]

The observations of settlement trends have also provided evidence of typical urbanization during the Early-Middle Bronze Age transition in the valley. As has been shown in Fig. 5, the number of sites in the transition has dramatically declined from ten to two settlements, but no significant change of hectarage of settled areas[2] between the two periods was confirmed. This indicates that medium and small-sized Early Bronze Age sites in the valley including Tell Ghanem al-'Ali were all abandoned, and that over 35 ha large-sized sites of Tell Bi'a[3] and Tell Thadayain were only settled during the Middle Bronze Age. Apparently, major part of the Early Bronze Age population in the valley migrated to the two urban centers in the subsequent period (Fig. 6).

An alternative migrating destination of the Early Bronze Age population in the valley might be the steppe on the Bishri Plateau. Archaeological features of such increased emphasis on steppe exploitation or nomadization in the Middle Bronze Age can also be presented (Fig. 6). For example, surveys by the Syrian-Japanese mission as well as K. Kohlmeyer have confirmed a site in which consists of a small Middle Bronze Age settlement and cemetery situated on the northern edge of the Bishri Plateau (Kohlmeyer 1986: 55 [site 115]; Nishiaki *et al.* 2009: 147-148 [site 23H]). It seems likely that the site was not simply a transitory camp of pastoral nomads, since agricultural and food processing tools (i.e., sickle blades, grinding slabs and pestles) were collected from the surface. Given the extremely high abandonment rate for Early Bronze Age settlements in the valley, however, the major habitation area of

1 Kohlmeyer's survey has also documented cemeteries of Hellenistic/Roman/Byzantine Periods (Kohlmeyer 1984).

2 Hectarage of several settlements was acquired from Google Earth through its measuring function, since little information on the size of the settlements was obtained from the literature (Kohlmeyer 1984).

3 Tell Bi'a is the only settlement that indicates settlement continuity between the Early and Middle Bronze Age in the region, but the occupied area apparently reduced in the Middle Bronze Age (Kohlmeyer 1996 cited in Schwartz 2007: 55).

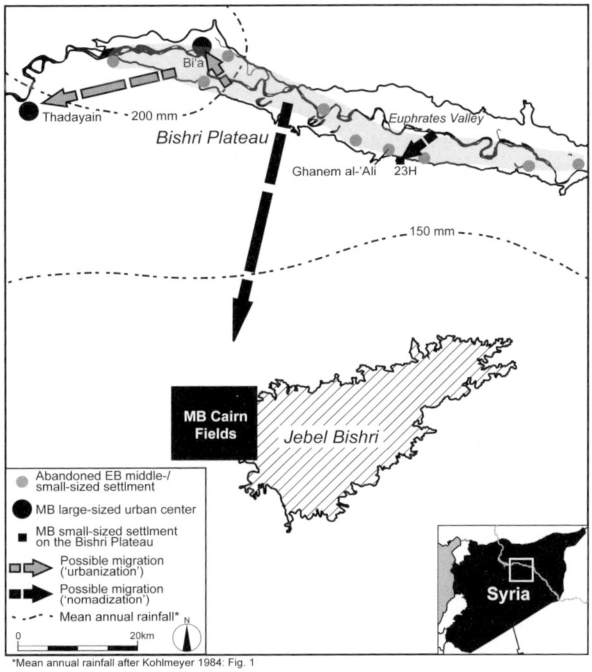

Fig. 6: Regional settlement abandonment, urbanization, and nomadization during the Early-Middle Bronze Age transition in the Syrian Middle Euphrates Valley.

the region apparently shifted from the valley to the steppe in the Early-Middle Bronze Age transitional period (Nishiaki this volume). Another evidence of nomadization in the Middle Bronze Age is massive cairn fields located in the Jebel Bishri, some 60km south from the Euphrates (Fujii and Adachi 2010; Fujii this volume). It appears that significant change of choice of burial place and burial practice itself also occurred between the two periods, suggesting nomadic adaptation in the Middle Bronze Age in the region.

The observed simultaneous processes of urbanization and nomadization are seemingly contradictory phenomena. Urban centers and nomads, however, might be an integral part of the Middle Bronze Age society in the region. As has been shown by several ethnographers, modern specialized pastoral nomads in the Near East have been completely involved in urban society not only in the sense of economic, but also in social and political (e.g. Marx 2012).

Recently, archaeologists have also added increased emphasis to the integration of nomadic and sedentary populations (Porter 2004; collected papers in Szuchman 2009). Then why did such urbanization and nomadization simultaneously occur during the Early-Middle Bronze Age transition in the region? Social responses to the 4.2ka BP abrupt climate change may be a key to the question (Weiss 2014). In that case, why was the adaptation of the mobile way of life in the region continued in the period of apparent climatic amelioration (Weiss *et al.* 1993) of the Middle Bronze Age? In the Upper Khabur region of northeast Syria, for example, survey results have provided evidence of the regional abandonment and temporal nomadic adaptation during only the cooling and aridification event (i.e., 2200-1900 cal. BC), while settlements were massively reestablished in the region in the subsequent period after increased precipitation (Ristvet 2012; Ristvet and Weiss 2013). Although the question remains to be solved, ecological conditions of mean annual rainfall between 150 mm and 200 mm in the region (Fig. 6), in which only vulnerable rain-fed cultivation can be operated because of year-to-year variability of mean annual rainfall (Wilkinson 2004: 41), will be part of the explanation. Alternatively, human agency like 'socio-political circumstances, chronic conflict, and sheer opportunism' as characteristics of the region (Wilkinson *et al.* 2012) should be considered as causes of the continuous nomadic adaptation in the Middle Bronze Age.

In any case, archaeozoological, archaeobotanical as well as lithic studies at Tell Ghanem al-'Ali and/or the steppe of the Bishri Plateau indicate that the inhabitants have exploited both the valley as habitation base/agricultural land and the steppe as temporary camping/livestock herding during the Early Bronze Age. On the other hand, the studies also suggest that trends of intensive steppe exploitation increased toward the end of the third millennium BC in the region (Akashi 2011, 2012; Hongo *et al.* 2010; Nishiaki 2010a, 2010b, this volume). Combined such evidence of increased steppe exploitation with the urbanization-nomadization trend suggested by settlement patterns in the region mentioned earlier, it seems that change of economic and social structure on the Syrian Middle Euphrates during the Early to Middle Bronze Age was essentially transition from agro-pastoral communities to seemingly dichotomized urban-nomad communities.

Concluding remarks

Attempts to answer the question of the deceased of the massive extramural cemeteries near Tell Ghanem al-'Ali in this paper provided insights into the nature of changing mobility patterns and economic and social structure during the Early to Middle Bronze Age on the Syrian Middle Euphrates. First, modeled mortality dynamics demonstrate that estimated thousands of graves in the nearby cemetery are more or less reasonable for the estimated number of deaths at the site, suggesting that the deceased are agro-pastoralists of the settlement, and that the existence of specialized nomadic pastoralists in the Early Bronze Age is an open question. Second, diachronic change of settlement trends and the appearance of new burial practices (i.e., cairns) in the Middle Bronze Age seemingly show the dichotomization of urban/steppe communities in the region. However, it is likely that the nature of the dichotomized populations was a differentiation of the Early Bronze Age agro-pastoral communities as social responses to environmental and socio-political changes in the Early to Middle Bronze Age transition. Although the apparent transformation of society should be associated with developments of political and cultural complexities, i.e., formation of tribal polities and the appearance of new ethnic groups or the Amorites in the Middle Bronze Age (e.g. Ristvet 2012; Schwartz 2012; Weiss 2014), the issues will be challenges for the future.

Acknowledgements

We are grateful to Professor Daniele Morandi Bonacossi (University of Udine, Italy), the organizer of the 8th ICAANE workshop 5 in Warsaw, and the participants of the workshop, for invaluable comments and suggestions. This study was produced within the framework of research programs directed by Professor Katsuhiko Ohnuma (Kokushikan University, Japan), which is financially supported by KAKENHI Grant-in-Aid for Scientific Research on Priority Areas from the Ministry of Education, Culture, Sports, Science and Technology, Japan (2005-2009), Sasakawa Scientific Research Grant from the Japan Science Society (2009), and the Science Research Promotion Fund from the Promotion and Mutual Aid Corporation for Private Schools of Japan (2010-2012).

Bibliography

Akashi, C.

2011 The subsistence and the plant use in Tell Ghanem al-Ali: Early Bronze Age Syria. *Al-Rafidan* 32: 105-110.

2012 The expansion of exploitation of the steppe in Tell Ghanem al-Ali (Early Bronze, Syria). *Bulletin of the Graduate Division of Literature of Waseda University* (4) 57: 83-95 (in Japanese).

Akkermans, P. M. M. G. and G. M. Schwartz

2003 *The archaeology of Syria: from complex hunter-gatherers to early urban societies ca. 16,000-300 BC)*. Cambridge University Press, Cambridge.

Aurenche, O.

1981 Essai de démographie archéologique. L'exemple des villages du Proche-Orient ancien. *Paléorient* 7/1: 93-105.

Buccellati, G.

2008 The origin of the tribe and of 'industrial' agropastoralism in Syro-Mesopotamia. In H. Barnard and W. Wendrich (eds.), *The archaeology of mobility: old world and new world nomadism*. Cotsen Institute of Archaeology, University of California, Los Angeles: 141-159.

Carter, E. and A. Parker

1995 Pots, people and the archaeology of death in northern Syria and southern Anatolia in the latter half of the third millennium BC. In S. Campbell and A. Green (eds.), *The archaeology of death in the ancient Near East.* Oxbow, Oxford: 96-116.

Chamberlain, A. T.

2006 *Demography in archaeology.* Cambridge University Press, Cambridge.

2009 Archaeological demography. *Human Biology* 81/2-3: 275-286.

Cooper, L.

2006 *Early Urbanism on the Syrian Euphrates*. Routledge, New York/London.

Falb, C., K. Krasnik, J.-W. Meyer, and E. Vila

2005 *Gräber des 3. Jahrtausends v. Chr. im syrischen Euphrattal: 4. Der Friedhof von Abu Hamed.* Saarländische Druckerei & Verlag, Saarwellingen.

Fujii, S. and T. Adachi

2010 Archaeological investigations of Bronze Age cairn fields in the northwestern flank of Mt. Bishri. In K. Ohnuma *et al.* (eds.), *Formation of tribal communities: integrated research in the Middle Euphrates, Syria (Al-Rafidan special issue).* The Institute for Cultural Studies of Ancient Iraq, Kokushikan University, Tokyo: 61-77.

Hasegawa, A.

2008 Surveying at Tell Ghanem al-'Ali. *Formation of tribal communities: newsletter* 10: 5-9 (in Japanese).

2010 Sondage at the site of Tell Ghanem al-'Ali. In K. Ohnuma *et al.* (eds), *Formation of tribal communities: integrated research in the Middle Euphrates, Syria (Al-Rafidan special issue).* The Institute for Cultural Studies of Ancient Iraq, Kokushikan University, Tokyo: 25-35.

Hassan, F. A.

1981 *Demographic archaeology.* Academic Press, New York.

Hongo, H., C. Akashi, L. Omar, K. Tanno and H. Nasu
2010 Zooarchaeology and ethnoarchaeobotany at Tell Ghanem al-Ali. In K. Ohnuma *et al.* (eds.), *Formation of tribal communities: integrated research in the Middle Euphrates, Syria (Al-Rafidan special issue)*. The Institute for Cultural Studies of Ancient Iraq, Kokushikan University, Tokyo: 97-104.

Kohlmeyer, K.
1984 Euphrat-Survey: Die mit Mitteln der Gerda Henkel. Stiftung durchgeführte archäologische Geländebegehung im syrischen Euphrattal. *Mitteilungen der Deutschen Orient-Gesellschaft zu Berlin* 116: 95-118.
1986 Euphrat-Survey 1984: Zweiter Vorbericht über die Mitteln der Gerda Henkel. Stiftung durchgeführte archäologische Geländebegehung im syrischen Euphrattal. *Mitteilungen der Deutschen Orient-Gesellschaft zu Berlin* 118: 51-65.
1996 The urban development of Tuttul/Tell Bi'a within the settlement system of the Euphrates Valley. *International colloquium "The Syrian Djezireh: cultural heritage and interrelations,"* Deir ez-Zor, April 1996, abstract.

Kume, S., Y. Miyata and S. Kadowaki
2012 Chemical approaches to funerary feasts in ancient Mesopotamia. *Mishima Kaiun Memorial Foundation Research Report* 2011(49): 150-157 (in Japanese).

Kume, S., A. Sultan, I. Ono and C. Akashi
2011 Sondage at Early Bronze Age Cemetery near Tell Ghanem Al-Ali. *Al-Rafidan* 32: 163-170.

Laneri, N., P. Pfälzner, S. Valentini (eds.)
2012 *Looking north: the socioeconomic dynamics of the northern Mesopotamian and Anatolian regions during the later third and early second millennium BC.* Harrassowitz Verlag, Wiesbaden.

Liverani, M.
1997 "Half-nomads" on the Middle Euphrates and the concept of dimorphic society. *Altorientaliche Forschungen* 24/1: 44-48.

Marx, E.
2012 The growth of a conception: nomads and cities. In H. Hazan and E. Hertzog (eds.), *Serendipity in anthropological research: the nomadic turn.* Ashgate Publishing, London: 31-46.

Meyer, J.-W.
2005 Auswertung und Zusammenfassung der Ergebnisse. In C. Falb *et al.*, *Gräber des 3. Jahrtausends v. Chr. im syrischen Euphrattal: 4. Der Friedhof von Abu Hamed.* Saarländische Druckerei & Verlag, Saarwellingen: 359-365.
2010 The cemetery of Abu Hamad: a burial place of pastoral groups?. In K. Ohnuma *et al.* (eds.), *Formation of tribal communities: integrated research in the Middle Euphrates, Syria (Al-Rafidan special issue).* The Institute for Cultural Studies of Ancient Iraq, Kokushikan University, Tokyo: 155-163.

Nakamura, T, M. Hoshino, T. Tanaka, H. Yoshida, T. Saito, K. Tsukada, Y. Katsurada, Y. Aoki, T. Ohta, A. Hasegawa, T. Kiuchi, K. Ohnuma, A. Al-Khabour and M. Al-Maqdissi
2010 Early Bronze Age strata at Tell Ghanem al-Ali along the Middle Euphrates in Syria: a preliminary report of 14C dating results. *Radiocarbon* 52/2-3: 383-392.

Nishiaki, Y.
2010a Archaeological evidence of the Early Bronze Age communities in the middle Euphrates steppe, north Syria. In K. Ohnuma *et al.* (eds.), *Formation of tribal*

communities: integrated research in the Middle Euphrates, Syria (Al-Rafidan special issue). The Institute for Cultural Studies of Ancient Iraq, Kokushikan University, Tokyo: 37-48.

2010b Early Bronze Age flint technology and flake scatters in the north Syrian steppe along the Middle Euphrates. *Levant* 42/2: 170-184.

Nishiaki, Y., S. Kadowaki and S. Kume

2009 Archaeological survey around Tell Ghanem Al-Ali. *Al-Rafidan* 30: 145-153, 160-163.

Numoto, H. and S. Kume

2010a Survey and sondage at the cemeteries near the site of Tell Ghanem al'Ali. In K. Ohnuma *et al.* (eds.), *Formation of tribal communities: integrated research in the Middle Euphrates, Syria (Al-Rafidan special issue)*. The Institute for Cultural Studies of Ancient Iraq, Kokushikan University, Tokyo: 49-60.

2010b Cleaning and survey of Early Bronze Age shaft graves at Wadi Daba cemetery near Tell Ghanem Al-'Ali. *Al-Rafidan* 31: 185-190.

Ohnuma, K., S. Fujii, Y. Nishiaki, A. Tsuneki, S. Miyashita, and H. Sato (eds.)

2010 *Formation of tribal communities: integrated research in the Middle Euphrates, Syria (Al-Rafidan special issue)*. The Institute for Cultural Studies of Ancient Iraq, Kokushikan University, Tokyo.

Orthmann, W.

1980 Burial customs of the 3rd millennium B.C. in the Euphrates Valley. In J.-C. Margueron (ed.) *Le Moyen Euphrate. Zone de contacts et d'echanges.* E.J. Brill, Leiden: 97-105.

Porter, A.

2004 The urban nomad: countering the old clichés. In C. Nicolle (ed.), *Nomades et sédentaires dans le Proche-Orient ancien.* Éditions recherche sur les civilisations, Paris: 69-74.

2007 The ceramic assemblages of the third millennium in the Euphrates region. In M. al-Maqdissi, V. Matoïan and C. Nicolle (eds.), *Céramique de l'âge du bronze en Syrie, II: L'Euphrate et la région de Jézireh.* Institut français de Proche-Orient, Beirut: 3-22.

2012 *Mobile pastoralism and the formation of Near Eastern civilizations: weaving together society.* Cambridge University Press, Cambridge.

Pruß, A.

2004 Remarks on the chronological periods. In S. Anastasio, M. Lebeau and M. Sauvage (eds.), *Atlas of preclassical Upper Mesopotamia*. Brepols, Turnhout: 7-27.

Ristvet, L.

2012 Resettling Apum: tribalism and tribal states in the Tell Leilan region, Syria. In N. Laneri, P. Pfälzner and S. Valentini (eds.), *Looking north: the socioeconomic dynamics of the northern Mesopotamian and Anatolian regions during the later third and early second millennium BC.* Harrassowitz Verlag, Wiesbaden: 37-50.

Ristvet, L. and H. Weiss

2013 The Habur region in the Old Babylonian period. In W. Orthmann, P. Matthiae and M. Al-Maqdissi (eds.), *Archéologie et histoire de la Syrie I: La Syrie de l'époque néolothique à l'âge du fer.* Harrassowitz Verlag, Wiesbaden: 257-272.

Roobaert, A. and G. Bunnens

1999 Excavations at Tell Ahmar-Til Barsib. In G. del Olmo Lete and J.-L. Montero Fenollós (eds.), *Archaeology of the Upper Syrian Euphrates, the Tishrin Dam area.* Editorial AUSA, Barcelona: 163-178.

Rowton, M. B.

1973 Urban autonomy in a nomadic environment. *Journal of Near Eastern Studies* 2/1-2: 201-215.

1974 Enclosed nomadism. *Journal of the Economic and Social History of the Orient* 17: 1-30.

Schwartz, G.M.

2007 Taking the long view on collapse: a Syrian perspective. In C. Kuzucuoğlu and C. Marro (eds.), *Sociétés humaines et changement climatique à la fin du troisième millénaire: Une crise a t-elle eu lieu en haute Mésopotamie?* Institut Français d'Etudes Anatoliennes, Istanbul/de Boccard, Paris: 45-67.

2012 Northern exposures: third to second millennium BC transformations in Upper Mesopotamia. In N. Laneri, P. Pfälzner and S. Valentini (eds.), *Looking north: the socioeconomic dynamics of the northern Mesopotamian and Anatolian regions during the later third and early second millennium BC.* Harrassowitz Verlag, Wiesbaden: 255-262.

Schwartz, G.M. and J. J. Nichols (eds.)

2006 *After collapse: the regeneration of complex societies*. University of Arizona Press, Tucson.

Szuchman, J. (ed.)

2009 *Nomads, tribes, and the state in the Ancient Near East: cross-disciplinary perspectives.* The Oriental Institute of the University of Chicago, Chicago.

Webb, J. M. and D. Frankel

2009 Exploiting a damaged and diminishing resource: survey, sampling, and society at a Bronze Age cemetery complex in Cyprus. *Antiquity* 83 (319): 54-68.

Weiss, H.

2014 The northern Levant during the intermediate Bronze Age: altered trajectories. In M. L. Steiner and A. E. Killebrew (eds.), *The Oxford handbook of the archaeologyof the Levant: c. 8000-332BCE.* Oxford University Press, Oxford: 367-387.

Weiss, H. (ed.)

2012 *Seven generations since the fall of Akkad.* Harrassowitz Verlag Wiesbaden.

Weiss, H., M. A. Courty, W. Wellerstrom, F. Guichard, L. Senior, R. Meadow and A. Currow

1993 The genesis and collapse of third millennium north Mesopotamian civilization. *Science* 291: 995-1007.

Wilkinson, T. J.

2004 *On the margin of the Euphrates settlement and land use at Tell es-Sweyhat and in the upper Lake Assad area*. The Oriental Insitutute of the University Chicago, Chicago.

Wilkinson, T. J., N. Galiatsatos, D. Lawrence, A. Ricci, R. Dunford and P. Graham

2012 Late Chalcolithic and Early Bronze Age landscapes of settlement and mobility in the Middle Euphrates: a reassessment. *Levant* 44/2: 139-185.

Wossink, A.

2009 *Challenging climate change: competition and cooperation among pastoralists and agriculturalists in northern Mesopotamia (c. 3000-1600 BC).* Sidestone Press, Leiden.

Zorn, J.R.

1994 Estimating the population size of ancient settlements: methods, problems, solutions, and a case study. *Bulletin of the American Schools of Oriental Research* 295: 31-48.

STEFAN L. SMITH, TONY J. WILKINSON and DAN LAWRENCE
Durham University

Agro-pastoral Landscapes in the Zone of Uncertainty

The Middle Euphrates and the North Syrian Steppe during the 4th and 3rd millennia BC

Sample archaeological surveys conducted within representative enclaves of the Middle Euphrates, the Khabur valley, and the north Syrian steppe demonstrate the existence of two distinct agro-ecological zones:

- A northern zone of relatively stable and long-term settlement
- A southern zone within an agro-ecological "Zone of Uncertainty" characterised by rapid phases of colonisation whose constituent settlements show short-term boom-and-bust life cycles.

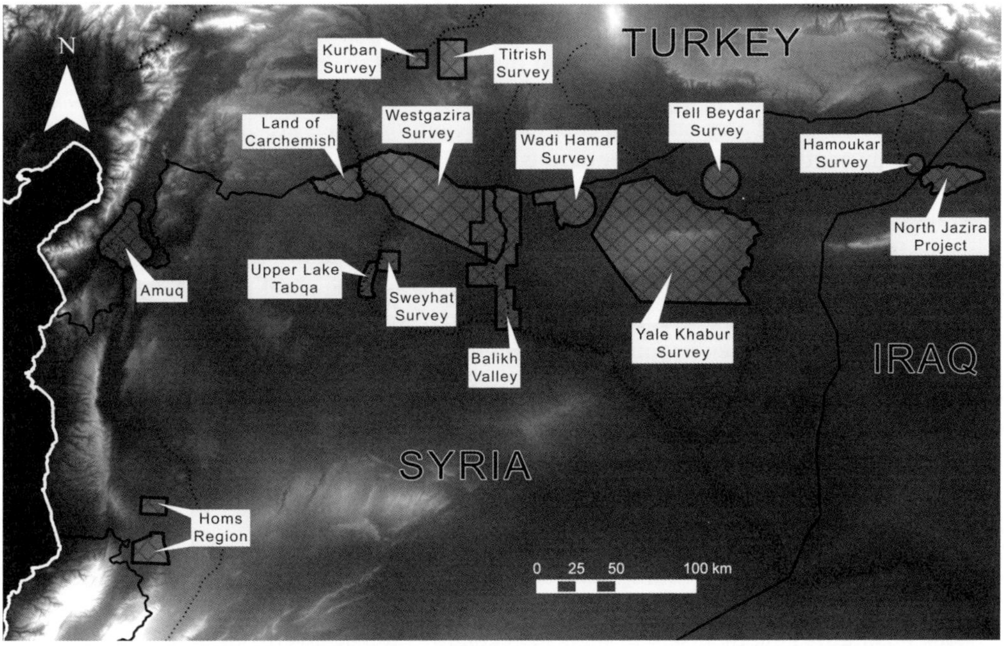

Fig. 1: ASTER GDEM elevation map [ASTER GDEM is a product of METI and NASA] of all the surveys that currently form the *Fragile Crescent Project* database.

Drawing on the databases of the Durham University Fragile Crescent Project (Fig. 1), settlement patterns from the fourth and third millennia BC are presented from the Euphrates to the western Khabur basin. These patterns enable us to compare the boom-and-bust trajectories of double-walled cities in the Western Jazira (within the Zone of Uncertainty) with more stable and arguably sustainable trends of cities, towns, and villages in the ecological zone to the north.

Studia Chaburensia 4 (2014), pp. 151–172.

We then assess long-term trends in settlement by the use of new methods of aggregating and combining disparate data from archaeological surveys conducted over the past thirty years.

Settlement trends and the growth of at least two distinct types of double-walled cities are discussed for these two ecological zones from the perspective of pastoral strategies as well as risk-taking opportunistic strategies such as trading and raiding. The discussion includes processes such as the commodification of textiles through pastoral production that may have operated within the margins of, amongst others, the large regional polities of Ebla and Mari. Evidence is drawn from archaeological surveys as well as palaeobotanical and faunal analyses to provide a basic model of settlement development within the climatically marginal zone of northern Syria and southern Turkey.

Introduction

In addition to the so-called "second urban revolution", northern Syria experienced a remarkable spread of settlement into areas that were climatically marginal for sedentary occupation. The aim of this paper is to contextualise this settlement, both in terms of the prevailing environment and the development of the agro-pastoral economy. Emphasis is upon what appears to be a remarkable co-evolutionary development of the pastoral and agricultural sectors, in association with the growth of major cities and their administrations. Any models that deal with this topic must grapple not only with the growth of the textile industries, but also the settlement of the steppe and the adoption of a relatively high-risk-high-return approach to economy. Examples of agro-pastoral communities and strategies are drawn from two parts of Durham University's Fragile Crescent Project (FCP) database: the Western Jazira in northern Syria and the Khabur basin in north-eastern Syria.

Following an outline of the significance of sheep rearing to the 3rd millennium economy, specifically at Ebla, we discuss modern-day strategies in the agro-pastoral zone, particularly models developed by agronomists working in the semi-arid region of Syria. We then examine the evidence for both Late Chalcolithic (LC) and Early Bronze Age (EBA) settlement, with particular emphasis on its often boom-and-bust nature as expressed in temporal models developed by the FCP (Lawrence 2012; Lawrence *et al.*2012). Finally, we suggest a qualitative comparison for the co-evolution of pastoral, barley-based agriculture and settlement development that may account for sedentary occupation within the so-called Zone of Uncertainty of Syria.

Ebla and the Bronze Age Pastoral Economy

It is well known that sheep-rearing formed a major pillar of the Eblaite economy. As stated by Gelb (1982, cited in 1986: 158),
"Sheep raising was the mainstay of the local economy; thousands of sheep were raised, supplying the wool for the production of textiles, the main export product of Ebla. Thus wool was the basis of Ebla's commercial prosperity and political power; ... [it] was an empire built on the backs of simple shepherds."

Estimates drawn from the Ebla texts of the actual number of sheep held by the city of Ebla range from 2 million to a more modest figure of some 670,000 (Milano 1995).[1]

1 Compare with Gelb 1986: 162 in Aleppo Province, 1975: 1,059,000 ovids and 44,000 bovids, i.e. 24 ovid to 1 bovid. Note, however, that in pre-tractor times there would have been more bovids.

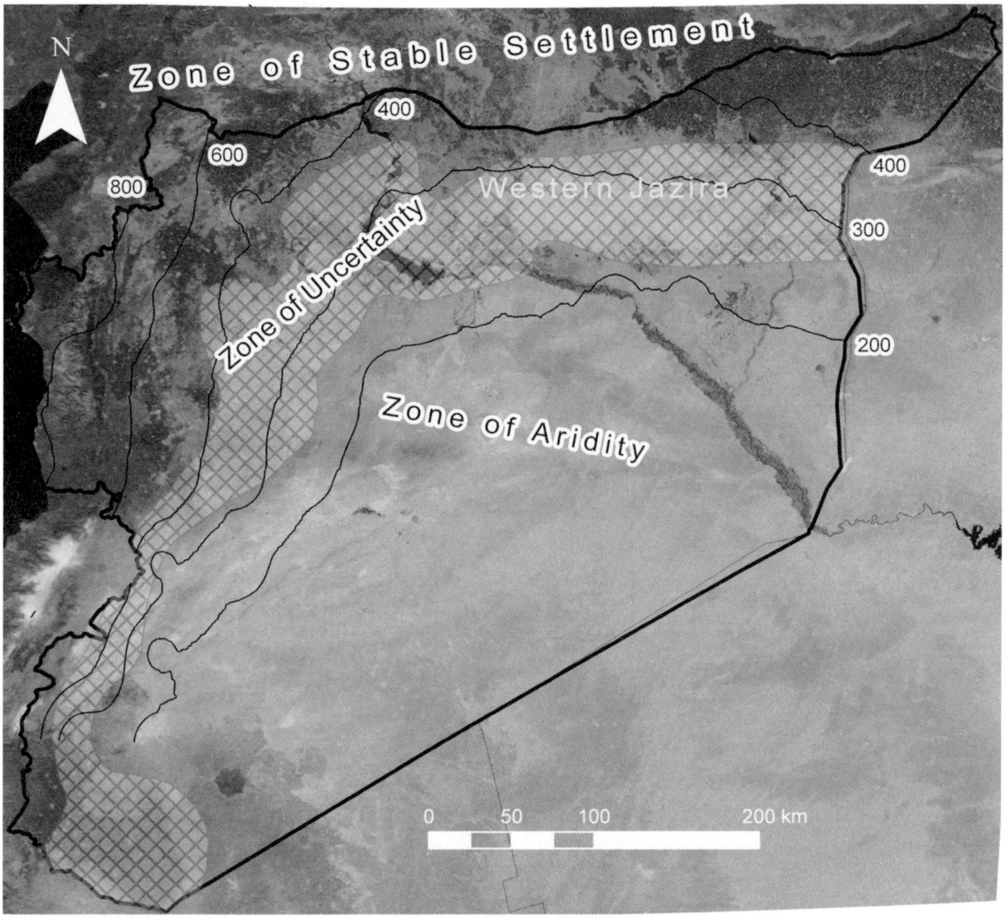

Fig. 2: Satellite image of Syria with average annual rainfall isohyets (from Meyer-Christoffer et al. 2011), and the Zone of Uncertainty (from Wachholtz 1996: 6) marked in crosshatching. The isohyet values are calculated from the mean monthly precipitation totals from 1951 to 2000, and were processed by Louise Rayne, Department of Geography, Durham University.

Meanwhile Archi estimates the number of animals directly controlled by the palace at Ebla to have fluctuated between 80,000 and 110,000, while emphasising that this does not include those herds kept by "the village communities scattered throughout the territory" (Archi 1993: 47). To graze such large numbers of animals it would have been necessary not only to utilise much of the land of Ebla, but also for these to be taken beyond the limits of the state's territory, according to Milano in, for example, the KUR (hill country) to the north and south of the city.

Clearly, the Eblaite administration would have required access to large areas of steppe grazing in order to raise such flocks. The appearance of these huge holdings of sheep provides the context for "a major transformation of ancient Mesopotamia" that entailed a replacement of plant fibres, particularly flax, by wool for the manufacture of textiles during the 3rd and perhaps part of the 4th millennium BC (McCorriston 1997: 517; Zeder 2003: 162).

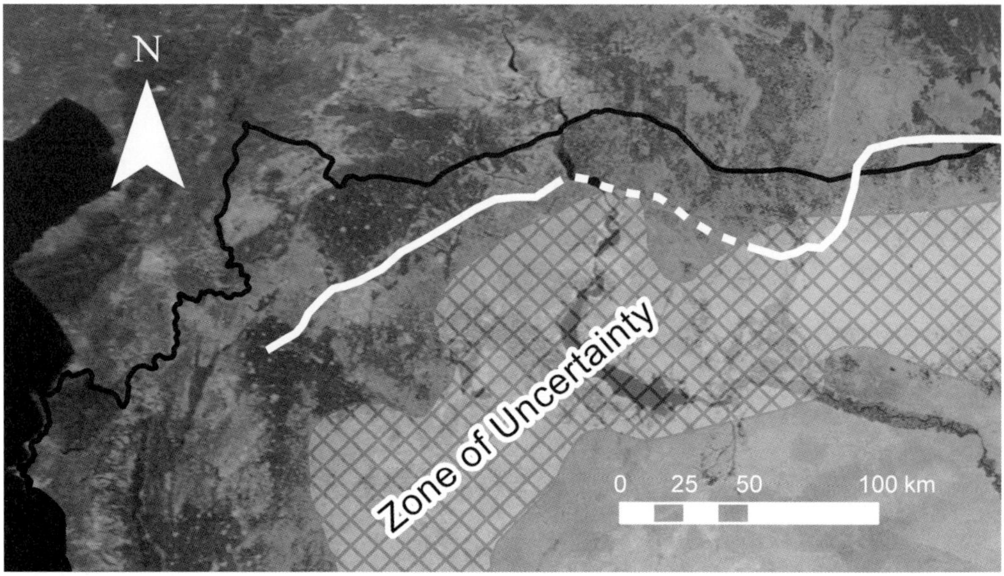

Fig. 3: Satellite image of northern Syria and southern Turkey with the Zone of Uncertainty and southern limit of large-scale Late Chalcolithic settlement (white line) indicated. Note that this boundary line does not take into account possible small LC occupations such as at Tell Chuera (Dohmann-Pfälzner and Pfälzner 2002: 13-14) and Tell Malhat (Kühne 1983)

Certainly the wool industry of Ebla was extraordinarily active and would not be matched until that of the Ur III state in southern Mesopotamia (Milano 1995). According to Zeder the ratio of sheep and goats to humans was 4 times that of today (Zeder 2003).

Overall, Gelb (1986: 160-161) considered agriculture (that is to say, cultivation) to have played a secondary role to the manufacture of textiles and metal. As will be shown in this paper however, if we consider agriculture to include both cultivation and pastures, and that animal feed and manure are crucial parts of an integrated economy, Ebla should surely be seen as an agro-pastoral society, the products of which contributed to significant economic production.

Sustainable Agriculture and the Zone of Uncertainty

The agricultural zones of northern Syria and southern Turkey range from a region with greater than 800 mm yearly rainfall in the north-west down to the arid south, characterised by less than 200 mm of rainfall per annum (Fig. 2). The latter zone is generally given over to pasture and has until recently been dominated by nomadic and semi-nomadic pastoralists, although today barley cropping is also prevalent (Cocks et al. 1988; Jones 1993; Wilkinson 2000). In the northern part of the region, where the mean annual rainfall is greater than 300 mm, crops are more diversified, including wheat, barley, lentils, grapes, and sometimes olives, and local village flocks can be grazed in the village catchments without the need to venture more widely in search of pasture.

In the zone of marginal cultivation, which falls between the rainfall isohyets of ca. 320 and 200 mm per annum, cereal cultivation is much riskier and thus agro-pastoral practices are the norm. Despite its climatically marginal location and the consequent high risk of crop failure, this zone is used for agriculture today, primarily for a closely integrated system of sheep and goat husbandry with rain-fed cereals (Wachholtz 1996: 5-9). Today, the sizes of flocks vary considerably; in 1991/1992, sedentary households in the Zone of Uncertainty south-east of Aleppo held flocks averaging 59 head of sheep and goats, whereas semi-sedentary and migratory households had an average of 189 and 160 head of sheep and goats, respectively (Wachholtz 1996: 49-51).

Archaeological data roughly defines a zone of what appears to be "sustainable settlement", distinct from the "Zone of Uncertainty" (Fig. 2). In the former zone a relatively abundant scatter of Late Chalcolithic sites (early to mid-4th millennium BC) gives way to a moderately dense scatter of smaller tells during the Early Bronze Age, which both cluster on the river terraces of the Euphrates as well as extending up its tributary valleys (Wilkinson *et al.* 2012). In addition, there are occasional major LC centres such as Samsat and, further west in the Afrin valley, Oylum Höyük. In the "Zone of Uncertainty", by contrast, pre-3rd millennium settlement is rare, and is often clustered as at the Uruk centres of Sheikh Hassan, Habuba Kabira and Jebel Aruda along the Euphrates (Fig. 4). This lack of significant LC occupation is especially noticeable in the Western Jazira, where only a handful of settlements from this period exist. In fact, when one draws a boundary line representing the edge of plentiful LC settlements in northern Syria and southern Turkey, it very closely follows the northern edge of the Zone of Uncertainty (Fig. 3). When 3rd millennium settlements occurred, they developed rapidly, in the Western Jazira into extensive double-walled cities dominated by central mounds with surrounding walls and lower towns. As indicated in Fig. 2, sustainable settlement occupies the zone north of the Syrian steppe, whereas the marginal "Zone of Uncertainty" (Wilkinson 2000) falls to the south within what is today the agro-pastoral zone.

The recovery of animal bones from excavated sites indicates that most Early Bronze Age communities within the zone of sustainability pursued relatively low-risk subsistence strategies of sheep/goat husbandry (Wattenmaker 1998: 184; Stein 1987).

Northern sites in the Middle Euphrates provide evidence for an economy dominated by domestic sheep, goat, pig and cattle, whereas southern sites like Sweyhat and Tell Chuera are characterized by increased numbers of sheep and goat, fewer cattle, virtually no pig and many more wild species (Weber 1997: 142; Vila 2010). These tendencies parallel the palaeobotanical record, with barley being less common in the north and dominant in the south (Table 1) (Miller 1997; Weber 1997).

For this paper, we concentrate upon settlement traces in the Western Jazira, with its remarkable concentration of double-walled cities, as well as other extensive sites which in aggregate appear to be as large as or even larger than those within the zone of sustainability. Noteworthy sites, from north to south within this zone, include Tell Khanzir (32 ha), Tell Chuera (65 ha), Tell Abu Shakhat (30 ha), Tell Mabtuh Sharqi (45 ha), Tell Mu'azzar (10 ha), and Tell Malhat (30 ha) (Fig. 4). As discussed later, although these sites all appear to have had upper and lower towns, and double (though not necessarily concurrent) surrounding walls, they fall into at least two distinct categories, belying their homogenous categorisation as "Kranzhügel".

However, it is Tell Sweyhat (40 ha), not in the Western Jazira, that provides the best starting point for evidence of the economy of EBA sites within the Zone of Uncertainty due to the wealth of available data on the outer town, settlement chronology, and palaeo-economy.[2]

2 The only site having undergone extensive archaeological investigations in the Western Jazira, Tell Chuera, has not yet had results of its palaeobotanical remains published

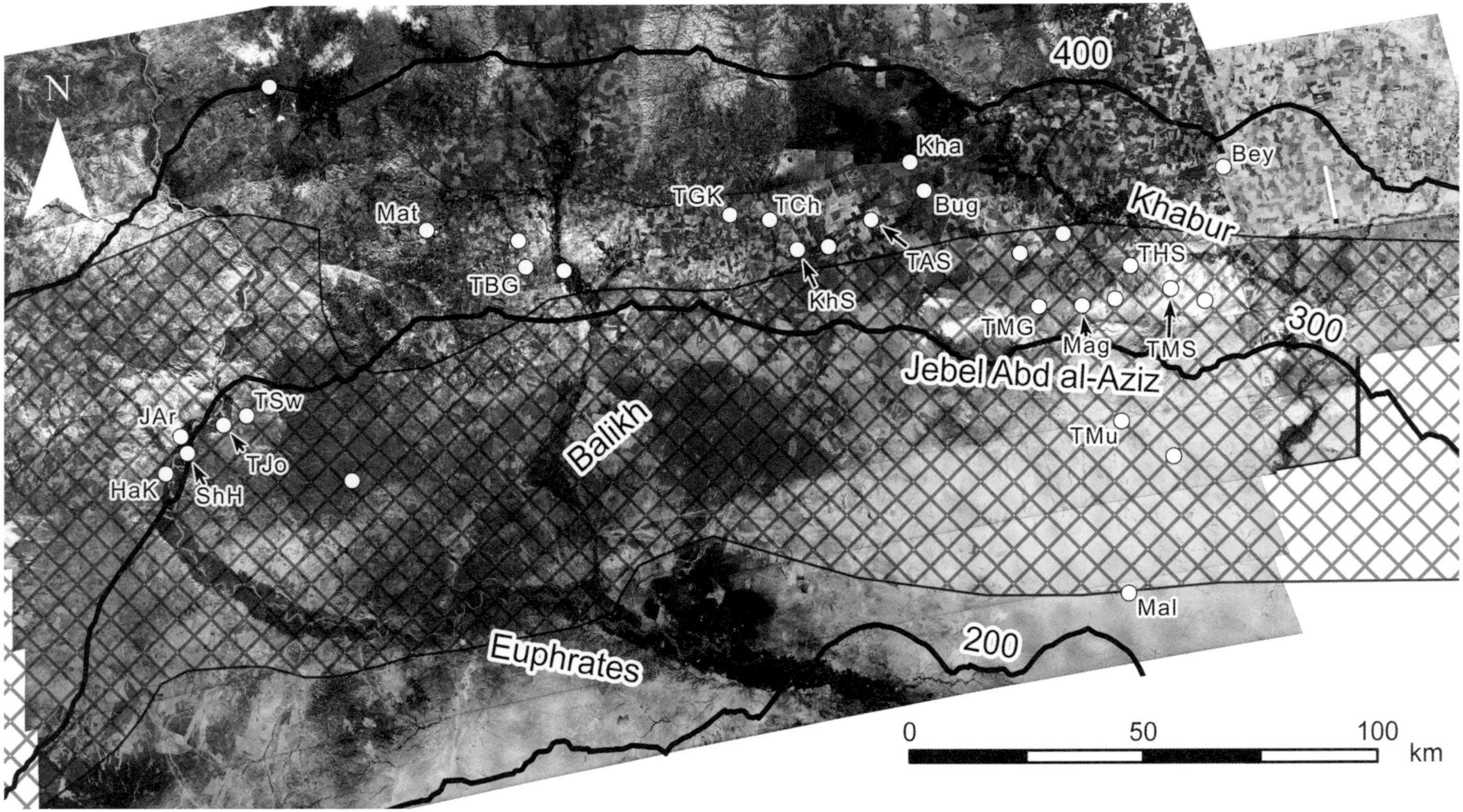

Fig. 4: Annotated CORONA satellite images (data available from the U.S. Geological Survey) of the Western Jazira with major sites labelled. Also includes isohyets (From Meyer-Christoffer et al. 2011; processed by Louise Rayne), and the Zone of Uncertainty in crosshatching.

TAS = Tell Abu Shakhat; TBG = Tell Barabra Gharbi; Bey = Tell Beydar; Bug = Tell Buga; TCh = Tell Chuera; TGK = Tell Ghajar el-Kebir; HaK = Habuba Kabira; THS = Tell Hammam Sharqi; JAr = Jebel Aruda; TJo = Tell Jouweif; Kha = Tell Khanzir; KhS = Tell Kharab Sayyar; TMG = Tell Mabtuh Gharbi; TMS = Tell Mabtuh Sharqi; Mag = Tell el-Magher; Mal = Tell Malhat; Mat = Tell Matin; TMu = Tell Mu'azzar; ShH = Sheikh Hassan; TSw = Tell es-Sweyhat

Table 1: Ratios of wild seed and percentages of barley and caprines from Umm el-Marra and Tell es-Sweyhat in the Zone of Uncertainty to the south contrasted with those from Kurban Höyük and Hacinebi in the zone of sustainable settlement to the north (Miller 1997). The figures that illustrate the increased appearance of caprine, barley, and wild cereal remains in the Zone of Uncertainty are in bold font.

wild seed: cereal Site:	Rainfall (mm)	LC: 3900 BC	early EB: 2800 BC	mid EB: 2500 BC	EB/MB: 2000 BC	MB: 1700 BC	LB: 1500 BC
Kurban Hoyuk	473	67	75	125	100		
Hacinebi	368	177					
Umm al-Marra	260		**1257**			**335**	**515**
Sweyhat	250				**950**		
percentage barley Site:	Rainfall (mm)						
Kurban Hoyuk	473	57	45	68	78		
Hacinebi	368	73					
Umm al-Marra	260		**93**			**90**	**88**
Sweyhat	250				**96**		
percentage caprines Site:	Rainfall (mm)						
Kurban Hoyuk	473	28	82	74	26		
Hacinebi	368	58					
Umm al-Marra	260		57			52	51
Sweyhat	250				**88**		

In contrast to sites further north in Turkey which exhibited a mixed economy, the plant remains of Sweyhat are dominated by barley (as were those of the neighbouring MBA site of Tell Jouweif). This, together with the high proportion of seeds from steppe plants to domestic cereals (mainly barley), argues for the significance of pastoralism (Table 1). However, the seemingly contradictory data, drawn from both palaeobotany and landscape features, argues both for the existence of a pastoral economy as well as cereal cultivation (Miller 1997; Wilkinson 2004). Therefore, as Danti has argued, agro-pastoral communities probably utilised a significant amount of barley and barley products as animals feeds, thereby increasing the demand on the cultivated landscape around the site (Danti 2000). Although Sweyhat was recognized as an agro-pastoral community in the report on the region (Wilkinson 2004), it remains necessary to explain why such an agro-pastoral economy should suddenly become viable in the 3rd millennium BC, when this did not appear to be the case before.

The present-day rural economy

Significantly, the rather sudden appearance of agro-pastoral communities in the Early Bronze Age (as well as perhaps during the Uruk, or LC 4-5) can be paralleled in modern-day Syria. In the 20th century AD, central Syria experienced a major expansion of the area of barley cultivation, accompanied by a similar increase in the number of sheep (Treacher 2000: 190; Cooper and Bailey 1990). As sheep production has become more dependent on food inputs from agricultural areas, farmers in these areas have responded by substituting barley for other crops as well as for fallow land. Moreover, flock owners and others have greatly extended barley-growing into the semi-arid rangelands where the average rainfall is less than 200 mm per annum (Treacher 2000: 191).

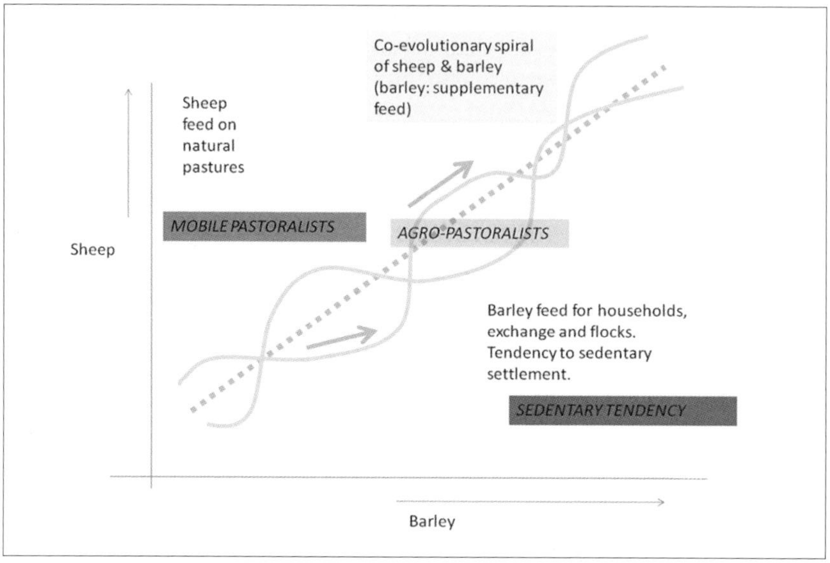

Fig. 5: Graph showing the co-evolution of pastoral sheep herding and agricultural fodder crop (barley) growing. The two central intersecting curvilinear lines represent the fluctuations in economic output type inherent in agro-pastoralism.

According to Métral (2000: 123), "this increase in barley cultivation, encroaching upon grazing lands, is both the result of and concomitant with an intensification of sheep-rearing and an increase in flock sizes." In other words, in these areas of climatically marginal land there appears to have been a co-evolutionary positive feedback spiral of pastoral activity and cultivation as sketched in Fig. 5. Ultimately, this also supported an increase in the human population, much of it sedentary or semi-sedentary. As discussed by Métral, such economic growth in a zone that is today extremely marginal in terms of "conventional" agricultural practices is accompanied by a wide range of risks, which are well dealt with by the local population. Moreover, as in modern financial markets, greater risk by farmers and herders may have led to higher average returns (Hazell *et al.* 2001: 16).

Archaeological applications

One must of course be extremely wary of applying a model derived from a modern context to the ancient world.[3] For example, although the hand of the early state (or perhaps royal policy) may have been a significant factor in the Ebla region and other parts of northern Syria, external demand for meat and the use of mechanical cultivation, which were significant factors in 20th century Syria, did not apply to the Bronze Age. Nevertheless, risk was hardly unknown to farmers of the ancient world. Usually however, risk is perceived in terms of subsistence societies, and therefore as something that would have been minimised as far as possible. Conse-

3 In the case of modern-day Syria, the effects of government policy and capitalist economies have played a significant role in allowing (or not enforcing limitations on) cultivation of the dry steppe. In addition, high demand for mutton from Saudi Arabia and the Gulf, as well as the introduction of tractors in the 20th century, have all played a role in the cultivation of the steppe (Jaubert 2000).

quently, strategies such as diversification of cropping or intensification (including irrigation) are regarded as ways of reducing such risk (Marston 2011: 191). However, in the ancient Near East, Bronze Age states such as Ebla that were involved in large-scale manufacture of textiles, metal trading, and other economic activities can hardly be regarded as subsistence economies. Instead, they had the potential capacity to invest in large-scale enterprises such as metal trade or sheep husbandry that could generate substantial returns, but also the capability to cope with the inevitable (sometimes substantial) losses inherent in such enterprises.

Rather than perceiving the land as a resource to be nurtured in the long-term, the herding strategies employed within the Zone of Uncertainty may have been seen as providing an immediate supply, perhaps for the "opportunistic stocking" of sheep and goats. For example, in semi-arid West Africa, development economists and geographers suggest that range land was not always perceived as a stable ecosystem that could be used up to a defined "carrying capacity". Rather, these were non-equilibrium pastoral systems that were opportunistically stocked with the result that livestock numbers were allowed to increase rapidly during verdant periods, but would then plummet when local conditions became drier and there was insufficient fodder (Mortimore 1998: 66, 73; also Stein 2005: 143).

Such "event-driven systems" tend not to become over-stocked because livestock are either taken outside the area or the animal population crashes, thereby relieving the stress on pastures. Economically, the pastoralists take advantage of the good times, but bear the losses during dry periods. The position of the Western Jazira within the Zone of Uncertainty thus provides the potential for great rewards, but also harbours significant levels of risk. This contrasts with the zone of stable settlement to the north, where more conservative strategies appear to have prevailed. Such challenges may have been acceptable either for some pastoral nomads or for the states of northern Syria such as Ebla.

It is clear then that several elements from the above model may be applied to the Early Bronze Age political and economic landscape. These include the following:

- The Zone of Uncertainty can form a productive economic niche as long as certain conditions are observed, and the scale of the economic system is such that there is sufficient "insurance" in place to absorb some of the risks and uncertainties. In other words, the size of the polity ameliorates the risk.
- Risk and perceptions of risk can play an important role in the development of agrarian practices – no single model can be applied to these landscapes through time. Herding practices can vary and be adopted in different ways depending upon circumstances.
- Any increase in the numbers of sheep may require a commensurate increase in the amount of feed. The large flocks that may have prevailed during certain periods of the EBA could not necessarily be supplied from pastures alone.
- Communities can occupy different areas of the cultivation / pastoral matrix (Fig. 5) at different times, fluctuating as necessary to accommodate the year-on-year variations in rainfall.

To conclude, when regional economies became sufficiently large in scale (as was arguably the case during the "second urban revolution" of the EBA), they could absorb risks in ways that more localised economies could not. Moreover, if the yields were sufficient (in this case in the form of large quantities of high-quality wool) and there was an institutional structure capable of absorbing the risk (e.g. the Ebla palace), then these otherwise marginal landscapes could become very productive, but only within a certain political-economic niche. This created a fragile system which collapsed very quickly when factors turned against the high returns of agro-pastoralism.

Settlement Trends in the 4th and 3rd Millennia

Comparison of data at the regional scale necessary to demonstrate these broad settlement trends requires the integration of different surveys into a single chronological framework. Differences in ceramic chronologies across northern Mesopotamia preclude direct comparisons between individual phases. In the Middle Euphrates Zone, for example, each of the four FCP surveys divided the millennium-long EBA into Early, Middle and Late phases, but none of these phases are of the same length. This can be contrasted with the Upper Khabur and Western Jazira, where surveys generally split the same period into only two phases, an Early and Late EBA. The solution developed in the FCP, and employed here, is to use published chronologies to relate each phase to calendar dates in years. Synthetic chronologies recently published for the Middle Euphrates (Porter 2007a, 2007b; Cooper 2006), the Jezirah (Lebeau 2000; Pfälzner 2010), and the Tell Chuera region (Meyer 2010a) provide the basis for this relation. By transforming the individual phases into a similar metric we can begin to model trends in settlement in a way which allows for direct comparison between surveys. In the following graphs, total settled area is displayed by one hundred year time blocks, reflecting the maximum degree of precision currently available in the date attributions for each phase.

Zone of sustainable settlement – Titrish-Kurban area

The neighbouring Kurban (see Wilkinson 1990) and Titrish (see Algaze *et al.* 1992; Algaze *et al.* 2001: 54-58) surveys in southern Turkey provide evidence that the area witnessed two periods of urbanisation during the 4th and 3rd millennia, the first in the latter part of the Late Chalcolithic centred on Samsat and the second in the mid-late EBA centred on Titrish Höyük. During the period when Samsat was the dominant site, reaching some 15 hectares, settlement was concentrated at small tells along the Euphrates river valley, with Titrish Höyük itself being uninhabited. This system was reasonably stable for almost 2000 years, from the end of the Ubaid to the beginning of the middle EBA. Sometime during the early EBA, Samsat appears to have declined slightly, and new sites, including the initially 5-hectare Titrish Höyük, were founded on a previously sparsely settled plain to the south of the Euphrates. Settlement exploded in this zone during the middle and late EBA, with Titrish growing to 40 hectares through the development of a substantial lower town, echoed with some temporal delay by the new establishment of numerous small settlements (Fig. 6). Agricultural intensification during this period is attested by the presence of low density sherd scatters surrounding the main site, interpreted as evidence of ancient manuring. Importantly, there was also an expansion of medium-sized "town" sites, including larger tells along the river such as Kurban and Lidar Höyüks. This massive expansion was short-lived however, and by the end of the EBA both Titrish and the medium-sized sites contracted significantly, coinciding with a rise in the number of small sites, perhaps indicative of the dispersal of the populations into the surrounding countryside. Ultimately these small sites were also abandoned as settlement declined still further into the MBA. Due to the sparse occupation during the early EBA (Fig. 6), it is unlikely that the entire population for growing centres such as Titrish came from the local region. Rather, the evidence suggests an influx of people to the region from elsewhere – whether they initiated habitation at Titrish and medium-sized tells or simply augmented the number of existing smaller settlements.

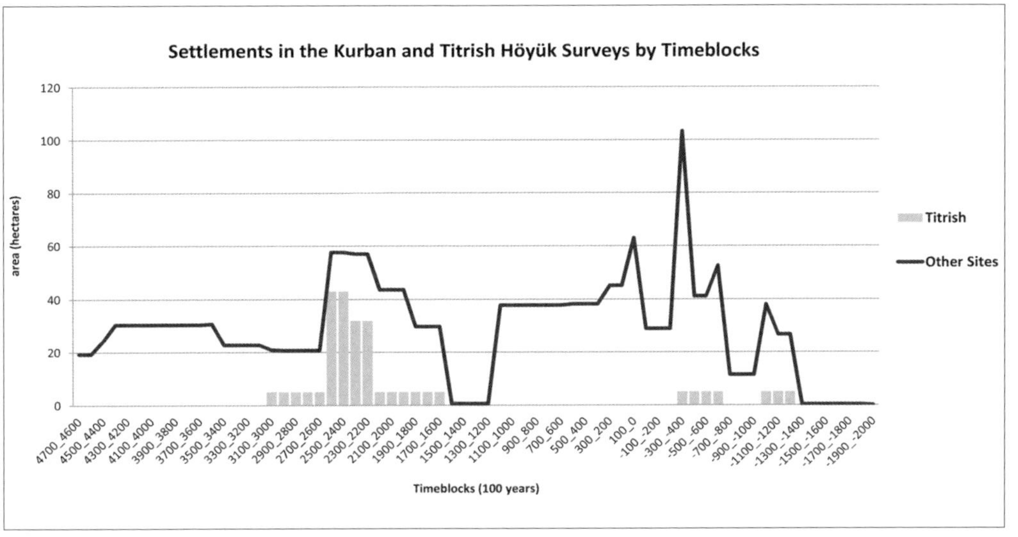

Fig. 6: Time block graph of the size of Titrish Höyük and other settlements in the Kurban and Titrish Höyük surveys, southern Anatolia.

The Zone of Uncertainty – Tell Chuera region

The only equivalent intensive survey in the heartland of the Western Jazira dry steppe is the Wadi Hamar Survey (see Pruß 2005; Kudlek 2006). This investigation of a 12 km radius around the small 3rd millennium tell of Kharab Sayyar also includes settlement data for a large regional centre: the 65-hectare double-walled Tell Chuera. Whereas settlement trends in the region during the 3rd millennium BC show a resemblance to the Kurban-Titrish region, its 4th millennium pattern is considerably different. This area saw a significant drop in human occupation in the mid-4th millennium (down to a total of three sites), a dearth of settlement that continued until the middle EBA (Fig. 7). From archaeological investigations elsewhere in the Western Jezirah,[4] we know that this was not a localised phenomenon, and that human occupation of the entire dry steppe region was either nomadic (to an extent as to leave no material remains) or simply non-existent during this time period. Notably, however, the mid-late 4th millennium did see occupation at Tell Chuera, albeit of very small proportions (Dohmann-Pfälzner and Pfälzner 2002: 13-14). By the end of the 4th millennium, even Chuera itself had been abandoned, representing an almost complete desertion of the area before the 3rd millennium settlement explosion.

This explosion of settlement has close parallels with the trajectory in the Kurban-Titrish region, with a rapid expansion in settlement during the mid-late EBA. Tell Chuera grew to 45 hectares, followed two centuries later by its expansion into a lower town to achieve a total of 65 ha. Simultaneously, the combined areas of smaller settlements in the surrounding landscape reached an unprecedented level (Fig. 7), a growth of occupation which would not be repeated until the Roman/Byzantine period. Furthermore, several medium-sized sites of between 20 and 40 hectares were newly settled, including Tell Ghajar al-Kebir, itself another

4 Most notably the Yale Khabur Survey around the Jebel Abd al-Aziz mountain range to the east; see Kouchoukos 1998.

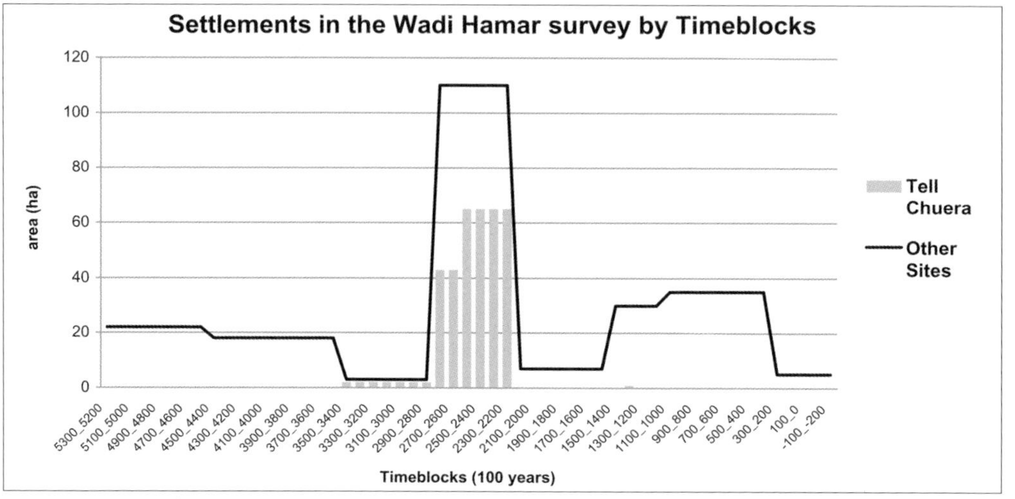

Fig. 7: Time block graph of the size of Tell Chuera and other settlements in the Wadi Hamar Survey, northern Syria.

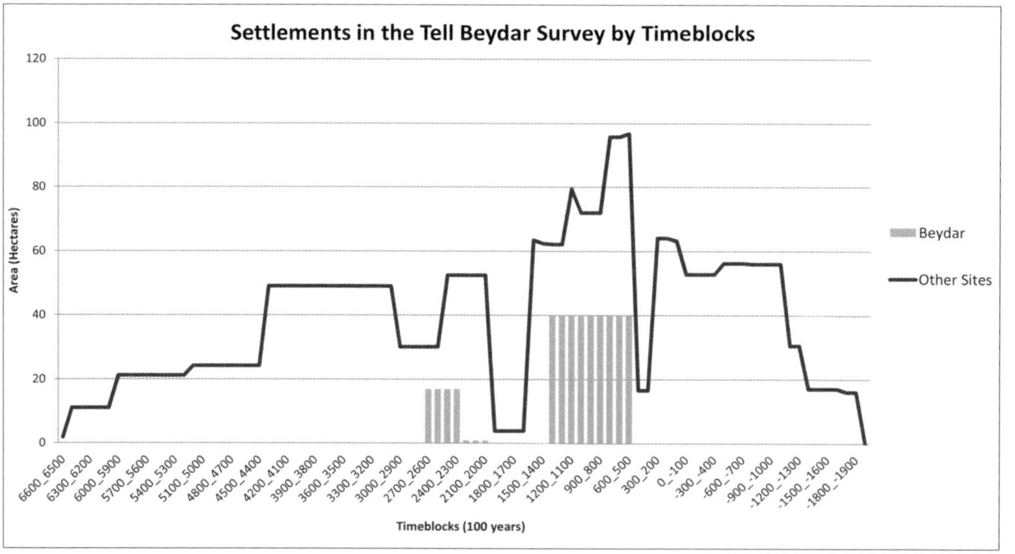

Fig. 8: Time block graph of the size of Tell Beydar and other settlements in the Tell Beydar Survey, north-eastern Syria.

double-walled city. The Chuera region saw a drastic collapse of settlement during the late EBA, and, as with the Kurban-Titrish region, became all but abandoned by the early MBA. This pattern is again reflected in the results of other surveys in the Western Jezirah,[5] suggesting the Wadi Hamar survey provides good proxy data for the dry steppe as a whole, especially the so-called "Kranzhügel zone".

5 The best examples being the Yale Khabur Survey (Kouchoukos 1998) and the Westgazira Survey (Einwag 1993).

Similar to the Kurban-Titrish region, the scale and rapidity of the expansion of settlement in the Western Jazira suggests the movement of external populations into the region. Here though, the complex and organised nature of the double-walled cities, as evidenced by the longstanding excavations at Tell Chuera, lend even greater credence to this hypothesis. The initial occupation of Chuera already shows evidence of a planned street network, a central administrative axis and town square, and the first town wall (a massive structure containing ca. 30 million bricks) (Meyer 2010b: 207). Furthermore, several of the features created during this initial period of building remained in use for the entire duration of Chuera's occupation, implying a construction that foresaw the growth of the city and incorporated this into its plans. This has led several recent interpretations to conclude that the double-walled settlements must have been founded by people with prior knowledge of urban planning, and not, as had previously been argued, by the sedentarisation of long-term nomadic groups (Meyer 2010b).

Agricultural core region – the Khabur basin

The expansion of settlement within the Zone of Uncertainty can be contrasted with the densely settled core area of the Khabur basin. Here, often only a few tens of kilometres from the region discussed above, a very different pattern can be observed. Surveys in this area provide evidence for long-term occupation of villages from at least the ceramic Neolithic period onwards, remaining fairly consistent until larger urban centres began to emerge from the beginning of the Late Chalcolithic, when the number of villages often increased also. However, this pattern shifted significantly in the early 3rd millennium, as shown by results from the Tell Beydar Survey (TBS) (Ur and Wilkinson 2008) in the western Upper Khabur drainage basin. In the region around this outlying ringwall settlement,[6] settlement drops significantly in the early EBA, with both villages and towns becoming abandoned shortly before the commencement of habitation at Tell Beydar itself, which quickly grew to its full size of 17 hectares (Fig. 8). Following this, although settlement at Tell Beydar remained stable until the start of the late EBA, occupation of smaller sites in its vicinity did not pick up again until the middle EBA. It therefore seems that the 3rd millennium occupation of Tell Beydar drew much of its population from local surroundings, prompting a drop in smaller settlements without significant overall population change in the area. Then, once the main tell had become established, the surrounding settlements returned to their former numbers.

A very similar pattern can be observed in northern Iraq, where the North Jazira Survey (NJS) (see Wilkinson and Tucker 1995) has shown that the early EBA saw a radical shift from a pattern of stable small settlements. During this period, the southern and western parts of the survey area become completely abandoned, the number of small settlements plummeting simultaneously with the large-scale occupation of the 66-hectare Tell al-Hawa (Lawrence 2012).

This is interpreted as providing an area of pastureland, whereas in the north and east of the region settlement nucleation and agricultural intensification prevailed (Wilkinson and Tucker 1995: 48-57). Most importantly, as in the TBS area, the overall population and total settled area appears to undergo very little change, if at all. Thus, in contrast to the expansion of settlement in the Middle Euphrates and Western Jazira, settlement in the north-eastern Jazira underwent a process of reorganisation during the EBA, rather than an influx of population from elsewhere.

6 See the next section for an explanation of this term.

Double-Walled Cities in the Western Jazira

The large, mostly circular double-walled cities of the Western Jazira had a profound influence on the landscape of the dry steppe. As has been established by several investigations (see Moortgat-Correns 1972: 25-52 for a summary of these), these sites are largely unique to this region, where they occur with relative abundance, scattered across the landscape. However, it is wrong to assume that all these double-walled cities are homogenous, both in terms of form and structure. Their distinctive form makes them easily identifiable on satellite imagery, enabling a rudimentary classification of these sites into two major categories (Fig. 9). The first might be termed "true" Kranzhügel to distinguish from the generic use of the term which has often been applied indiscriminately. This group is identified by a strong similarity with Tell Chuera; a circular upper mound with a central depression and a massive surrounding wall, and a concentric circular lower town on a terrace with its own surrounding wall (see Fig. 10). It is important to note that at Tell Chuera, there is ample evidence that the upper town wall fell into disrepair following the establishment of the lower town and construction of the second wall (Meyer 2010b: 204). This would indicate that, despite modern-day appearances (especially from aerial views), the site was never actually doubly fortified at any given time, and it is not unreasonable to assume the same might be true of other sites in the true Kranzhügel group.

The second category, termed "ringwall settlements",[7] appears distinct on satellite imagery (Fig. 9). The majority of sites in this group are clustered around the Jebel Abd al-Aziz, but a few exist west of the Balikh, and at least one example, Tell Beydar, is located beyond the Western Jazira in the Khabur drainage basin. These are defined by a circular inner mound with no central depression and a barely visible, at times seemingly nonexistent, surrounding wall. Around this, a "lower town" with no visible structural remains on an extremely low (if any) terrace is enclosed by a very clear outer wall. However, the only intensively investigated ringwall settlement, Tell Malhat, showed evidence that its lower town was occupied, but that modern traces of it are all but invisible (Quenet 2012). The shape of the lower town wall of ringwall settlements is not always circular; Tell Mu'azzar, for example, is pentagonal, while Tell el-Magher is almost square (Moortgat-Correns 1972: 30-31). This, together with the lack of both a substantial lower town terrace and a central depression in the upper town indicates a significant structural difference from true Kranzhügel. Indeed Meyer considers upper-town depressions to be a crucial factor in the primary construction phase of many Kranzhügel, connected with the initial establishment of a planned street network (Meyer and Hempelmann 2006: 25).

This classification is not limited to physical differences between sites however, as it is also reflected in the spread of double-walled cities across the Western Jazira. As can be seen on the map of their distribution (Fig. 11), the aforementioned categories divide into two clear geographical zones. With few exceptions, the Kranzhügel in the mould of Tell Chuera form a northern band located both east and west of the Balikh, on the border of the

7 Not to be confused with the term "citadel city", which was first used by A. L. Oppenheim (1964: 130-132) – for discussion and further references see Wilkinson *et al.* (2012: 175-176) – to describe certain Middle Euphrates sites such as Titriş Höyük and Sweyhat; small LC tell sites that in the EBA became walled lower town settlements with an elite acropolis on the original tell. There is no evidence to suggest such acropoleis in the double-walled settlements of the Western Jazira (see Meyer 2010c: 199).

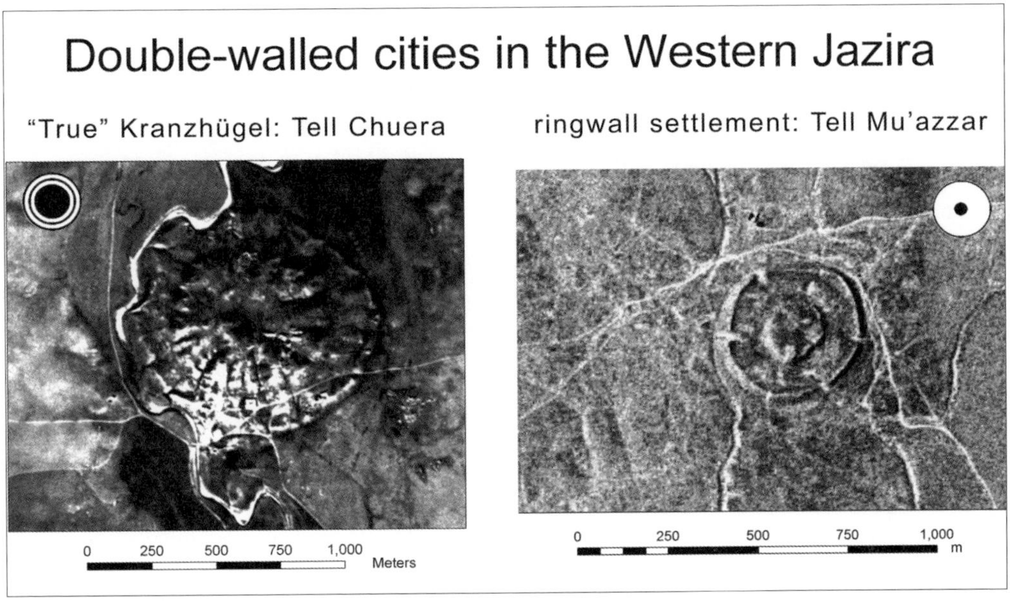

Fig. 9: CORONA satellite images (data available from the U.S. Geological Survey) illustrating the typical variations between Kranzhügel and ringwall settlements, as well as the symbols used to denote them for Fig. 11.

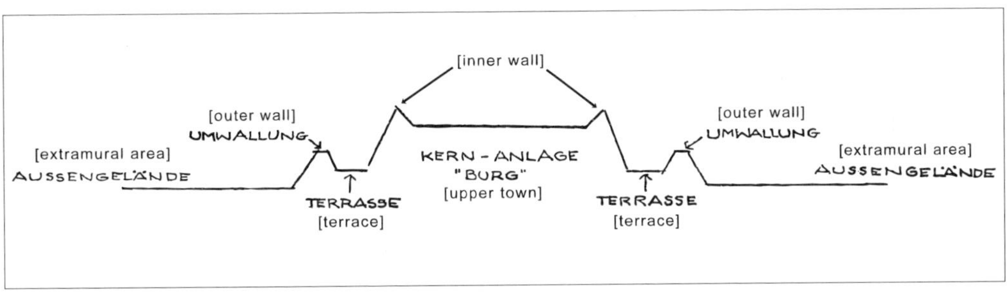

Fig. 10: Cross-section of a typical Kranzhügel (based on Tell Abu Shakhat) (Moortgat-Correns 1972: 28, bracketed annotations by the authors).

Zone of Uncertainty. Ringwall settlements on the other hand are mostly clustered in the eastern half of the region. Unlike the true Kranzhügel, they exist in locations encompassing a broad range of average rainfall per annum, stretching from the 400 mm to the 220 mm isohyets; the entire north-south breadth of the Western Jazira. Thus the form and structure of double-walled cities appear, to some extent at least, to be related to their location within the Zone of Uncertainty. This can be regarded as further evidence of a deliberate colonisation of this region, either by multiple groups of people each bringing their own methods of construction, or by polities which purposefully selected different types of settlement to best suit different climatic landscapes.

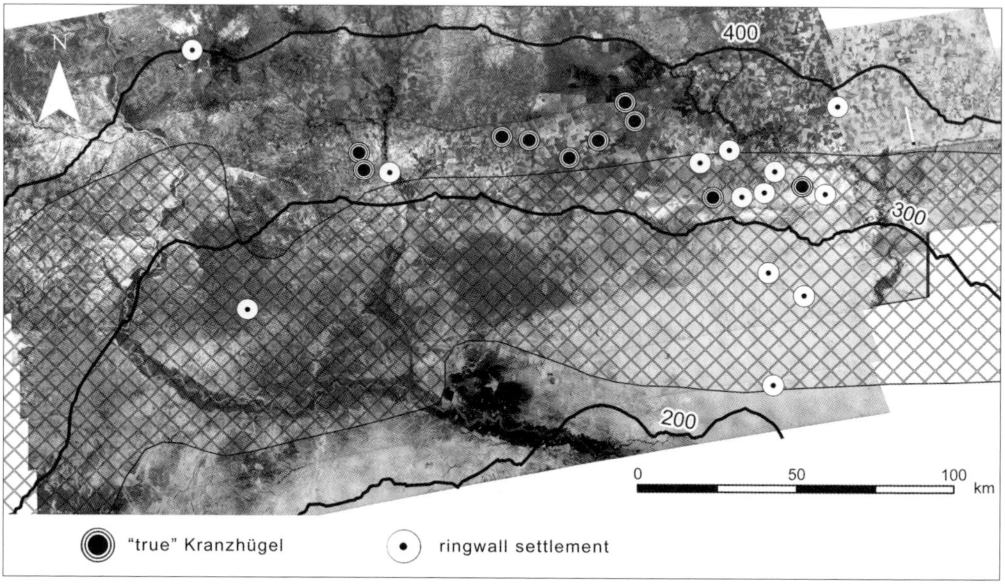

Fig. 11: CORONA satellite images (data available from the U.S. Geological Survey) of the Western Jazira, with the locations of Kranzhügel and ringwall settlements marked. Also includes isohyets (from Meyer-Christoffer et al. 2011), and the Zone of Uncertainty in crosshatching.

The Zone of Uncertainty: Settlements and Agro-pastoral Strategies

As has been pointed out for the area of Tell Sweyhat (Danti 2000), it is likely that a significant amount of the agricultural land around large settlements in the Zone of Uncertainty would have been devoted to barley and/or fodder crops. It is therefore notable that, according to the palaeobotanical and faunal record, the numerical growth of flocks of sheep appears to parallel the growth of barley (McCorriston and Weisberg 2002; Zeder 2003); in fact the two appear to exhibit a co-evolutionary development. Moreover, once barley-based agriculture became significant, sedentary communities would have been necessary both to tend fields and to service the management of the huge flocks that appear to have existed at the time of the Ebla archives. In other words, extensive pastoralism of large flocks, the cultivation of barley, and the development of settlement would have been co-related in the steppe lands of the Zone of Uncertainty.

During the second half of the 3rd millennium BC, the Middle Euphrates, the Western Jazira steppe, the Middle Khabur, and large areas of central Syria witnessed a dramatic expansion of settlement out of "core regions". These cores include regions such as the Orontes valley corridor, Harran Plain (southern Turkey), and Khabur basin (Fig. 12), which apparently sees a decline in settlement (and possibly in social complexity) during the Ninevite V period, at the start of the 3rd millennium BC (Weiss 2003).

This speaks for a shift in population focus towards newly established settlements, most of which occurred within the Zone of Uncertainty. At approximately this time, or slightly earlier, evidence exists for a massive switch in the manufacturing of textiles, from plant fibres to the use of wool (McCorriston 1997). Because the development of settlement in the steppe occurred over the 3rd millennium BC, and may have even commenced in the late 4th millennium, it can be suggested that these two processes were part of the same phenomenon.

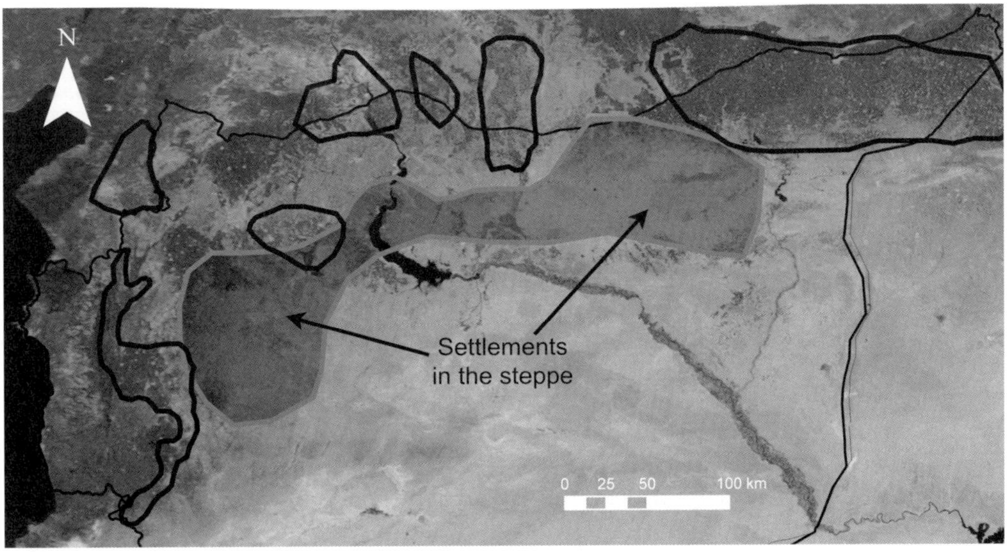

Fig. 12: Satellite image of Syria with core agricultural regions outlined in black, and the general location of settlements in the dry steppe indicated.

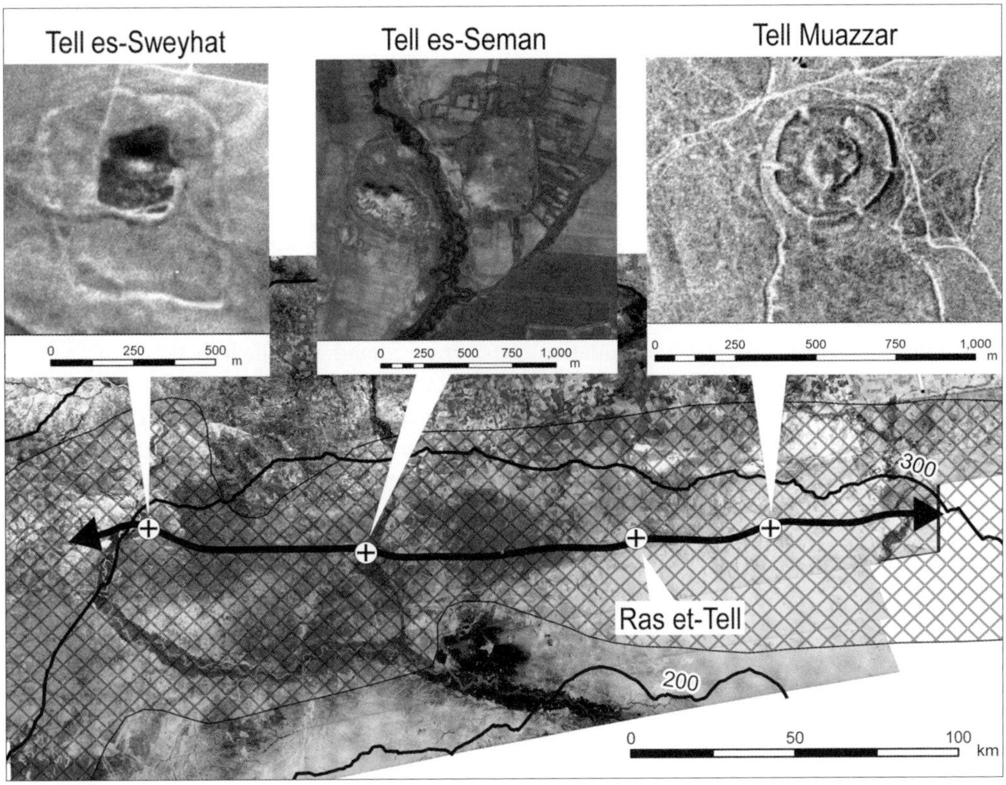

Fig. 13: CORONA satellite images (Data available from the U.S. Geological Survey) of the Western Jazira, with the proposed east-west itinerary marked and important stopping points along the route illustrated.

At this time, states such as Ebla appear to have attained a level of institutional organisation and economic power capable of exploiting the niche of the Zone of Uncertainty for agro-pastoral purposes, thereby investing in high-risk-high-gain ventures.

By contrast, the pasture lands that developed in the densely populated region of northern Iraq represented somewhat marginal lands that had formerly been occupied by settlements, and which were subsequently abandoned. Rather than being outside the zone of long-term settlement, these enclaves of enclosed pasture appear to represent a different, more localised form of pastoral production within the cultivated zone.

A further factor that affected the boom-and-bust trajectories of settlements in the Zone of Uncertainty that cannot be overlooked is connectivity. For the Western Jazira, trade and transport routes have long been interpreted from textual evidence (e.g. Goetze 1953); pathways across the landscape that encouraged the growth of "stopping stations" which grew into large cities. This has notably been suggested as the possible origin of Tell Malhat (Kühne 1983), explaining its location at the very southern boundary of the Zone of Uncertainty, a region that sees a mere 220 mm average annual rainfall (see Fig. 4).

Another similar ringwall settlement, Tell Mu'azzar, can be regarded as the eastern continuation of a route passing from Sweyhat via Tell es-Seman (a large site straddling both banks of the Balikh) and the unique religious site of Ras et-Tell (see Moortgat-Correns 1972), drawing a near-perfect east-west line (Wilkinson 2004: 186-187) (Fig. 13).

Thus, while the establishment of widespread occupation may have been affected by, and indeed given rise to, transport links across the dry steppe, the collapse of these routes would subsequently have further contributed to the demise of such settlements.

To conclude, while states exploiting the Zone of Uncertainty needed to be large enough to deal with the risks involved and absorb the inevitable occasional losses, the fragile economies within the dry steppe would only have persisted as long as rainfall remained sufficient (and sufficiently stable) for flocks to be fed and barley to be grown. In addition, if trade patterns were disrupted or weakened, and/or the resources of the state were diminished, high-risk agro-pastoralism would have become less profitable, and the area would be abandoned except for potential seasonal visits. At Tell Chuera, there is good evidence for a relatively abrupt decline in the late EBA, with no significant archaeological layers of destruction (Meyer 2009: 56). This evidence against a violent end to occupation goes hand-in-hand with the hypothesis of a voluntary withdrawal from the Zone of Uncertainty. In contrast, in the sustainable zone to the north, which may not have participated in high-risk-high-return endeavours to the same degree, economies would have been less disrupted by the above factors, and, in the event of them occurring, life could have returned to earlier levels of subsistence without significant effect on settlement.

Acknowledgements

The Fragile Crescent Project was funded by the Arts and Humanities Research Council (UK), grant no. AH/FO10095/1. We wish to thank the AHRC, The British Academy and CBRL for funding. Thanks also go to Durham University and its Department of Archaeology for providing funding for the Informatics Laboratory which was crucial for the analyses undertaken. Our colleagues from the project also deserve thanks, specifically Louise Rayne for processing rainfall data. From Frankfurt University, we thank Prof. Jan-Waalke Meyer and Veronika Kudlek for providing survey data for the Wadi Hamar region. In Syria, the Land of Carchemish Project, Homs, and Sweyhat Surveys benefited significantly from advice and help from the Directorate of Antiquities and Museums in Damascus, Homs, and Aleppo.

Bibliography

Algaze, G., T. J. Wilkinson and A. Mısır
1992 Şanlıurfa Museum/University of California Excavations and Surveys at Titriş Höyük, 1991: A Preliminary Report, *Anatolica* 18: 33-60.

Algaze, G., G. Dinckan, B. Hartenburger, T. Matney, J. Pournelle, L. Rainville, S. Rosen, E. Rupley, D. Schlee and R. Vallet
2001 Research at Titriş Höyük in southeastern Turkey: The 1999 Season, *Anatolica* 27: 23-106.

Archi, A.
1993 Trade and Administrative Practice: The Case of Ebla, *AoF* 20/1: 43-58.

Cocks, P. S., E. F. Thomson, K. Somel and A. El-Moneim
1988 Degradation and Rehabilitation of agricultural land in North Syria, *ICARDA* 119, Aleppo.

Cooper, L.
2006 *Early urbanism on the Syrian Euphrates,* London.

Cooper, P. and E. Bailey
1990 An assessment of risk associated with increasing sheep and feed supplies in Syria. Current and alternative strategies, *ICARDA annual report* 1989: 289-316.

Dohmann-Pfälzner, H. and P. Pfälzner
2002 Untersuchungen zur Urbanisierung Nordmesopotamiens im 3. Jt. v. Chr. am Beispiel des Tall Chuera, Syrien. In R. Eichmann (ed.), *Ausgrabungen und Surveys im Vorderen Orient I*, OrA 5, Rahden: 1-22.

Danti, M. D.
2000 *Early Bronze Age Settlement and Land Use in the Tell es-Sweyhat Region, Syria.* Ph.D. dissertation, Department of Anthropology, University of Pennsylvania.

Einwag, B.
1993 Vorbericht über die archäologische Geländebegehung in der Westgazira, *DaM* 7: 23-43.

Gelb, I. J.
1986 Ebla and Lagash: Environmental Contrast. In H. Weiss (ed.), *The Origins of Cities in Dry-Farming Syria and Mesopotamia in the Third Millennium B.C.,* Guilford CT: 157-167.

Goetze, A.
1953 An Old Babylonian Itinerary, *Journal of Cuneiform Studies* 7/2: 51-72.

Hazell, P., P. Oram and N. Chaherli
2001 *Managing droughts in the low-rainfall areas of the Middle East and North Africa,* EPTD Discussion Paper 78, Washington DC.

Jaubert, R.
2000 Government policy and productive systems in the arid regions of Syria since since the 1930s. In M. Mundy and B. Musallam (eds.), *The Transformation of Nomadic Society in the Arab East*, Cambridge: 99-109.

Jones, M. J.
1993 Barley-based farming systems of the Mediterranean. In M. J. Jones, G. Mathys and D. Rijks (eds.), *The Agrometeorology of Rainfed Barley-based Farming Systems,* Aleppo: 129-144.

Kouchoukos, N.
1998 *Landscape and Social Change in Late Prehistoric Mesopotamia*, Ph.D. dissertation, Department of Anthropology, Yale University.

Kudlek, V.
2006 *Ein Beitrag zur Siedlungsgeschichte im Bereich von Tell Chuera*, M.A. Dissertation, Institut für Archäologie und Kulturgeschichte des Vorderen Orients, Goethe-Universität Frankfurt am Main.

Kühne, H
1983 Tall Malhat ed-Deru: Eine Station auf dem Wege nach Kappadokien?. In R. M. Boehmer and H. Hauptmann (eds.), *Beiträge zur Altertumskunde Kleinasiens: Festschrift für Kurt Bittel*, Mainz: 299-308.

Lawrence, D.
2012 *Early Urbanism in the Northern Fertile Crescent: A Comparison of Regional Settlement Trajectories and Millennial Landscape Change*, Ph.D. dissertation, Department of Archaeology, Durham University.

Lawrence, D., J. Bradbury and R. Dunford
2012 Chronology, Uncertainty and GIS: A Methodology for Characterising and Understanding Landscapes of the Ancient Near East. In W. Bebermeier, R. Hebenstreit, E. Kaiser and J. Krause (eds.), *Landscape Archaeology. Proceedings of the International Conference Held in Berlin, 6th-8th June 2012*, Topoi Special Volume 3, Berlin: 353-359.

Lebeau, M.
2000 Stratified Archaeological Evidence and Compared Periodizations in the Syrian Jezirah During the Third Millennium BC. In C. Marro and H. Hauptmann (eds.), *Chronologies des pays du Caucase et de l'Euphrate aux IVe-IIIe Millenaires*, Istanbul: 167-192.

Marston, J. M.
2011 Archaeological markers of agricultural risk management, *JAA* 30: 190–205.

McCorriston, J.
1997 The Fiber Revolution, *Current Anthropology* 38/4: 517-535.

McCorriston, J. and S. Weisberg
2002 Spatial and temporal variation in Mesopotamian agricultural practices in the Khabur Basin, Syrian Jazira, *JAS* 29: 485-498.

Métral, F.
2000 Managing risk: sheep-rearing and agriculture in the Syrian steppe. In M. Mundy and B. Musallam (eds.), *The Transformation of Nomadic Society in the Arab East,* Cambridge: 123-144.

Meyer, J.-W.
2009 Zur Entstehung urbaner Zentren in Nordostsyrien – Tell Chuera. In J.-W. Meyer (ed.), *Zwischen Euphrat und Rhein: Ergebnisse des Graduiertenkollegs "Archäologische Analytik"*, FarS 7, Wiesbaden: 55-59.
2010a Versuch einer historischen Einordnung von Tell Chuera in die politisch-historische Entwicklung Nordostsyriens im 3. Jt. v.Chr. In J.-W. Meyer (ed.), *Tell Chuera: Vorberichte zu den Grabungskampagnen 1998 bis 2005*, VFMOS 2 II, Wiesbaden: 11-34.
2010b Tribal community and State: The Change of Settlements and Settlement Patterns in Upper Mesopotamia During the 3rd Millennium B.C. – A Re-evaluation, *al-Rafidan special issue "Formation of Tribal Communities"*: 203-211.
2010c Überlegungen zur Siedlungsstruktur. Eine erste Analyse der geomagnetischen Prospektion. In J.-W. Meyer (ed.), *Tell Chuera: Vorberichte zu den Grabungskampagnen 1998 bis 2005*, VFMOS 2 II, Wiesbaden: 199-221.

Meyer, J.-W. and R. Hempelmann
2006 Bemerkungen zu Mari aus der Sicht von Tell Chuera – Ein Beitrag zur Geschichte der ersten Hälfte des 3.Jts.v.Chr., *AoF* 33/1: 22-41.

Meyer-Christoffer, A., A. Becker, P. Finger, B. Rudolf, U. Schneider and M. Ziese
2011 *GPCC Climatology Version 2011 at 0.5°: Monthly Land-Surface Precipitation Climatology for Every Month and the Total Year from Rain-Gauges built on GTS-based and Historic Data*, DOI: 10.5676/DWD_GPCC/CLIM_M_V2011_050.

Milano, L.
1995 Ebla: a third millennium city-state in ancient Syria. In J. Sasson (ed.), *Civilizations of the Ancient Near East,* Vol. 2, New York: 1219-1230.

Miller, N. F.
1997 Farming and herding along the Euphrates: Environmental constraint and cultural choice (fourth to second millennia B.C.). In R. Zettler (ed.), *Subsistence and Settlement in a Marginal Environment: Tell es-Sweyhat, 1989-1995 Preliminary Report,* Philadelphia: 123-132.

Moortgat-Correns, U.
1972 *Die Bildwerke vom Djebelet el Beda in ihrer räumlichen und zeitlichen Umwelt,* Berlin.

Mortimore, M.
1998 *Roots in the African dust: Sustaining the sub-Saharan Drylands*, Cambridge.

Oppenheim, A. L.
1964 *Ancient Mesopotamia: Portrait of a Dead Civilization*, Chicago.

Pfälzner, P.
2010 Introduction and Synthesis: Urban Development and Ecology at Tell Mozan. In K. Deckers, M. Doll, P. Pfälzner and S. Riehl (eds.), *Development of the Environment, Subsistence and settlement of the city of Urkesh and its region,* Wiesbaden: 1-12.

Porter, A.
2007a You say Potato, I say...Typology, Chronology and the Origin of the Amorites. In C. Kuzucuoğlu and C. Marro (eds.), *Sociétés Humaine et Changement Climatique à la Fin du Troisième Millénaire.* Istanbul: 69-115.
2007b The Ceramic Assemblages of the Third Millennium in the Euphrates Region. In M. Al-Maqdissi, V. Matoian and C. Nicolle (eds.), *Céramique de l'Âge du Bronze en Syrie II: L'Euphrate et le region de Jezireh.* Beirut: 3-21.

Pruß, A.
2005 Wadi Hamar-Survey: Zur Besiedlungsentwicklung der Region um Tell Chuera, *Alter Orient Aktuell* 6: 12-14.

Quenet, P.
2012 Preliminary Results of the Khirbet Malhat Survey, North Syria, paper presented at the 8th ICAANE, Warsaw, 30 April - 4 May.

Stein, G.
1987 Regional economic integration in early state societies: third millennium BC pastoral production at Gritille, Southeast Turkey, *Paléorient* 13/2: 101-111.
2005 ‚Invisible' Social Sectors in Early Mesopotamian State Societies. In: V. L. Scarborough (ed.), *A Catalyst for Ideas. Anthropological Archaeology and the Legacy of Douglas Schwarz*, Santa Fe: 121-148.

Treacher, T. T.
2000 Linkages between livestock and agricultural systems in Syria. In M. Mundy and B. Musallam (eds.), *The Transformation of Nomadic Society in the Arab East*, Cambridge: 187-196.

Ur, J. A. and T. J. Wilkinson
2008 Settlement and Economic Landscapes of Tell Beydar and its Hinterland. In M. Lebeau and A. Suleiman (eds.), *Beydar Studies 1*, Subartu XXI, Turnhout: 305-327.

Vila, E.
2010 Etude de la faune mammalienne de Tell Chuera. In J.-W. Meyer (ed.), *Tell Chuera: Vorberichte zu den Grabungskampagnen 1998 bis 2005,* VFMOS 2 II, Wiesbaden: 223-291.

Wachholtz, R.
1996 *Socio-economics of Bedouin Farming Systems in Dry Areas of Northern Syria,* Kiel.

Wattenmaker, P.
1998 *Household and State in Upper Mesopotamia: Specialized Economy and the Social Uses of Goods in an Early Complex Society,* Washington DC.

Weber, J.
1997 Faunal remains from Tell es-Sweyhat and Tell Hajji Ibrahim. In R. Zettler (ed.), *Subsistence and Settlement in a Marginal Environment: Tell es-Sweyhat, 1989-1995 Preliminary Report,* Philadelphia: 133-168.

Weiss, H.
2003 Ninevite 5 Periods and Processes. In E. Rova and H. Weiss (eds.), *The Origins of North Mesopotamian Civilization: Ninevite 5 Chronology, Economy, Society,* Subartu IX, Turnhout: 593-624.

Wilkinson, T. J.
1990 *Town and Country in Southeastern Anatolia, Volume 1: Settlement and Land Use at Kurban Höyük and Other Sites in the Lower Karababa Basin*, OIP 109, Chicago.
2000 Settlement and land use in the zone of uncertainty in upper Mesopotamia. In R. Jas (ed.), *Rainfall and Agriculture in Northern Mesopotamia,* Leiden: 3-35.
2004 *On the Margin of the Euphrates: Settlement and land use at Tell es-Sweyhat and in the Upper Lake Tabqa Area, Syria,* OIP 124, Chicago.

Wilkinson, T. J. and D. J. Tucker
1995 *Settlement development in the North Jazira, Iraq,* Iraq Archaeological Reports 3, Warminster.

Wilkinson, T. J., N. Galiatsatos, D. Lawrence, A. Ricci, R. Dunford and G. Philip
2012 Late Chalcolithic and Early Bronze Age Landscapes of Settlement and Mobility in the Middle Euphrates: A Reassessment, *Levant* 44(2): 139-185.

Zeder, M.
2003 Food provision in urban societies: a view from Northern Mesopotamia. In M. L. Smith (ed.), *The Social Construction of Ancient Cities,* Washington DC: 156-183.

OLIVIER BARGE, CORINNE CASTEL and JACQUES ÉLIE BROCHIER
CNRS, Université Lyon 2, CNRS, Université Lyon 2 and CNRS, Aix-Marseille Université

Human impact on the landscape around Al-Rawda (Syria) during the Early Bronze IV

evidence for exploitation, occupation and appropriation of the land

The work undertaken between 2002 and 2010 by the Franco-Syrian team[1] on the site of Tell Al-Rawda has led to confirmation that this was a new city[2] constructed in the steppe in the middle of the third millennium and abandoned only a few centuries after its foundation, about 2200 BC.[3] The work of Bernard Geyer's team ("Mission des Marges Arides") had already shown that Al-Rawda was positioned on the eastern edge of a front of colonisation that developed in the Early Bronze IV, before these places were abandoned by sedentary populations to be later resettled for a relatively brief period in the Byzantine period (Geyer, Besançon and Rousset 2006: 56-59 and 57, fig. 1). The arid context gives rise to questions concerning the subsistence conditions of a population that, although difficult to estimate, would appear high in relation to the potential resources offered by this environment.[4] This population would certainly have had an impact in terms of exploitation and modification of a fragile environment; today we observe a slow degradation (Debaine and Jaubert 2006) over the long term, although this is not possible to define it temporally.

Can we learn how the population of the second half of the third millennium in this region organised and exploited this territory? What were the reasons for this relatively brief occupation? What were the environmental constraints that the population had to face and that required adaptation, and what led to the abandonment of this territory?

More than 4000 years later, the study of the relation of these humans to their environment is based on data with many gaps, and there are more questions than certainties. However, the thorough survey of the microregion of Al-Rawda combined with data from the excavations and from the paleo-environmental studies has permitted some observations concerning the population and their exploitation of the environment. This work has also led

1 The Franco-Syrian archaeological Mission of Tell Al-Rawda is co-directed by Corinne Castel, CNRS, and Nazir Awad, Director for Heritage at the General Directorate of Antiquities and Museums of Syria.

2 The meaning of a "new city" in this case is a regular city, conceived as a whole before being built, and thus the result of a planned urbanism. This does not exclude the possibility that it could have been founded on a previous settlement, as will be seen below.

3 These chronological elements rest in particular upon radiocarbon determinations calibrated through Bayesian analysis. For a presentation of the method and results obtained at Tell Al-Rawda, cf. chapter III.3 « Fondation, occupation et abandon de la ville d'Al-Rawda; analyse bayésienne des déterminations radiocarbone » written by J.É. Brochier and C. Castel *in* Castel, Barge, Awad and Al-Maqdissi, (forthcoming).

4 It is considered that climatic conditions have not fundamentally changed since the 3rd millennium, which does not exclude the possibility that slight variations could have had repercussions on human activities and on processes of sedimentation.

Studia Chaburensia 4 (2014), pp. 173–185.

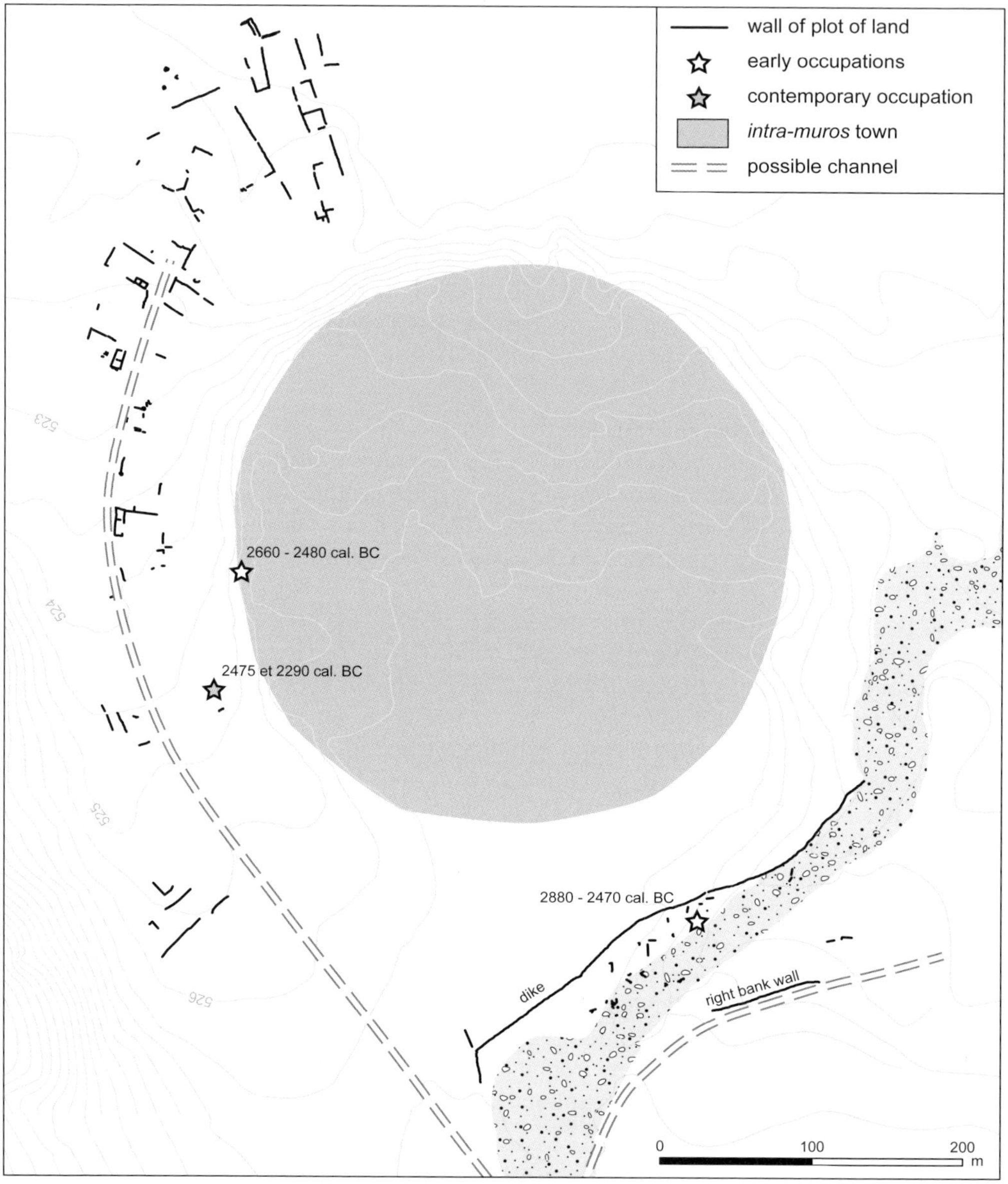

Fig. 1: The land divided into plots, the city and the hypothetical hydraulic constructions.

us to propose a few hypotheses on the subsistence modes and on the relation of the population to its environment in the Early Bronze IV.

Three points clarify these relations: the existence of indices for irrigated agriculture in areas near the city, the possible relation of the site with the kites found in the surroundings, and the configuration of the network of settlement sites near the Very Long Wall or TLM (Très Long Mur).

The *fayda* and its agricultural transformations

During excavations in different areas of the city, forty-one complete or fragmentary swing plough heads were found in stratigraphic contexts. These stone objects, often wrongly considered to be hoes, shovels or ploughshares[5], are all symmetrical in shape and generally triangular. They show bevelled wear of the cutting edges on one side, of the point on the other side, and present many striations related to use. These swing plough heads are evidence that agricultural activity occurred near the city, as the archaeobotanical data indicate. The latter show that diversified farming was practised at Al-Rawda, both dry and most probably irrigated (Herveux 2004), as seeds from weeds of irrigated crops (*Phalaris*) were found in stratigraphy. Moreover, the *fayda*, the flat surface area that surrounds the tell, presents today a silty fill, decisive for favourable pedological conditions; according to present-day inhabitants, the barley harvest can reach, in humid years, 20 to 40 hundredweight (100 kg) per hectare.[6] This *fayda* is quite large (about one hundred hectares) and it is crossed by the Wadi Al-'Amur which skirts the tell on the east. In its breadth, flatness, and proximity to the city, it constitutes in principle an obvious surface area for cultivation, like other *faydas* near Al-Rawda.

However, the combined information gleaned from different studies shows that the topography of the fayda was different at the beginning of the occupation of the city.

There exist at several points under the present surface of the *fayda*, anthropic deposits contemporary or anterior to the occupation of the city (Fig. 1):

- an exceptional flood in 2007 locally cut into the bed of the Wadi Al-'Amur, about 150 m south/south-east of the tell, brought to light ashy anthropic deposits. They had accumulated a few decimetres under the surface of the present bed of the *wadi* during a short period between 2880 and 2470 cal. BC, before the foundation of the city;
- a trench dug in sector 2a, west of the tell, showed that two levels of occupation existed that pre-date the first rampart, the oldest situated at almost 3 m below the present surface of the *fayda*. The narrowness of the trench at its bottom does not allow determination of the nature and extent of these early occupations, but two radiocarbon dates (between 2660 and 2480 cal. BC) situate them a little before the foundation of the city;
- a borehole sample taken in the *fayda*, some thirty meters to the west/south-west of the tell, crossed a thick ashy level of between 1.20 and 1.80 m below the surface, accumulated between 2475 and 2290 cal. BC, thus contemporary to the first years of occupation of the new city.

These three indications thus closely frame the foundation of the city which took place between 2560 and 2430 cal. BC. They show that the fill that we observe today, thicker to the west than to the east, and thicker to the north than to the south, began after the foundation of the city (Fig. 2). At that moment, the flat surface of the *fayda* did not exist. Instead, in certain places, especially to the west, there existed stony terraces with chaotic topography, typical of alluvial formations of the *wadi*, and probably silty areas, fairly well-developed in the flare of the valley, which enabled partial cultivation. The dating of the end of the fill accumulation is more problematic. It rests on the presence, on the surface of the *fayda* and around the tell, from the north to the south-west, of long walls that appear to de-

5 A study was made of these objects, written by C. Castel, in Castel, Barge, Awad and Al-Maqdissi (eds.), forthcoming.

6 For comparison, the FAO has estimated that the average yield for barley on a world scale was 27.2 q/ha in 2004.

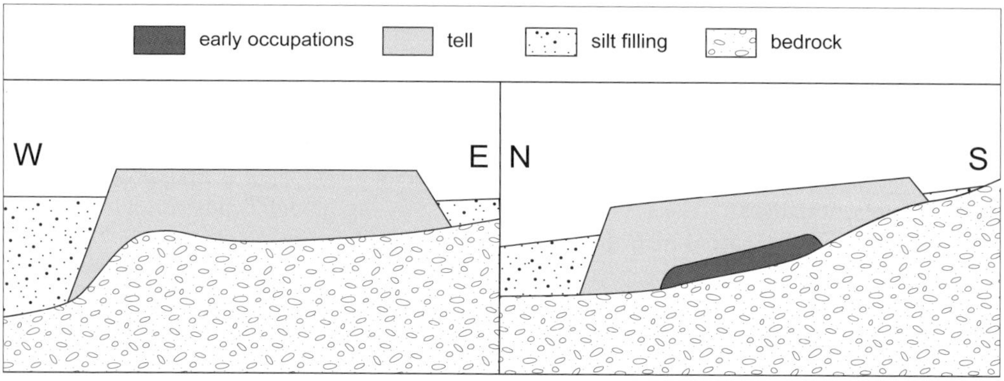

Fig. 2: Sketch in section of the sedimentary fill of the *fayda*.

fine division of the land into plots. They take up the urban fabric in radial lines, and extend it into the *fayda* beyond the fortifications (Fig. 1). It is thus probable that this land division was contemporary to the last period of the occupation of the city. In this hypothesis, the fill appears to have occurred particularly rapidly, three centuries at the most, during which time the city was occupied.

Several phenomena could account for what appears to have been a major sedimentary event, but they will not be discussed in this article.[7] It is evident that, whatever the origin and quantity of the materials that make up the fill, the latter could have been favoured by the presence of dividing walls that slowed down the rapid flow of floodwaters. It is plausible that a division of the land around the city could have evolved during the occupation of the city and not only existed at the very end of this period.

With the raising of the surface level, the *fayda* was levelled progressively, up to the moment when, upriver, the wadi became destabilised at its entrance into the *fayda*; it then changed course towards the east and adopted the channel that we observe today.

This could have occurred naturally; it could also have been influenced by the construction of a dike, towards the end of the occupation of the city, to protect the tell, threatened by floods because of the raising of the level of the *fayda*. A long wall may be observed over 340 meters, south of the tell, against which the *wadi* bed appears to lie (Fig. 1). It is situated on the left bank. In this hypothesis, the part of the fayda to the west and north-west of the tell is free of the *wadi* and thus a planned division by walls could constructed, which brought a continuation of the urban fabric onto the land that had become completely cultivable.[8]

Other elements are coherent with this interpretation; the setting up of a hydrological model, based on the present topography, has shown that a slight adaptation could easily enable control of the *wadi* waters (Barge and Moulin 2008). A simple low wall or an earth levee of less than a meter in height, built perpendicular to the *wadi* at the entrance to the *fayda* and at the level of the dike, would have enabled deviation of the waters, when needed, towards the

7 There is a discussion of this subject in the chapter "L'occupation d'Al-Rawda, la mise en place de la *fayda* et l'événement de 4200 BP" written by Jacques Élie Brochier in Castel, Barge, Awad and Al-Maqdissi (eds.), forthcoming.

8 This hypothesis is only acceptable if the constructions protruding at the surface of the fayda, north-west of the tell, date to the Early Bronze.

western part or eastern part of the *fayda*. The differences in slope at the entrance to the *fayda* are in fact very slight, which explains 1) the ancient shift in the course of the wadi, 2) that the highest floodwaters flow to either side of the tell, and 3) the possible presence of channels, supported by several indications.

In fact, two channels, on either side of the tell, could have originated farther upstream in this hypothetical control of the water flow of the *wadi* (Fig. 1). To the east, the presence of a wall 75 meters long on the right bank could be understood as the wall of a channel that originated a little farther upstream and that supplied the eastern part of the *fayda*. A line was revealed on the geomagnetic map in this place and a model intended to evaluate the depth of the vestiges shows that this channel would have had a depth beneath the ground surface of 1 to 2 m and would have corresponded to an incision about 2 m wide.[9] On the west side, the lines of the land division show the possible presence (topographically and in the mosaic of plots) of a canal that would have occupied the ancient bed of the *wadi*. It would have thus enabled irrigation of the plots surrounded by the low walls that are observed, even their possible extension more to the north and more to the east. This channel also appears to be confirmed by examination of the geomagnetic map. Several sausage-like shapes appear at the end and in the continuation of the presumed channel. The main one is subdivided at several points. It is similar to sausage shapes that surround the tell, which we know through excavation are ditches, the external part of the system of fortifications.

Thus, the examination of the adaptations made next to the *wadi* a little upstream from the tell, the analysis of the topography (hydrological model), the disposition of the plots and the shapes revealed by the geophysical map are not formal proof, but make up a body of evidence that could indicate irrigated agriculture, also supported by the archaeobotanical data.

Al-Rawda and the kites: utilisation in the Early Bronze Age?

In general, in spite of a number of studies, the question of dating kites has not been satisfactorily answered. The propositions for dating cover millennia, from the Neolithic to the sub-contemporary. The age of these constructions, whether relative or absolute, remains very difficult to establish in contexts where archaeological material is either absent or rare and without any clear stratigraphic relation to the construction. This is the case in the region of Al-Rawda, where seventeen kites have been discovered (Fig. 3), but without the possibility of obtaining chronological evidence concerning the period of construction or use of these constructions.

The most reliable data available today show that these kites were in use mainly in the Chalcolithic and the Bronze Age (Échallier and Braemer 1995; Nadel *et al.* 2010 and Nadel *et al.* in press.). Moreover, in the immediate environs of Al-Rawda, a higher density of these constructions around the city can be observed (Fig. 4a). If the scale is changed, it is clear that the aggregate is not unique, but remains nevertheless observable (Fig. 4b).[10] As there are no chronological data, it may be asked whether the kites of the region were in use at the time of occupation of the city, and in this hypothesis, what could have been the function of these structures in the context of the territory around the city.

9 This geophysical survey was carried out in 2010 by Rozan Al-Kontar, doctoral student at the University of Strasbourg I, under the direction of Marc Munschy, professor at the University of Strasbourg I, École et Observatoire des Sciences de la Terre.

10 We have been able to verify the absence of kites in the zones between these aggregates.

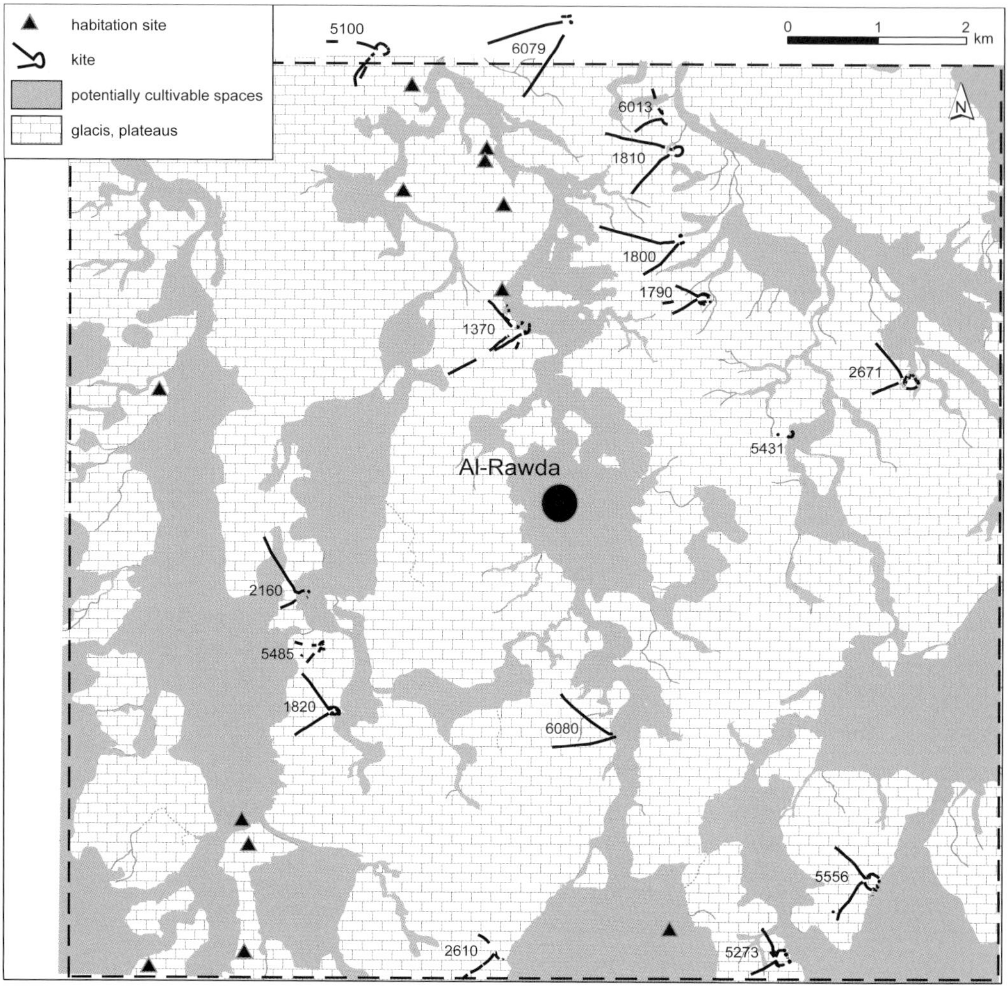

Fig. 3: Distribution of kites and potentially cultivable zones.

Near Al-Rawda, the kites are distributed according to meridians that correspond to the interfluves between the valleys of the Wadi Qastal, the Wadi Al-Amur and the Wadi Al-Zarub (Fig. 3). They consist of long branches (between 100 and 900 m) that are generally well preserved. However, the enclosures are poorly preserved; their trace is often only partial or even totally absent, with no indication as to whether this is a problem of preservation or whether they were absent to begin with.

Where they can be examined, their shape is circular. Some kites present what could be pens, but the relation to the enclosure wall is never clear enough to identify them as such, except for one case (kite 1820). All the kites present a break in the slope near the entrance to the enclosure; the branches lie in the plateaus with their crusty surfaces and discontinuous vegetation, while the enclosures are located on the slopes that link the plateaus with the valleys. All are oriented to the west.

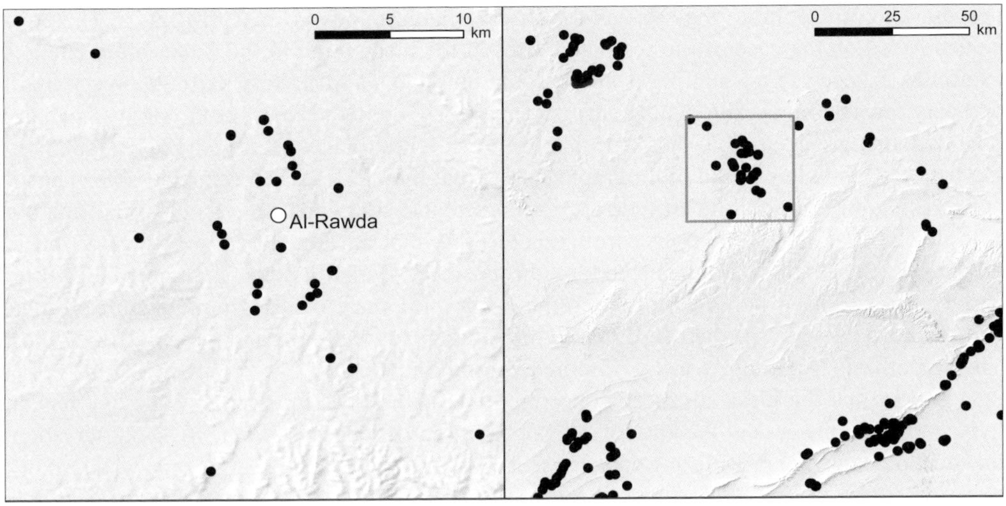

Figure 4: The kites on a regional scale.

These last two characteristics are strong arguments for the identification of these kites as traps, for the collective hunting of gregarious animals. However, several elements render this interpretation difficult:

- Certain kites possess no enclosure and these are often discontinuous or present partitions;
- The possible pens lie directly on the limestone crust, in such a way that no pit could have been dug. The others are positioned either on the line of the enclosure, or inside it, or outside it, and it is impossible to know whether they were part of the kite or simply little pastoral enclosures of a late period, constructed with the materials of the enclosure.

These pens certainly do not play the role of traps or of hides for the hunters.

There are gaps and off-centre parts in the branches. In several cases there are divisions into two, forming corridors, as though this adaptation served to direct the animals. In kite 1370, a corridor ends towards a series of enclosures, of which some are placed side by side with little buildings (site RW19). A trench opened in one of these buildings (RW19.4117) showed that it had been probably occupied over a long period, and probably in the Bronze Age.

Moreover, the study of the bone remains found in stratigraphy at Al-Rawda shows that traces of wild fauna are rare but those of sheep and goats are abundant. Hunting had a negligible role in the food economy of the city (Vila and El Besso 2006). These elements taken together, if accepted for the Early Bronze, point towards herd management rather than hunting. For the kites, certain interpretations see the branches of the kites as fencing the valleys or are thought to be intended for wild animals. This is not true for Al-Rawda; on the contrary, the kites are situated on uncultivable plateaus (and the enclosures on the first slopes along their edges), so that the kites avoid, without exception, the zones most potentially favourable for agriculture (Fig. 3). In the use of kites on the territory around the city, those situated beyond its immediate environs could have defined a zone of defence of the cultivated areas.

This interpretation is very hypothetical, and brings into question the idea that the kites had a role in hunting. This idea is based on the accounts of travellers between the end of the 16th and the beginning of the 20th century who described hunting activities that did not necessarily correspond to the kites as we observe them, in any case at Al-Rawda. This idea is also based on the case of the restricted kites in the Negev and in the Sinai, often cited as examples, whose use for hunting is not in doubt but which appear very particular, very small and very few in comparison to the many others discovered in the Near East. Finally, the many new data that are accessible thanks to the use of satellite images (Kennedy 2012), added to the list of all the kites in the bibliography, show that their extension is considerable. Thus it appears that the temporal incertitude corresponds to a spatial expansion that is broad and not completely established. Nothing permits us to think that all is identical everywhere and in all periods; the cautious and nuanced hypothesis of J.-C. Échallier and F. Braemer (Échallier and Braemer 1995) proposes use for herding or for more than one use (hunting and herding), and was refuted on bases that were more virulent than solid (Rosen and Perevolotsky 1998). This hypothesis is perhaps worthy of more exploration.

It is clear that the ideas advanced concerning the use of kites in relation to the city remain hypothetical. In the case of the kites of Al-Rawda, essentially because of their bad conservation, they will probably not provide any data that will enable clarification of general or local questions.

A network of settlement sites

A thorough survey intended to record Bronze Age developments of all kinds (settlements, funerary groups, vestiges of agro-pastoral activities…) was carried out in the environs of the city. The method of spatial sampling used[11] guaranteed that the condition of exhaustiveness could be met satisfactorily in a squared-off area 10 km per side centred on the city, and in a partial manner[12] in areas with the same surface to the south and south-east. Within this perimeter, some forty vestiges of the Early Bronze IV were discovered, drawn and studied (Fig. 5).

Many are too modest, simple buildings built against an enclosure, to correspond to sedentary occupations. Moreover, for a number of them it was not possible to find elements that could enable attribution of a date that is certain.

Besides these two categories, there are more than twenty sites whose characteristics (in particular the nature of the architectural remains and the proportion of sherds collected on the surface) indicate that they can be reasonably thought to have been occupied throughout the year,[13] and for which the chronological attribution is sure. Their cartography shows that the eastern part of the zone is empty of sedentary occupations; the team of Bernard Geyer has already emphasized this unusual situation by fixing the boundary of the colonisation front at the valley of the Wadi Al-ʻAmur. In detail, it can even be said that south of Al-Rawda, with the exception of one modest site, this boundary follows more the valley of the Wadi Qastal. However, the western part of the zone studied presents a high density of sites, in particular in the valley of the Wadi Al-ʻAmur north of Al-Rawda and in the valley of the Wadi Qastal in the south.

This distribution was matched to the ecological zoning that includes precipitation, and thus the potential for exploitation decreasing from west to east. The topographical location of the sites also obeys this logic as all are situated at the bottom of the valley. Given the probably

11 *Cf.* chapter I « Méthodologie » *in* Castel, Barge, Awad and Al-Maqdissi, forthcoming.

12 The survey of the first microregion around the city was completed in 2008; the second microregion open to our investigations was explored in 2009 and 2010. Unfortunately, it has not been able to be completed at the present time.

13 *Cf.* in chapter IV.4, « Les sites d'habitat autour d'Al-Rawda », the part « L'interprétation des vestiges d'habitat: critères et limites » written by C. Castel in Castel, Barge, Awad and Al-Maqdissi, forthcoming.

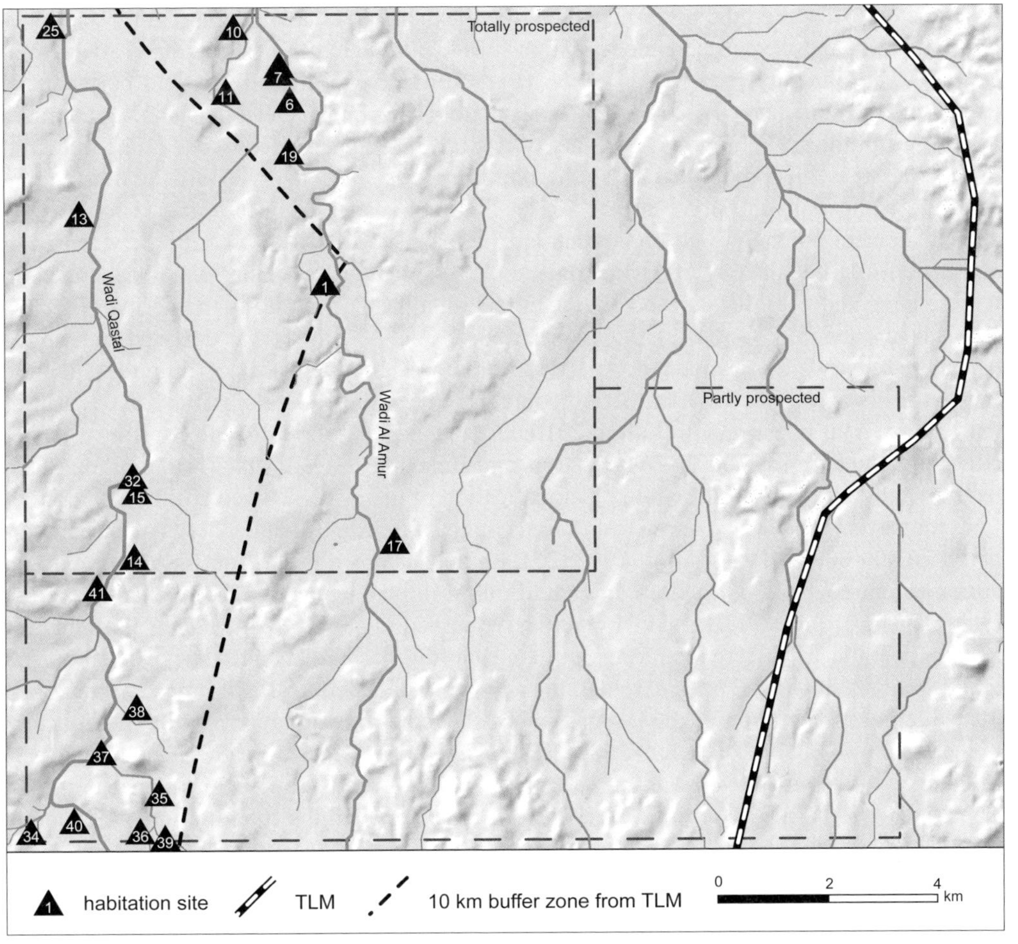

Fig 5: The network of settlement sites and the Very Long Wall.

severe environmental conditions that started in the 3rd millennium, the east-west gradient of aridity is obviously the first factor of explanation.

However, the Very Long Wall (TLM, Très Long Mur), found over more than 220 km by the Mission des Marges Arides (Geyer 2009) and which passes only 10 km east of Al-Rawda, possibly explains this distribution. This wall has not been dated with certainty, but different elements indicate that it probably dates to the Early Bronze IV (Geyer *et al.* 2010: 64).

Too modest, in spite of its remarkable length, to be considered a true defence wall (Geyer *et al.* 2010: 67), the TLM would nevertheless have been an impressive feature in the landscape, in that its regular orientation flouts the reliefs and the *wadis* that it crosses to the point of crossing locally the same thalweg many times.[14] The line that it followed would have thus probably been a symbolic boundary and its function would appear to have been the definition

14 This point could be observed exactly in the second microregion of Al-Rawda. The TLM follows a straight line independently of the topography, crossing seven times the Wadi es Slima and its tributaries (*cf.* in Castel *et al.* in press, the paragraph written by S. Sanz and C. Castel, « Le TLM »).

of a territory. However, the map shows that there existed an area having a regular width between the wall and the permanent occupation sites identified in the microregion of Al-Rawda. The internal boundary of a buffer zone[15] (Fig. 5) of 10 km around the wall corresponds roughly to the boundary of the sedentary occupation. This line even stretches along the main aggregates of the sites; moreover, the city of Al-Rawda is exactly positioned at the inflexion of this line. It is thus probable that the distribution of the sites observed and the outstanding situation of the main city of the region are related to the very existence of the TLM.

This element points to a chronological attribution of the TLM to a period between 2500 and 2200 BC, when the city of Al-Rawda and the secondary occupations which depended on it were inhabited. It is difficult to believe, in fact, that the TLM dates to the period of Ur III, as has been proposed (Geyer 2010: 69), when the city of Al-Rawda was deserted and almost all the secondary sites abandoned. At the same time, the hypothesis according to which the kings of Ebla of the proto-Syrian period could have been the instigators of the TLM or have participated in its construction (Lafont 2010: 86), in a context of rivalry and confrontation between the kingdoms of Ebla and Mari (Archi and Biga 2003) stands out. The definition of the territory marked by the TLM could be an indication of the increased need of the kingdom of Ebla for pasture land, as we know from the accounts of Palace G that the herds of sheep numbered 70,000-80,000 head (Archi 1992: 27). Rather than as a protection for an agricultural zone (which remained divided up and could only fill local needs) against nomads from the east (Geyer 2010: 67, 71), the TLM would have served to define a zone newly exploited for its high potential as pasture land for part of the year (this has already been suggested in Castel and Peltenburg 2007: 613), while remaining a territory under control. This appears even more plausible in consideration of the importance of sheep herding and wool textile manufacture in the economic life of Ebla (Gelb 1986: 162) in which "wool has the same standard value of reference as barley in Mesopotamia" (Biga 2002: 279), in the period of the palace archives. The "fiber revolution" (McCorriston 1997) could have played the role of a major economic stimulus invested in these territories newly exploited by sedentary populations.

As for when the TLM was constructed, during the three centuries corresponding to the occupation of the city, the answer is difficult to find at this time, unless the city of Al-Rawda, situated far to the east in a steppic zone, established in an environment that was on the face of it unfavourable to settlement and at the very limit of occupation for settlers, took the reason for its existence from its location. Its particular character of a "new city", based on a regular pre-established plan, reveals that it was not the result of progressive growth, but of a political will expressed in a context of territorial conquest.

We will not go back over the political power likely to have been at the origin of this urban foundation, but we have seen that Ebla, which extended, during the second half of the 24th century BC, its influence over this region of arid margins of the Fertile Crescent, is a possible candidate.[16] We may thus propose as a hypothesis that the TLM dates to the foundation of the city of Al-Rawda, about 2500 BC, at the moment of the beginnings of territorial states. Al-

15 A buffer zone defines an area of 10 km on either side of the wall. The internal boundary on the map corresponds to the projection of all points of the TLM perpendicular to the latter. For this question and that of the tribal confederation of the Ib'al mentioned in the Ebla texts.

16 For this question and that of the tribal confederation of the Ib'al mentioned in the Ebla texts of the 24th century, *cf.* Castel and Peltenburg 2007: 613 and the pertinent bibliography, as well as the chapter written by A. Catagnoti "Tell Al-Rawda: une ville de la région d'Ib'al dans les archives d'Ebla?" in Castel, Barge, Awad and Al-Maqdassi, forthcoming.

Rawda, whose defensive character is clearly confirmed[17], would then have played the role of a fortress at the boundary of a territory. The discovery of towers dated to the Early Bronze IV around Al-Rawda suggest watchtowers[18] related to the TLM and the fortified city.

By way of a conclusion

The three points brought up in this article deal with subjects that are independent of each other: evidence for agricultural practices, hypotheses concerning hunting and/or herding, occupation of the land. However, they concern a time-space that is particularly unusual, that of a relatively short-lived city established in a difficult environment.

Through the series of material evidence presented, the city of Al-Rawda appears to have left its mark on the environment. These indications are the material manifestation of a territory,[19] which could be considered to be an "*agencement de ressources matérielles et symboliques*" (Debarbieux 2003). With these three points, we touch on several symbolic dimensions of this territory: the dimension of the *appropriated*, seen in the boundary demarcation of the Very Long Wall, as well as the *developed and managed*, seen in the agricultural developments. These dimensions, and possibly the kites, refer also to the dimension of the *perceived*, representations of the environment as resources.

Bibliography

Archi, A.

1992 The City of Ebla and the Organization of its Rural Territory, *Altorientalische Forschungen* 19: 24-28.

Archi, A. and M.G. Biga

2003 A Victory over Mari and the Fall of Ebla, *Journal of Cuneiform Studies* 55: 1-44.

Barge, O. and B. Moulin

2008 The Development of the Syrian Steppe during the Early Bronze Age. In H. Kühne, R. M. Czichon, F. J. Kreppner (eds.) *Proceedings of the 4th International Congress on the Archaeology of the Ancient Near East,* (Berlin, 29th March to April 3rd, 2004), Wiesbaden, Harrassowitz Verlag, Vol. 1: 19-28.

Biga, M.G.

2002 Les foires d'après les archives d'Ebla. In *Mémoires de Nabu* 7, Florilegium Marianum VI, Recueil d'études à la mémoire d'A. Parrot, Paris (SEPOA): 277-288.

Brochier, J.

Forthcoming L'occupation d'Al-Rawda, la mise en place de la fayda et l'événement de « 4200 cal. BP ». In C. Castel, O. Barge, N. Awad and M. Al-Maqdassi (eds.), *Des villes neuves aux franges du désert de Syrie à la fin du 3e millénaire : Travaux de la mission archéologique franco-syrienne d'Al-Rawda 2002-2008 ; Travaux de la mission syrienne de Tell Sha'irat 2007-2008 et Tell Es Sour 2009,* Travaux de la Maison de l'Orient et MAEE, Lyon.

17 The city is surrounded by four lines of fortification. A good number of the sites of the microregion are defended by an enclosing wall.

18 *Cf.* the paragraph entitled « Un système de défense regional? » written by Quenet in Castel *et al.*, in press..

19 The necropolis that extends on the plateau to the west of the *fayda* can also be mentioned.

Castel, C., N. Awad, R. Al-Kontar, T. Babour, M. Bano, O. Barge, J. É. Brochier, C. Calastrenc, B. Chiti, A Cuny, A. Emery, K. Hammad, L. Herveux, V. Kudlek, F. Laliberte, M. Munschy, S. Munos, A. Pelissier, B. Perello, P. Quenet, E. Regagnon, S. Sanz, A. Wild

in press Rapport préliminaire sur les activités de la mission archéologique franco-syrienne d'Al-Rawda, travaux 2007-2010, *Akkadika.*

Castel, C., O. Barge, N. Awad and M. Al-Maqdassi (eds.)

Forthcoming *Des villes neuves aux franges du désert de Syrie à la fin du 3e millénaire : Travaux de la mission archéologique franco-syrienne d'Al-Rawda 2002-2008 ; Travaux de la mission syrienne de Tell Sha'irat 2007-2008 et Tell Es Sour 2009,* Travaux de la Maison de l'Orient et MAEE, Lyon.

Castel, C and E. Peltenburg

2007 Urbanism on the Margins: Third Millennium BC Al-Rawda in the Arid Zone of Syria, *Antiquity* 81: 601-616.

Debaine, F. and R. Jaubert

2006 La dégradation de la steppe: hypothèses et évolution du couvert végétal. In R. Jaubert and B. Geyer (ed.), *Les marges arides du Croissant Fertile: Peuplements, exploitation et contrôle des ressources en Syrie du Nord*, TMO 43/2, Lyon: 149-165.

Debarbieux, B.

2003 Territoire. In J. Lévy et M. Lussault (eds.), *Dictionnaire de la géographie et de l'espace des sociétés*, Belin, Paris: 910-912.

Échallier, J. C. and F. Braemer

1995 Nature et fonctions des « desert kytes »: données et hypothèses nouvelles, *Paléorient* 21/1, CNRS Editions : 35-63.

Gelb, I.J.

1986 Ebla and Lagash: Environmental Contrast. In H. Weiss (ed.),*The origins of cities in dry-farming Syria and Mesopotamia in the Third Millenium B.C.*, Four Quarters Publishing Co., Guilford, Connecticut: 157-167.

Geyer, B.

2009 Die Syrische Mauer. In A. Nunn (ed.), *Mauern als Grenzen,* Verlag Philipp von Zabern, Mainz am Rhein: 38-45.

Geyer, B., N. Awad, M. Al-Dbiyat, Y. Calvet, M.-O. Rousset

2010 Un « Très long Mur » dans la steppe syrienne, *Paléorient* 36/2: 57-72.

Geyer, B., J. Besançon and M.-O. Rousset

2006 Les peuplements anciens. In R. Jaubert and B. Geyer (ed.), *Les Marges arides du Croissant fertile. Peuplements, exploitation et contrôle des ressources en Syrie du Nord,* Travaux de la Maison de l'Orient 43, Lyon, Maison de l'Orient et de la Méditerranée Jean Pouilloux: 55-69.

Herveux, L.

2004 Étude archéobotanique préliminaire de Tell Al-Rawda, site de la fin du Bronze ancien en Syrie intérieure, *Akkadica* 125/1: 79-91.

Kennedy, D.

2012 Kites – new discoveries and new type, *Arabian archaeology and epigraphy* 2012/23: 145-155.

Lafont, B.
2010 Contribution de la documentation cunéiforme à la connaissance du „Très Long Mur“ de la steppe syrienne, *Paléorient* 36/2: 73-89.

McCorriston, J.
1997 The Fiber Revolution: Textile Intensification, Alienation, and Social Stratification in Ancient Mesopotamia, *Current Anthropology* 38: 517-549.

Nadel, D., G. Bar-Oz, U. Avneri, E. Boaretto and D. Malkinson
2010 Walls, ramps and pits: the construction of the Samar Desert kites, southern Negev, Israel, *Antiquity* Vol. 84 n° 326: 976-992.

Nadel, D., G. Bar-Oz, U. Avneri, D. Malkinson and E. Boaretto
In press Ramparts and walls: Building techniques of kites in the Negev Highland », *Quaternary International.*

Rosen, B. and A. Perevolotsky
1998 The function of 'desert kites' - hunting or domestication? », *Paléorient* 24: 107-111.

Vila, E. and M. El Besso
2006 Résultats préliminaires de l'étude de la faune d'Al-Rawda (Campagnes 2002 à 2004), *Akkadica* 126/2: 111-119.

BERND MÜLLER-NEUHOF
Deutsches Archäologisches Institut – Orientabteilung, Berlin

Desert irrigation agriculture

Evidences for Early Bronze Age Rainwater-Harvesting Irrigation Agriculture at Jawa (NE-Jordan)

Jawa is located in the western part of the basalt desert (*al-harra*) in the northern Jordanian Badia at an altitude of c. 1035 m a.s.l. The site is situated on a basalt hummock bordered by the deeply incised Wadi Rajil in the north and a smaller wadi in the south, which was another arm of the Wadi Rajil in the distant past, but due to its higher level today, is no longer connected to it. Jawa was discovered from the air by the aviator A. Poidebard in 1931 and was partly excavated by S. Helms from 1973 to 1976, followed by several minor investigations until 1986 by S. Helms and A. Betts. The earliest occupations on the site date back into the Late Chalcolithic/Early Bronze Age I (LC/EBA I) (2nd half of the 4th millennium BC). During that time Jawa had its largest areal extent of 0.12 km^2 (12 ha). A much smaller reoccupation took place at the end of the Early Bronze Age IV/early Middle Bronze Age I (EBA IV/MBA I) (beginning of the 2nd millennium BC) (Helms 1981; Betts 1991). Since then the site was never occupied again.

However, occupation evidence in the immediate vicinity of Jawa date to the Lower Palaeolithic (Late Acheulian), Epipalaeolithic (Natufian), PPNB, Late Neolithic, LC/EBA I, EBA IV/MBA, the Roman-Byzantine and Islamic Periods (Umayyad, Mamluk, Abbasid) (Müller-Neuhof forthcoming). Most of the datable finds hint towards temporal presence of diverse groups in the area (campsites) or derived from Jawa itself during its main occupation phases. The only exception is a small Roman tower on a summit north of Jawa, documenting the permanent presence of a small group of people in the region in the Roman period.

In 2010 and 2011 the direct vicinity of Jawa was surveyed within the framework of the archaeological survey project „*Aride Lebensräume im 5. bis frühen 3. Jahrtausend v. Chr.: Mobile Subsistenz, Kommunikation und Ressourcennutzung in der Nördlichen Badia (Nordostjordanien)*“[1], which focused on the documentation of Chalcolithic/Early Bronze Age socio-economic activities in the arid region of the Jordanian Northern Badia.[2] During these surveys evidence for water harvesting technologies was identified, which partly had already been observed by Jawa's excavator S. Helms, namely the flood water harvesting system, which was used for storing wadi water in several pools located at the foot of Jawa, and which presumably irrigated terraced wadi gardens west of Jawa (Helms 1981:135-198). During our surveys in the past several years additional gardens with remains of check dams,

1 This project (Arid habitats in the 5th to the early 3rd millennium B.C.: mobile subsistence, communication and key resource use in the Northern Badia [NE-Jordan]) has been carried out since spring 2010 and is funded by the *Deutsche Forschungsgemeinschaft* (MU-3075/1-2). Furthermore, this project is affiliated with the Council of British Research in the Levant (CBRL) and to the excellent-cluster TOPOI of the Berlin universities.

2 See http://www.dainst.org/de/project/badia-jordan, Müller-Neuhof 2012.

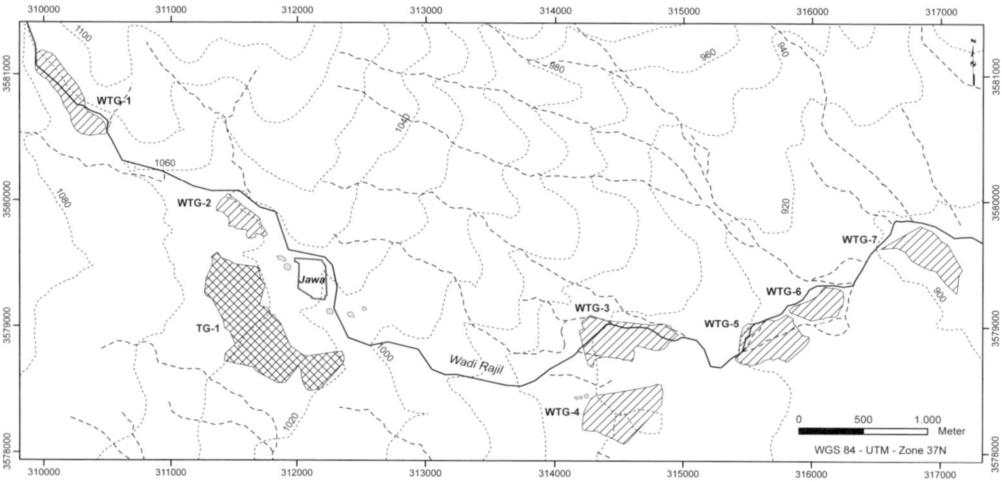

Fig. 1: Location of the wadi terrace garden structures and the terraced gardens in the vicinity of Jawa. Legend: TG = Terraced Gardens (rainwater harvesting); WTG =Wadi Terrace Gardens (Floodwater harvesting) (© DAI, Orientabteilung / J. Meister).

overflows, garden walls, and channels were identified on wadi terraces east of Jawa. These sites will be explored in detail in spring 2013.

During our survey around Jawa in spring 2011 the discovery of large terraced gardens on an adjacent plateau south of Jawa was a great surprise. These gardens are located on a much higher level than the gardens on the wadi terraces and had therefore no connection to the Wadi Rajil (Fig. 1). These terraced gardens obtained their water exclusively from local precipitation. A detailed exploration and documentation of these terraced gardens was carried out in autumn 2011 in a three weeks survey campaign.

Jawa's water procurement systems

The major water source for Jawa is the Wadi Rajil, on the south bank of which Jawa is located. The Wadi Rajil originates in the summit areas of the Jebel Druze (Jebel Arab), the highest elevation (1,803 m a.s.l.) of the volcanic mountain region called Hawran in southern Syria. Here the annual precipitation averages more than 500 mm per year. The entire water catchment area of the Wadi Rajil covers an area of 270 km^2, mainly on the SE flanks of the Jebel Druze (Whitehead *et al.* 2008: 518).

Prior to the construction of several dams within this catchment area on Syrian territory beginning in the late 1960s, the Wadi Rajil was a large contributor of the Azraq wetlands where the wadi finally discharges after its S-shaped course through the western part of the basalt desert. The comparably high precipitation rate in the higher regions of the Hawran resulted in water flowing in the wadi from October onwards. The fact that winter precipitation accumulates in form of snow, prolongs the wadi flow period beyond the end of the rainy season (which ends around mid-April), presumably until mid- to late May. Thus the Wadi Rajil was a very reliable water source for the settlers in Jawa at the end of the 4th millennium BC. Especially under these circumstances, wetter climate conditions can be assumed for the 4th millennium until the end of the Early Bronze Age in the 3rd millennium BC (Rambeau and Black 2011: 95-98).

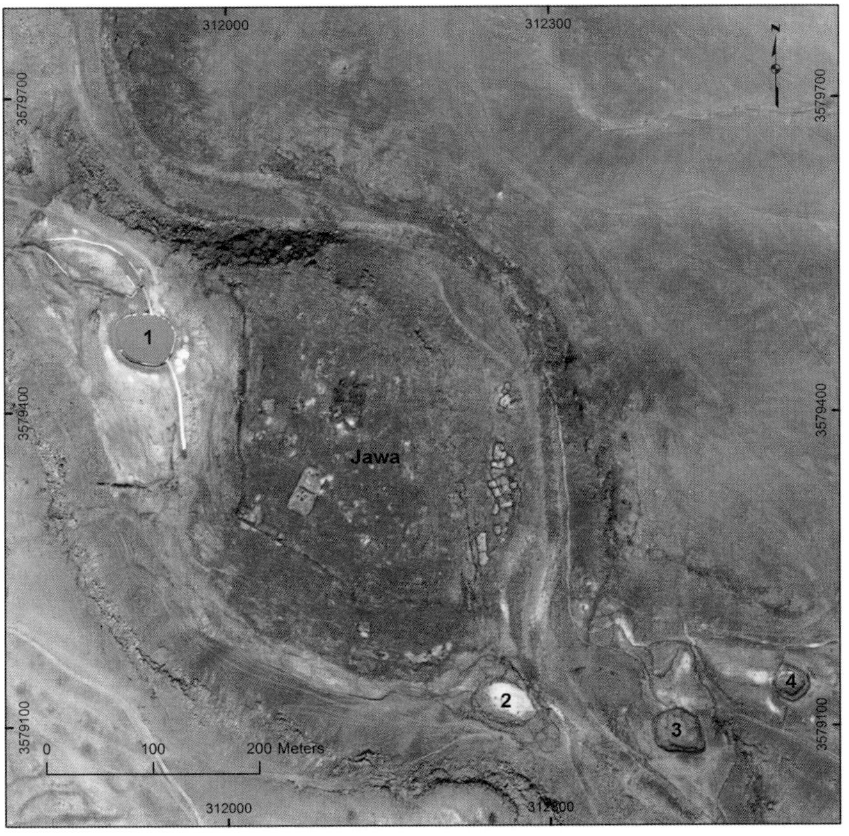

Fig. 2: The pools around Jawa (Note: pool no. 1 is a modern construction replacing three ancient pools). (World-View 2, 17. March 2011)

Fig. 3: Pool near Jawa. (Photo: DAI, Orientabteilung / B. Müller-Neuhof)

At the end of the 4th millennium BC Jawa's inhabitants developed two different strategies of wadi water harvesting. First and best known is water storage in pools, which are situated at the foot of the elevation on which the settlement is located (Helms 1981: 171-180; Whitehead *et al.* 2008) (Figs. 2 and 3). The water was directed to these installations via check dams in

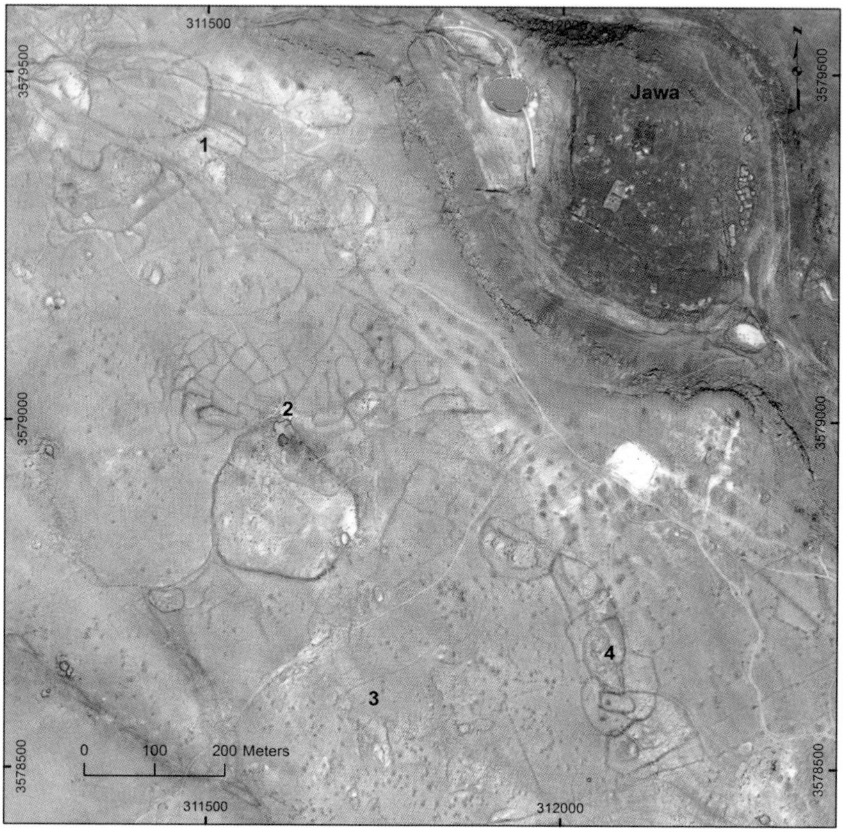

Fig. 4: Satellite image of the terrace garden clusters on the plateau opposite of Jawa. (World-View 2, 17. March 2011)

Fig. 5: Collapsed garden stone wall. (Photo: DAI, Orientabteilung / B. Müller-Neuhof)

Fig. 6: Overview on the garden cluster No. 2. (Photo: DAI, Orientabteilung / B. Müller-Neuhof)

the wadi and channels carrying the water to the pools. In addition, gardens were constructed on several wadi terraces, irrigated with wadi water that was directed via check dams, channels, sluices and overflows into these gardens, which are enclosed by remains of garden walls. Preliminary surveys, especially in the previously unidentified gardens on wadi terraces east of Jawa, revealed some C/EBA artefacts, mostly lithics. However, a later reutilisation of some of these structures, particularly in the 1930s, cannot be excluded (Lancaster and Lancaster 1997:371-372). An abandoned farm house close to one of these gardens hints at later agricultural activities here. Planned detailed survey in spring 2013 will generate more information on these gardens on wadi terraces.

Nevertheless, it seems that the water supply of the Wadi Rajil did not entirely meet the demand of Jawa's residents. This might be the reason for the existence of a second water harvesting system, located on the adjacent plateau south of Jawa, which relied entirely on local precipitation exclusively used for irrigation agriculture.

The garden terraces of Jawa – layout and rainwater harvesting technology

Four garden clusters define the terraced gardens south of Jawa, covering an area of almost 0.33 km^2 within a c. 6 km^2 precipitation catchment system (Fig. 4). The irrigation of these gardens relied entirely on local precipitation, which probably did not exceed 100 mm per year.[3]

The orientation of the slope of the garden area is towards south-east according to the general topographic situation. The single gardens are enclosed by basalt stone walls, which are collapsed today. The basalt boulders for the construction of the walls were collected from the surface of these gardens (Fig. 5).

The sizes of the single gardens within the garden clusters differ according to the topographic situation. Gardens located on comparably steep slopes are much smaller than gardens in nearly flat areas.

However, the steepest parts of the area did not exceed a decline of 4.01%. This makes it difficult to identify these structures as terraced gardens while standing on the ground (Fig. 6),

3 Current precipitation rates are between 50 mm in the southeast of the Northern Badia and more than 150 mm in the northwest of the Northern Badia, with an average of 70-80mm (Al-Ayyash *et al.* 2012: 128). The closest weather station to Jawa is located in Safawi, c. 19km SSE of Jawa. The average recorded annual precipitation in Safawi between 1943 and 2003 was 73 mm (Whitehead *et al.* 2008: 519).

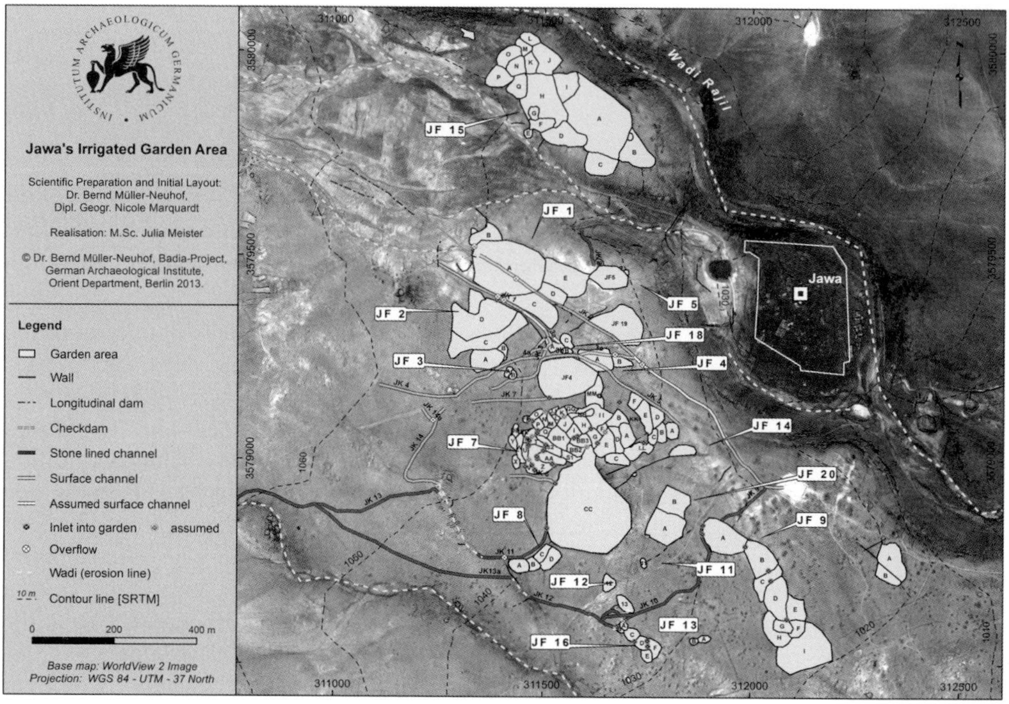

Fig. 7: Map of the terrace gardens with indication of garden walls, channels and inlets/overflows.

although prior consultancy of aerial photos and satellite images give a clear picture of these garden structures. Walls, channels, and in some cases even inlets and overflows, are visible on these images. With these images as a base the components of the garden system and further structural features are easily identifiable on the ground.

As already mentioned, these gardens depended entirely on local precipitation, which falls directly on their surfaces and was additionally distributed into the gardens from the adjacent areas. This surface water distribution into the gardens was carried out by channels, which tapped the precipitation directly from adjacent summits and small wadis nearby, directing the water into the gardens. These small wadis originate nearby and are part of the same local precipitation catchment area (Fig. 7).

Some of the channels directed the water in large, virtually flat gardens, but most of them first supplied the highest located gardens. After the sufficient saturation of the soil of these gardens, the water was distributed to the lower lying gardens of the same terraced gardens system in a cascade manner via overflows, which could have been easily being sealed with stones and mud.

The channels were constructed in different ways. We've called one channel type a "surface channel". These channels are comparably short and tap the adjacent elevation summits. The main characteristic of surface channels is that they are not sunk into the ground. They were constructed by removing the basalt boulders from the surface of their course. Some of these boulders were placed on the sides of the channel banks to stabilize them and to prevent stones and larger debris from rolling or sliding into the channels. Simultaneously, these channel banks were permeable for surface water flow in from the adjacent slopes.

Fig. 8: View on a cleared inlet of a surface channel into a garden, with surface channel in the background. (Photo: DAI, Orientabteilung / B. Müller-Neuhof)

The course of these surface channels is characterised by slight curves which enabled the collection of additional surface water from the slopes of the elevation and probably reduced the flow velocity in order to prevent erosion. Surface channels have a constant width between 6 and 8 m. The point of entry of the surface water into fields or field clusters is defined by inlets as well as overflows (Fig. 8).

Another channel type is slightly sunken into the ground and is single or double stone-lined (Fig. 9). A sounding in one single stone-lined channel revealed probable remains of a clay sealed channel bed that would have prevented the water from draining.

These single or double lined channels are usually much longer than the surface channels. These channels mostly run at an angle between c. 45° and 60° along the contour line of the slopes with an almost constant decline of 3.8 % (channel 13/13a), which allowed sufficient velocity while at the same time permitted the collection of surface water running

Fig. 9: Stone lined channel for water distribution from a small wadi to the terrace gardens. (Photo: DAI, Orientabteilung / B. Müller-Neuhof)

Fig. 10: Collapsed terraced gardens walls. Note the white greyish coloured lichen cover on the stone walls indicating the west / north-west faces of the wall and the dark grey to black colour on the opposite (dry) face.
(Photo: DAI, Orientabteilung / B. Müller-Neuhof)

Fig. 11: Basalt boulder with lichen cover.
(Photo: DAI, Orientabteilung / B. Müller-Neuhof)

down the slopes into these channels. This especially applies to a very long channel (channel 13/13a), which tapped a wadi close to its head with a check dam, directing the water over a distance of 1006 m to one garden cluster.

Date of the rainwater harvesting irrigation system and the terraced gardens

Several of the single gardens were randomly surveyed. Pottery artefacts mostly dating into the Roman/Byzantine and Islamic periods (Umayyad, Mamluk and Abbasid period) were discovered in the plain large garden areas, referring to nomadic campsites in these areas, comparable to modern Bedouin camps in the same areas. Lithic artefacts however, which were discovered in all surveyed garden areas, are mostly dating into the Chalcolithic/Early Bronze Age, among these remains of cortical scrapers and blades with broad platforms have to be mentioned, which gives one hint to an early (C/EBA) date of the construction of this system.

The possible time span in which these gardens were constructed is additionally narrowed by the lichen cover on the surface of the basalt boulders, which were used for constructing the garden walls, which are today collapsed. Lichen cover was especially observed on the basalt boulders on the north-western and western faces of the collapsed walls, which is the weather side (Fig. 10). These faces are entirely covered with lichen, which partly extends to parts of the boulder surfaces on the dry side, which is the south-eastern side of the boulders and walls respectively (Fig. 11). A comparison with the Roman tower just north of Jawa, where the north-western edge of the lower part of the tower has been exposed to the weather since its construction, shows just minimal lichen coverage. Therefore the construction of the garden walls was definitely carried out prior to the Roman period. Since we can fully exclude such large scale construction activities in the Late Neolithic and earlier periods, the remaining possible periods in which the construction of the rainwater harvesting irrigation system and the terraced gardens was organised, are the two occupation periods of Jawa itself, the Late Chalcolithic/Early Bronze I or the late Early Bronze IV/Early Middle Bronze Age.

The large size of the terraced garden system refers to a task that was carried out in a limited time frame, possibly just a few years. Such a task required a large workforce not only for construction, but also for maintenance of the entire system and the gardening itself. Therefore it is highly possible, that these gardens were constructed in the major occupation phase of Jawa, the Late Chalcolithic/Early Bronze I period, in the second half of the 4th millennium BC, when Jawa possibly had 5,000 inhabitants or more (Helms 1981; Whitehead 2008: 520).

Therefore we can ascertain that Jawa's terraced garden system, artificially irrigated by rainwater harvesting, is the hitherto oldest known system of that kind. However, water harvesting for water storage and irrigation, but focussing on seasonal wadi floods, are much older and the earliest known examples are dating into the PPNB. Terrace walls constructed in the wadis perpendicular to the wadi flow direction, were used as barrages for storing the water to saturate the sediments and to direct the water into cisterns. Such structures have been identified in PPNB sites in the Wadi Ruweishid ash-Sharqi, Wadi Abu Tulayha and Wadi Quweir in southern Jordan (Fujii 2010).

Summary and Conclusion

The discovery of the Late Chalcolithic/Early Bronze Age terraced garden structures close to Jawa, which were artificially irrigated in a highly sophisticated manner by rainwater harvesting, has several far reaching consequences for our current knowledge about Jawa itself and the Late Chalcolithic and Early Bronze Age cultures of the Southern Levant and adjacent regions.

The entire water harvesting system of Jawa, which consists of floodwater harvesting for storage in pools, floodwater harvesting for irrigation agriculture in gardens on wadi terraces, and rainwater harvesting for irrigation agriculture on the terraced gardens, shows a highly sophisticated adaption to the local environmental conditions. The water supply for Jawa and its population relied on two different and independent water resources: the local precipitation for irrigating the terraced gardens, and the (much higher) precipitation in the Jebel Druze (Jebel Arab) region, which delayed arrival in the form of floodwater in the Wadi Rajil. Furthermore, the terraced gardens at Jawa are the earliest evidence so far for rainwater harvesting irrigation in connection with artificial terraced gardens in all of SW Asia and maybe even worldwide. The highly sophisticated system probably points to long experience in the construction of such systems elsewhere that have not yet been identified. This system probably originally derived from terrace walls, which already had been constructed in the PPNB and in the Pottery Neolithic (PN) in wadis perpendicular to their flow directions for generating plots for agriculture in terms of soil collection, preventing soil from erosion, and moisture storage (see Fujii 2010; Kuijt *et al.* 2007). That means that the early settlers of Jawa already possessed an intellectual "irrigation technology package".

The rainwater harvesting terraced garden system especially reflects a detailed planning and management of all individual construction tasks, which have to be organised by specialists, which also implies the management of a large workforce. Additionally, management and a larger workforce were needed for maintaining the system and operating the terraced garden system in times of local precipitation. With the discovery of these irrigation systems it becomes obvious that the population of Jawa was able to support itself with agricultural products grown in the immediate vicinity of the site, which simultaneously enlarged the range of products of Jawa's inhabitants in trade networks. It becomes clear that Jawa's location in an arid and inhospitable environment is not as odd as assumed by several authors (e.g. Helms 1981:59-68). In fact, Jawa was able to provide itself with all necessary products for daily life; furthermore, it most probably produced an adequate surplus for exchange or trade.

However, the advantages of Jawa's existence in such an "odd" location are not only explained by the infrastructure that supported a sufficient subsistence for Jawa's population. In fact, there are more important regional if not supra-regional reasons to explain the town in this location. This can only be done when the common geographical perspective towards Jawa, which is a western perspective, changes to an eastern perspective. During the surveys in the hinterland east of Jawa, abundant evidence for intensive animal pastoralism dating into the C/EBA was discovered. Additionally, at least one permanent settlement with fortifications, which is much smaller than Jawa, was identified in the eastern part of the basalt desert. Finally, the large LC/EBA flint mines in the greater Wadi Ruwayshid region east of the basalt desert have to be mentioned, where cortical scraper blanks were produced on an almost industrial scale for export in other regions. These mines were probably a component of the Jawa socio-economic network (Müller-Neuhof 2013a; 2013b; Müller-Neuhof forthcoming). Such an eastern perspective on Jawa, which includes its eastern hinterland, shows clearly the rich and diverse economic aspects (agriculture, pastoralism, mining of raw material and trade) Jawa benefited from and probably exercised control on.

Bibliography

AlAyyash S., R. Al-Adamat, H. Al-Amoush, O. Al-Meshan, Z. Rawjefih, A. Shdeifat, A. Al-Harahsheh and M. Al-Farajat

2012 Runoff estimation for suggested water harvesting sites in the Northern Jordanian Badia. *Journal of Water Resource and Protection* 4: 127-132.

Betts, A.V.G. (ed.)

1991 *Excavations at Jawa 1972-1986, Stratigraphy, Pottery and Other Finds*. Edinburgh.

Fujii, S.

2010 Domestication of runoff surface water: current evidence and new perspectives from the Jafr Pastoral Neolithic. *Neo-Lithics* 2/10: 14-32.

Helms, S.W.

1981 *Jawa Lost City in the Black Desert.* London.

Kuijt, I., B. Finlayson and J. MacKay

2007 Pottery Neolithic landscape modification at Dhra', *Antiquity* 81: 106-118.

Lancaster, W. and F. Lancaster

1997 Indigenous resource management systems in the Bâdia of the Bilâd ash-Shâm, *Journal of Arid Environments* 35: 367-378.

Müller-Neuhof, B.

2012 Northern Badia. In D.R. Keller, B.A. Porter and C.A. Tuttle (eds.), Archaeology in Jordan, 2010 and 2011 Seasons. *American Journal of Archaeology* 116: 700-702.

2013a The Wādī ar-Ruwayshid mining complex: Chalcolithic / Early Bronze Age cortical tool production in NE Jordan. *Annual of the Department of Antiquities of Jordan* 56.

2013b Southwest Asian Late Chalcolithic / Early Bronze Age demand for "big-tools": specialized flint exploitation beyond the fringes of settled region. *Lithic Technology* 38.3: 220-236.

forthcoming The Jawa Hinterland: Studies on Chalcolithic / Early Bronze Age Mobile Subsistence, Communication and Key Resource Use in the Northern Badia (NE Jordan).

Rambeau, C. and S. Black

2011 Paleoenvironments of the southern Levant 5,000 BP to present: linking the geological and archaeological records. In S. Mithen and E. Black (eds.), *Water, Life and Civilisation, Climate, Environment and Society in the Jordan Valley*: 94-104. International Hydrology Series, Cambridge.

Whitehead, P.G., S.J. Smith, A.J. Wade, S.J. Mithen, B.L. Finlayson, B. Sellwood and P.J. Valdes

2008 Modelling of hydrology and potential population levels at Bronze Age Jawa, Northern Jordan: a Monte Carlo approach to cope with uncertainty. *Journal of Archaeological Science* 35: 517-529.

MARIA GIOVANNA BIGA
La Sapienza University, Rome

The Syrian steppes and the kingdom of Ibal in the third millennium B.C.

new data from the Ebla texts[1]

Ibal and its location

Several years ago it was proposed that Ibal was a confederation of small towns created by seminomadic tribes living in the steppe south of Ebla, not too far from Homs, to the east of the Orontes, in an extensive region that may have included Tell al Rawda and Qatna further south.[2] Ibal was the capital of the confederation.[3] It was composed of different areas that were named according to their location close to particular features; in fact the texts refer to Ibal, Ibal of the steppes, Ibal of the canal, Ibal of the water reservoir etc (Bonechi 1993:188; ARES II: 296-297).

From the texts dealing with relations between Ebla and Ibal we also learn many toponyms which must certainly have been in the region of Ibal;[4] none can be identified with certainty and they are not mentioned very frequently because the scribes generally record only relations with the most important cities of the confederation; usually they mention the relationships between Ebla and the capitals of other states. Exceptions are the sites in the Jezira around Nagar cited in some texts (Archi 1998: 1-15; Biga in press). Some new texts quoted in this article which deal with the military campaigns against Ibal conducted by vizier Ibbi-zikir allow the addition of more toponyms to the already long list regarding the territory of Ibal. In fact the number of toponyms concerning Ibal is surprising; is it possible that there were so many important sites in an arid steppe region which had difficulty ensuring the supply of water?

Ibal is not named in the letter of Ennadagan, king of Mari, to the king of Ebla because Ibal was probably not on the Euphrates and thus not on the itinerary of the kings of Mari who all followed the Euphrates to penetrate into north Syria.

In the campaign against Mari, which is the best documented, several sites on the itinerary of vizier Ibbi-zikir and his army are cited. Ibal is not mentioned because it seems to have been on another itinerary. For this reason it also seems more probable that Ibal was located towards

1 I would like to thank Daniele Morandi Bonacossi for inviting me to participate in the workshop "Settlement Dynamics and Human-Landscape Interaction in the Steppes and Desert of Syria" organized by him in Warsaw. The following abbreviations are used: ARET=Archivi Reali di Ebla.Testi; ARES=Archivi Reali di Ebla. Studi; AAM=Annual Account of Metals; MAT=Monthly Account of Textiles.

2 For nomadism and pastoralism in third millennium Syria s. recently Porter 2012. For the location of Ibal s. Bonechi 1993: 187-188; ARES II: 298-299; Pettinato 1988: 253-256 placed it in Palestine. More recently Morandi Bonacossi 2009: 56-68; Catagnoti in press.

3 In third millennium Syria kingdoms were called by the name of the capital s. Biga 2012: 25, and n.26.

4 For a list of these toponyms s. Bonechi 1993: 188; ARES II: 298.

Studia Chaburensia 4 (2014), pp. 199–207.

Qatna, which was certainly already a site of some importance (Morandi Bonacossi 2009: 56-68), on a lake, with access to spring water as excavations have demonstrated. The confederation of Ibal may have extended into the region of Qatna, as far north-east as Tell al-Rawda.[5]

Therefore it may have been an important zone that permitted Mari to reach Qatna, and from there the south as far as Egypt.

The heads of the confederation were defined in some texts as "en" "king", using the same title as the king of Ebla, and in a single text five kings (en) of Ibal are recorded. But in other texts – and more often – the maximum political authority seems to be a superintendent (ugula), a term that quite probably indicates the head of a tribal organization, a sheikh, in areas with partial urbanization. In some regions, such as the region of Ibal, the two terms "en" and "ugula" were used to indicate the maximum political authority.

Several names of the "en" and "ugula" of Ibal in different periods are already known and others are in the texts still unpublished (TM.75.G.1419 obv. VII 2-3: *Ki-da*-NI ugula *Ib-al* ¾); a prosopographic study of them needs to be done on all the texts in chronological order.

Reconstruction of the history of relations between Ibal and Ebla

Ibal is referred to in all the periods covered by the Ebla archives. There is already much published data on Ibal and the organization of the confederation, but much is still unpublished.

Now that the Ebla texts have been put in chronological order and many texts have been joined and studied in recent years at the Idlib museum, it is possible to trace the history of the complex relations between Ibal and Ebla during the whole period of the archives. It will be a lengthy task to compile the complete history of the relationships between these two political entities from all the texts that mention Ibal. I would like to outline here some new aspects of this history, using mostly the texts dealing with textile deliveries, which are very useful because they are full of events.

Relationships between Ebla and Ibal are already attested in the earliest times and also from texts of the period of king Irkab-damu (the penultimate king of Ebla) and his vizier Arrukum (ARET XV voll. 1-2). As it is often the case with these kingdoms, relations were sometimes good and sometimes conflictual. At the time of king Irkab-damu a treaty between Ibal and Ebla is recorded, and representatives of Ibal came to swear allegiance in the temple of KUra. But Ibal was never included in the short list of Ebla's best allies, which were small kingdoms nearby, quite probably all around its borders, reaching the Euphrates in the east. These are the kingdoms of Nirar, Ra'ak, Dub, Burman, Lumnan, Emar and Kakmium. With some of them interdynastic marriages are attested, but no interdynastic marriage has so far been verified between Ebla and Ibal. A queen of Ibal, Maza'adu, is mentioned but there is no proof that she was an Eblaite princess and her marriage is not referred to with the usual formulas.

Ibal and some cities in its territory are in the list of around 30 kingdoms of Ebla quoted in the texts ARET I 1-8, which regularly received Eblaite textiles.

From the text ARET XIII 11 it seems evident that Ibal was under the political influence of the kingdom of Manuwad. The text reports a letter written to the king of Manuwad by the heads of Ibal. The text is difficult to interpret, but it is evident that four chiefs of Ibal, referred to by their personal names – Igalim, Puzurhal, Abail, Binimani – went to swear allegiance

5 The site of Tell al-Rawda, according to the reconstruction of C.Castel and her mission, was in a region where the rainfall level was at the lower limit necessary for survival

(lit. to offer oil) to the king of Ebla,[6] the king of Nirar and the king of Gakam. They came to the city of Šulanu, a town close to the regions of Gakam and Ibal.

Ibal is often mentioned in relation to the kingdom of Manuwad, which was certainly south of Ebla, towards the kingdom of Mari.

Many years ago, Pettinato pointed out that Manuwad (and Gakam) must have been states in the same region as Ibal (Pettinato 1988: 257-258).

Ibal and Manuwad were both within the political sphere of Mari. From text ARET XIII 13 it emerges clearly that Ibal at that time was under the influence of Mari and had relations with it. There was no news of a conflict between Ibal and Mari, probably because Mari controlled Ibal. Text ARET XIII 13 demonstrates that Ibal was allied with Mari. Then some tribes of the confederation stipulated an alliance with Ebla, at the time of the last king Išar-damu and vizier Ibrium.

The military campaigns of vizier Ibbi-zikir and the last king Išar-damu against cities of the Ibal confederation[7]

Despite the swearing of allegiance and the regular deliveries to at least 4 chiefs of Ibal documented in texts ARET I 1-8 (during Ibbi-zikir's first years as vizier), relations between Ebla and Ibal were ambiguous and remained so for a long period.

For this reason, after several military campaigns in the north and in the Euphrates region,[8] the last vizier of Ebla in the reign of king Išar-damu (Ibbi-zikir, who had possibly already decided to attack Mari) tried to defeat the powerful confederation of Ibal. I do not propose to quote all the texts related to these military expeditions; some of them have been joined thanks to the presence of Ibal.[9]

Between the years 7/8 and 12 that Ibbi-zikir was vizier he was involved in military expeditions against cities of Ibal. The military campaigns that took place for 4 years in the region of Ibal against various towns of the confederation have left a vivid trace in several texts that refer to them. Only the military campaign against Mari is better documented, and from a larger number of texts.

Already in Ibbi-zikir's 7th year of office the first military campaign against a town of Ibal is mentioned.

In the text MAT TM.75.G.2622+ARET III 185+ (dated to Ibbi-zikir's 7th year as vizier) the queen gives a precious gift of gold to Raizu, who brought to Ebla the news that the city of Durdu of the confederation of Ibal had been conquered.

For Ibbi-zikir's 8th year as vizier we have information about the military campaign against Ibal from three different types of text: AAM TM.75.G.2428 (=MEE XII 35), MAT

6 This interpretation of the text (different from Fronzaroli's interpretation) is based on the fact that the king of Ebla never went to swear allegiance in another city, s. Biga in press.

7 In ARES II: 299 some texts relating to military campaigns against Ibal are quoted, but at that time it was not evident that the word níg-kas was used by Eblaite scribes to indicate a military campaign; for this reason the authors of ARES II considered them as journeys. Only some years later did it become evident to me, after also having spoken with the epigraphists working on the Mari texts, that we were dealing with military campaigns, s. Biga 2008: 311-319 (publication of the paper presented at RAI 1997, Venice). Already Pettinato 1988: 264 had supposed that the word níg-kas might indicate not only a journey, but also a war.

8 For the wars of vizier Ibbi-zikir in the first three years s. Biga 2003: 79-87; Biga 2010: 35-57.

9 For some texts related to the military campaigns against Ibal s. Biga 2008: 321-322.

TM.75.G.1419 and the contemporary mu-DU text TM.75.G.10236, which also mentions the military campaign (*in* níg-kas *Ib-al*°¾).

The text AAM MEE XII 35 (22a) registers that the queen gave a precious gift to Iti-Ilam, who brought to Ebla the news that Ibal had been conquered; gifts of weapons, precious daggers for Ibbi-zikir who entered Ibal and took up residence there (22b); gifts to other people including the son of Ibbi-zikir, Dubuhu-Ada, and the *badalum* of Harran and of Ursaum who participated in the military campaign (23-26). The participation of two kings of two kingdoms north of Ebla as Ebla's allies should be noted. Apparently they were not interested in what happened in the south, but evidently Ebla needed soldiers from other armies and the king of Harran was a natural ally because he was a son-in-law of the king of Ebla, having married Eblaite princess Zugalum.[10]

The text is parallel in some parts to MAT TM.75.G.1419, which is therefore contemporary with it.

Much of the text MAT TM.75.G.1419 is devoted to registering the textiles given to various people involved in the military campaign against Ibal. The organization documented by this text is very similar to that of the military campaign against Mari that took place a few years later.

At the beginning linen textiles are given to Anenadu for the king to prepare some items of clothing destined for the king (en) for the military campaign against Ibal. It is clear that on this occasion Ishar-damu went to war together with vizier Ibbi-zikir, whereas against Mari Ibbi-zikir had many allies but was without the king! I previously wrote that sometimes the king, who was still young but no longer a child, went to war with vizier Ibbi-zikir. This is another case (Biga 2010: 45, 50).

Immediately afterwards, Puzurri receives linen textiles to prepare clothing for Ibbi-zikir. In this case it is not stated explicitly that these were for the military campaign against Ibal, but it is clear from the context. Then, as usual, various types of deliveries for other purposes are registered at the Ebla court. Among them, one aktum-textile to divide to make equipment for 4 donkeys of the king and Ibbi-zikir, possibly used during the military campaign although this is not made explicit.

Immediately after, 3 textiles ('adaum, aktum and ib coloured and of good quality) and 2 sicles of gold for two earrings are delivered to Iti-¼Ilam, who brought to Ebla the news that Ibal had been taken and Ibbi-zikir resided there. Then the text registers several textiles, 1 mina of gold for various weapons, daggers for Ibbi-zikir who is in Ibal. The queen gives Ibbi-zikir more textiles and 2 gold earrings.

People who helped in the war receive precious gifts: Guša son of Iram-malik of the city of Zik, a person from Galatum and to Iti-¼Kamiš, maškim-functionary of Đarani who captured Kidani, chief of Ibal. Textiles are given to Natiru, a person of Ibal, because they surrendered.

Then gifts from the queen to Dubuhu-Ada, son of Ibbi-zikir, are registered, possibly because he accompanied his father in the military campaign. Then several deliveries not related to the war are recorded.

But again in obv. XI 23: 1 linen textile to prepare two felt for the king and 1 felt for Ibbi-zikir because they went on the military expedition against Ibal.

In the text MAT TM.75.G.1898+2447 of the same period (month isi) the military campaign against Ibal is cited again.Textiles to cut to make equipment for 2 donkeys (IGI.NITA)

10 On this marriage and the relations between Ebla and Harran s. Biga 2010a: 47-48; Biga 2010b: 159-165.

of the king and 2 donkeys of Ibbi-zikir, plus 2 linen textiles for a donkey of Ibbi-zikir, who went to the military campaign against Ibal.[11]

In the next year, Ibbi-zikir's 9th, again the vizier went to war against Ibal as documented by a number of texts: AAM TM.75.G.2508=MEE XII 37 (news of the military campaign against Ibal), mu-DU text TM.75.G.10236 (rev. IV 26-V 10: linen textiles *in* níg-kas *Ib-al*°¾) and by the MAT TM.75.G.1442 (month igza), MAT TM.75.G.1942 (month giNI).

MAT TM.75.G.1442 is parallel in several passages to AAM MEE XII 37 (TM.75.G.2508) and is therefore contemporary with it. In both texts the gift is registered of two gold earrings given by the queen and by princess Tište-damu to Ibgi of the family of Gigi, who brought to Ebla the news that Ibal had been taken. AAM MEE XII 37 obv. IV 28-46 registers the gift of gold earrings to Ibgi of the family of Gigi who brought the news of the defeat of Budan (of the confederation of Ibal). MAT TM.75.G.1442 registers the same gift and the textiles given to Ibgi, the messenger of the defeat of Budan.[12] The text continues registering the deliveries of more textiles given by princess Tište-damu to Iram-malik of the family of Ugura, who participated in the military campaign against Ibal, and textiles given to the king and to Ibbi-zikir who went to the war against Ibal. The king therefore also participated in this military campaign.[13]

In fact in AAM MEE XII 37 two military campaigns are mentioned: one against Zabu which gave the name to the year (MEE XII 37 rev. IV 11-13: DIŠ mu níg-kas *Za-bù*¾) and another one against Ibal. It is probable that Zabu was in the same region as Ibal and that Zabu must be added to the list of toponyms of the Ibal region. But there is no clear indication of this in either the AAM or its parallel MAT TM.75.G.1942 (same year, month giNI).

In Ibbi-zikir's 10th year a military campaign against Sidau of Ibal is documented by AAM TM.75.G.2429 (=MEE XII 36) and MAT TM.75.G.2418+3810 (same year, month isi). In MEE XII 36 a gift to Dubuhu-malik of the family of Dadud, who brought the news that Sidau of Ibal had been taken, and gifts to vizier Ibbi-zikir who resided in Sidau, are registered. In the parallel text TM.75.G.2418+3810 the same gifts are given to Ibbi-zikir who resides in Sidau of Ibal and to Dubuhu-malik who was the messenger of the victory against Sidau of Ibal. Linen textiles are presented to the king and to Ibbi-zikir who went to the military campaign against Sidau. A person of the city of Sitiaga entered Sidau: perhaps Sitiaga was close to Sidau and is another city of the region to add to the already long list. The queen thanks the goddess Išhara, sending textiles to the three sanctuaries of Mane, Zuramu and Uguaš; Ilzi takes these textiles to the temples. After the military campaign there was an allegiance ceremony between Ebla and Sidau of Ibal, with the offering of oil.

In the fragment ARET XII 701, not yet joined to the pertinent tablet, precious gifts to Abu-Kura, another messenger of the victory against Sidau of Ibal, are registered too. There are other examples of two messengers coming to Ebla with the news of the same victory (Biga 2008: 298-300).

11 This text is a join I did some years ago; then the cleaning of the tablet permitted me to discover that the word (nig-kas) for military campaign was there. TM.75.G.1898+2447 obv. V 2-15: 1 aktum-TÚG TAR *ma-da-lu* 1 aktum-TÚG TAR *bù-a-lu* 2 IGI.NITA en 2 IGI.NITA *I-bí-zi-kir* lú ‹è› [níg]-kas *Ib-al*°¾..

12 TM.75.G.1442 obv. II 8-III 5: 1 *'à-da-um*-TÚG-II [1 a]ktum-TÚG [1 í]b+III-TÚG-[sa°]-gùn 3+2 [+x] ‹NI› kù-gi 2 geštuμ-lá *Ib-gi* lú *Gi-gi ma-lik-tum* ì-na-sum 1 *'à-da-um*-TÚG-II 1 aktum-TÚG 1 íb+III-TÚG-sa°-gùn 2 geštuμ-lá 1 1/2 kù-gi *‹Ti› -iš-te-da-mu* ì-na-sum níg-AN.AN.AN.AN *Ib-al*°¾ šu-ba-ti.

13 TM.75.G.1442 obv. III 6- IV 2: 1 'à-da-um-TÚG-II 1 aktum-TÚG 1 íb+III-TÚG-sa°-gùn *‹Ìr›-am-‹ma›-‹lik›* lú *Ù-gú-ra Ti-iš-te-da-mu* ì-na-sum šè [ì]-ti níg-kas 2 íb+V-TÚG-sa°-gùn an- šè túg-túg é ti-túg *en wa I-bí-zi-kir in* ud ì-ti níg-kas *Ib-al*°¾.

AAM MEE XII 36 registers the military campaign against the city of NIligau and another against Sidau of Ibal. The mu-DU text TM.75.G.2073 is of the same year and registers a military campaign against Ibal.

In Ibbi-zikir's 11th year as vizier (in AAM TM.75.G.2507), the queen and Tište-damu gave gifts to Ibgi of the family of Gigi who brought the news that Ibal had been conquered; Dubuhu-Ada received gifts too. He is the same messenger who came to Ebla with the news of another victory against Ibal several years before.

The text MAT TM.75.G. 2372 refers to a military campaign of vizier Ibbi-zikir against another city of the confederation, Budan. A textile was offered to Ibbi-zikir on the day that Budan of Ibal was defeated and its sheep taken. The king and queen gave textiles to Ibte-damu of the family of Ahar-damu, who brought the news that Ibbi-zikir defeated Budan of Ibal and took its sheep (Biga 2008:321)

In Ibbi-zikir's 12th year there was another war against a city of the confederation, Wazaru (ARET VIII 534; Biga 2008: 322).

The fact that for four or five years vizier Ibbi-zikir, sometimes (at least twice) with the king and with Dubuhu-Ada, his son, was compelled to return to war in the same region suggests that the confederation covered a huge area. It was impossible to subdue the area in one single campaign; to pacify this extensive territory took four years. Only against Ibal are several military campaigns attested, and it seems to have been of great importance to Ibbi-zikir to obtain Ibal's submission before undertaking the great expedition against Mari.

The military campaign against Mari; Ibal's alliance with Ebla

It seems therefore that a careful expansionist policy of Ebla can be seen, which was completely planned. Ebla tried to isolate Mari, to remove its allies, such as Adu, Manuwad and Ibal and perhaps others that we can no longer identify.

After many victorious military campaigns, and probably after having destroyed a large number of villages and occupied some cities of Ibal, Ibbi-zikir undertook his great expedition against Mari. Ebla had an impressive number of allies against Mari. Mari itself will certainly have had allies as well, but we are not permitted to know them from the texts of Ebla!

We know about Ebla's allies because after the war Ebla sent precious gifts, registered in several texts, to all the allies who had contributed to the Eblaite victory.[14]

Among these allies there were several cities of the confederation of Ibal.

In MAT TM.75.G.2335, parallel to the metal text TM.75.G.2426, there is a list of kingdoms who received gifts from Ebla because they were allied with her in the war against Mari. We find 9 sheiks of Ibal and several other people (perhaps also sheiks) of other toponyms of the Ibal confederation.[15]

14 According to the interpretation of Biga, s. Archi and Biga 2003: 13-44; Biga 2008: 322-331. Recently Durand 2012: 117-132 proposed another interpretation: Ebla and its allies went to help Mari, instead of against Mari. Among the allies, according to a recent proposal of Pomponio 2013: 71-83 there was not the Mesopotamian city of Kiš but another one with the same name, closer to Ebla and in the region of the Khabur.

15 Part of this text has already been published in Archi-Biga 2003: 40-44. The text will be published, together with all those relating to the military campaign against Mari, in a volume of ARET by Biga and Chambon. TM.75.G.2335 rev. VII 14-VIII 14: 4 *'à-da-um*-TÚG-II 5 gu-dùl-TÚG 10 lá-1 aktum-TÚG 4 íb+IV-TÚG sa° gùn 5 íb+III-TÚG gùn ugula *Ib-al*°¾ 1 *'à-da-um*-TÚG-II 3 gu-dùl-TÚG 2 aktum-TÚG 2 sal-TÚG 1 íb+IV-TÚG sa° gùn 2 íb+IV-TÚG gùn *Ib-al*°¾ lú EDEN¾ 3 *'à-da-um*-TÚG-II 5 gu-dùl-TÚG 3 aktum-TÚG 5 sal-

The confederation seems completely pacified and dominated by Ebla and under total Eblaite control, as documented by some texts of the last years of Ebla, like for example ARET IV 1 (65)-(70) which refers to all cities of the Ibal confederation receiving textiles.

Conclusions

Despite the fact that there is no certainty, the unanimously accepted location of the confederation of Ibal south of Ebla, in the region between Qatna and Tell al-Rawda, is confirmed by the Ebla texts studied in this paper. Unfortunately the numerous texts regarding the complex relations between Ebla and Ibal do not allow the identification of an itinerary between the two states. However it is certain that the confederation included a large number of centres, some of which were important; more cities of the confederation can be identified from the texts studied here. The region was therefore densely inhabited.

There are very few areas for which the scribes mention as many sites as the Jezira around Tell Brak/Nagar and that around Ibal. But while for Nagar identification is certain, for Ibal there are only hypotheses, however probable these may appear.

Tell al-Rawda shows no signs of destruction; it may well have been a city that quickly accepted Eblaite superiority and where people from Ebla were installed, as its material culture would suggest. In the microregion around Tell al-Rawda numerous inhabited sites have been found, but these were not true tells because the sites that can be dated were only occupied for a short period during the BAIV (perhaps BAIVb). None of these sites exceeds 4.5 hectares. About 20 of them were probably inhabited by settlers (Castel and Peltenburg 2007: 601-616; Castel *et al.* 2005: 51-96; 2008: 5-54 and in press).[16]

In the region around Qatna several tells have been identified (Morandi Bonacossi 2009: 56-68).[17]

Given the several qualifications of Ibal, the location was proposed in a steppe region with some lakes or reservoirs, from which a canal might have been constructed.

We do not know the word used by Eblaite scribes to indicate a "lake"; perhaps the word "water reservoir" did so. The lake of Qatna certainly existed in the third millennium. A canal would probably have started from a lake; thus we have also "Ibal of the canal".

As has already been pointed out (Bonechi 1993: 187-188), we must look for Ibal in the region where the various trade routes – attested later, in the II millennium – passed. In the period of the royal archives of Mari (XVIII cent. B.C.) there were three roads to Qatna from Mari. Therefore the Ibal region could have been that which extended from Qatna towards the middle Euphrates. The control of this territory perhaps also guaranteed the passage of Eblaite caravans towards the south-west of Syria and then on to Egypt.

We must therefore leave to archaeologists the task of identifying the many tells in this steppe region, suited to nomadic sheep-rearing and densely inhabited.

TÚG 3 íb+IV-TÚG sa° gùn 5 íb+III-TÚG gùn *Lá-sa-nu*¾ 1 *'à-da-um*-TÚG-II 1 gu-dùl-TÚG 1 aktum-TÚG 1 sal-TÚG 1 íb+IV-TÚG sa° gùn 1 íb+IV-TÚG gùn *Da-ù*¾ lú 'à-mu 10 lá-3 gu-dùl-TÚG 10 lá-3 sal-TÚG 10 lá-3 íb+II-TÚG gùn *Da-ù*¾ lú IGI:KEŠDA a 3 gu-mug-TÚG 3 sal-TÚG 3 íb+III-TÚG gùn *Si-da-ù*¾.

16 I am most grateful to Corinne Castel for bibliography regarding and information about Tell al-Rawda.

17 I am most grateful to Daniele Morandi Bonacossi for bibliography on and information about Qatna.

Bibliography

Archi, A.

1998 The Regional State of Nagar According to the Texts of Ebla, *Subartu* 4, 2: 1-15.

Archi, A and M. G. Biga

2003 A Victory over Mari and the Fall of Ebla, *JCS* 55: 1-44.

Biga, M.G.

2003 Las Guerras de Ebla. In M. Alonso Baquer *et al.* (eds.), *La Guerra en Oriente Próximo y Egipto*, Madrid: 79-87.

2008 Au-delà des frontières: guerre et diplomatie à Ebla, *Orientalia* 77 (2008): 289-334.

2010a War and Peace in the Kingdom of Ebla (24th Century BC) in the First Years of Vizier Ibbi-zikir under the Reign of the Last King Ishar-damu. In M.G. Biga and M.Liverani (eds), *ana turri gimilli, Fs W.Mayer, Vicino Oriente Quaderni* V, Roma: 39-57.

2010b More on Relations between Ebla and Harran at the Time of the Eblaite Royal Archives (24th Century BC). In S. Dönmez (ed.), *Veysel Donbaz'a Sunulan Yazilar DUB.SAR É.DUB.BA.A. Studies presented in Honour of Veysel Donbaz*, Istanbul: 159-165.

2012 Tra Egitto e Siria nel III millennio a.C., *Atti dell'Accademia delle Scienze di Torino* 146: 17-36.

in press Inherited Space- 3rd Millennium Political and Cultural Landscape. In *Proceedings of a TOPOI Workshop, Constituent, Confederate, and Conquered Space in Upper Mesopotamia: The Case of the Mittani Transition*, Berlin, 13-15 July 2009.

Biga, M.G. and A. Catagnoti

in press The contribution of the Ebla textual sources to the study of circular cities of the third millennium BC. In *Proceedings of the International Workshop, Circular Cities of Early Bronze Syria*, Lyon, 21-23 November 2013.

Bonechi, M.

1993 *I nomi geografici dei testi di Ebla*, RGTC 12/1, Wiesbaden.

Castel, C. *et al.*

2005 Rapport préliminaire sur les activités de la mission archéologique franco-syrienne dans la micro-région d'al-Rawda (Shamiyeh): deuxième et troisième campagnes (2003 et 2004), *Akkadica* 126: 51-96.

2008 Rapport préliminaire sur les activités de la mission archéologique franco-syrienne dans la micro-région d'Al-Rawda (Shamiyeh): quatrième et cinquième campagnes (2005 et 2006), *Akkadica* 129/1: 5-54.

in press Rapport préliminaire sur les activités de la mission archéologique franco-syrienne d'Al-Rawda, travaux 2007-2010", *Akkadica.*

Castel, C. and E. Peltenburg

2007 Urbanism on the margins: third millennium B.C. al-Rawda in the arid zone of Syria, *Antiquity* 81: 601-616.

Catagnoti, A.

in press Tell Al-Rawda: une ville de la région d'Ib'al dans les archives d'Ebla ? In C. Castel *et al.* (eds.), *Des villes neuves aux marges du désert de Syrie à la fin du 3è millénaire. Travaux de la mission archéologique franco-syrienne d'Al-Rawda 2002-2008,Travaux de la mission syrienne de Tell Esh-Sha'irat 2005-2008.*

Durand, J.-M.
2012 Sargon a-t-il dèstruit la ville de Mari?, *RA* 106: 117-132.
Morandi Bonacossi D.
2009 Tell Mishrifeh and its Region During The EBA IV and the EBA-MBA Transition. A First Assessment. In P.J. Parr (ed.), *The Levant in Transition, Palestine Exploration Fund Annual Series* IX, London & Leeds: 56-68.
Porter, A.
2012 *Mobile Pastoralism and the Formation of Near Eastern Civilizations*, Cambridge.
Pomponio F.
2013 Further considerations on Kiš¾ in the Ebla texts, *RA* 107: 71-83.

NELE ZIEGLER and HERVÉ RECULEAU
CNRS (UMR 7192, Paris) / HIGEOMES[1]

The Sutean Nomads in the Mari Period

Suteans are one of the nomadic tribes that the archives from Mari and other contemporaneous sites mention frequently. In our contribution, we would like to give a quick overview of the documentation already published and present the binational research programme on historical geography HIGEOMES and the database developed by it. The focus will be on the geographical aspects of the Sutean presence close to the Euphrates.

Suteans in the Mari Archives According to the ARCHIBAB/HIGEOMES Database

In order to elaborate a comprehensive database of textual indications on the historical geography of Upper Mesopotamia during the second millennium BCE, project HIGEOMES relies on the indexing capacities of the ARCHIBAB database, developed in Paris since 2008 by D. Charpin and his team.[2] This database offers an overview of the archival contexts in which Suteans are mentioned in the Mari documentation, while the enriched indexes developed within the database by project HIGEOMES make possible a refined analysis of the themes and objects associated with Suteans in these documents.

Mentions of Suteans according to the ARCHIBAB database[3]

In the present state of documentation, Suteans are mentioned in 70 published texts from Mari, including letters, administrative documents and legal texts.

1 The present article is the result of research developed by French and German cooperation project HIGEOMES (ANR / DFG), which brings together data on the historical geography of Upper Mesopotamia from textual and archaeological sources, and analyzes it with the help of modern tools such as geo-informatics and web semantics. The project joins forces with Paris (UMR 7192) and the University of Dijon (for web semantics), but also with members belonging to other institutions or research groups in France (universities of Brest UBO, Lille 3 UMR 8164 or UMR 7041 Nanterre), as well as several German institutions: the Freie Universität Berlin (with a focus on Middle-Assyrian texts), the Johannes-Gutenberg Universität Mainz (working on archaeological data), and the Fachhochschule Mainz (for geo-informatics) – for more information, see www.digitorient.com. The project is currently in its final phase of development and will be followed in April 2014-2016 by the French-German cooperation project TEXTELSEM (ANR / DFG); the results will be available in 2016 on the internet. Given that this paper focuses on a nomadic group for which archaeological evidence is almost completely absent, only the textual material will be taken into consideration here. We are indebted to M. Puntigam and D. Charpin for their corrections and remarks.

2 For a presentation of the ARCHIBAB project, see Charpin in press a and b. The project homepage can be accessed at : http://www.archibab.fr. The bibliographical abbreviations used for texts in the present article follow the norms of ARCHIBAB, which can be consulted on the site.

3 All texts can be found in the database ARCHIBAB (http://www.archibab.fr); for this reason, bibliography for the text editions is not given below.

Letters (26 texts)
A.1289+; A.2801; A.2993+; A.3297+; *ARM* 1 100; *ARM* 3 12; *ARM* 5 23; *ARM* 6 15; 44; 51; 57; 58; *ARM* 13 44; 106; *ARM* 14 78; 79; *ARM* 26/1 225; *ARM* 26/2 311; 482; 483; 497; *ARM* 27 161; *ARM* 28 8; *FM* 8 31; M.6159; M.11009+.

Administrative documents (42 texts)
Expenditures: *ARM* 7 19; 110; 133; 165; 169; 210; *ARM* 18 61; 64; *ARM* 21 336; *ARM* 23 241; 446; 448; *ARM* 31 199; D. Charpin, *MARI* 3 68; M.6731 (= *ARM* 30, p. 251). **Accounting summary**: *ARM* 22 150; 153; 327; *ARM* 23 449; 450; 451; *FM* 4 44; M.6654+ (*ARM* 30, p. 253). **Inventory**: D. Charpin, *Mél. Veenhof* 8; **Receipt (šu.ti.a)**: *ARM* 23 67; T.178; T.240; T.259; T.260. **Delivery**: *ARM* 23 341; T.179. **Contribution (mu.DU)**: *ARM* 7 225; *ARM* 21 2; 4; 10; *ARM* 24 32; *FM* 10 68. **List of people**: *ARM* 21 407; *ARM* 22 27. **Amount + personal name**: *ARM* 9 244; *ARM* 23 243. **Memorandum**: *ARM* 23 592.

Legal texts (2)
Official statement: A.2279. **Purchase**: *ARM* 8 9.

The presence of Suteans around Mari is already attested during the reign of Yahdun-Lim (six texts) and the period of the Kingdom of Upper Mesopotamia (six texts), even if, unsurprisingly, most known examples date from the reign of Zimri-Lim, the last king of Mari (fifty-six texts) – two texts being impossible to date.

- **Yahdun-Lim (unknown year)**: D. Charpin, *Mél. Veenhof* 8; T.179; T.259 (8/iii); T.240 (17/iii); T.260 (12/iv). **Yahdun-Lim 4**: T.178 ([…]/vii).
- **Kingdom of Upper Mesopotamia (unknown year)**: A.3297+; M.11009+; *ARM* 1 100. **Ep. Awiliya**: *ARM* 8 9 (21/ii°). **Ep. Addu-bani**: D. Charpin, *MARI* 3 68 (28/iii); **Ep. Ṭab-ṣilli-Aššur**: *ARM* 7 19 (22/v).
- **Zimri-Lim (unknown year)**: A.2279; A.2993+; *ARM* 3 12; *ARM* 5 23, *ARM* 6 15; 44; 57; 58; *ARM* 7 210; 225; *ARM* 9 244; *ARM* 13 44; 106; *ARM* 14 78; 79; *ARM* 23 243; *ARM* 26/1 225; *ARM* 26/2 311; 482; 483; 497; *ARM* 27 161; *ARM* 28 8; *FM* 8 31; M.6159; *FM* 10 68 (16/x); *ARM* 23 241 (28/xi). **Zimri-Lim 1**: A.1289+; A.2801; *ARM* 23 341 (24/xi). **Zimri-Lim 2**: *ARM* 23 67 (5/xi). **Zimri-Lim 3**: *ARM* 21 407; *ARM* 22 327; M.6654+ (= *ARM* 30, p. 253) ([…]/[viii+]); *ARM* 22 27 (26/ix); *ARM* 23 446([…]/[x]); *ARM* 18 61 (14/xi); *ARM* 23 448 (15+/xi); 449 (20/xi); 450 (21/xi); 451 ([21]/[xi]); *ARM* 22 150 (24/xi); M.6731 (*ARM* 30, p. 251) (9/xii). **Zimri-Lim 5**: *ARM* 21 336 (5/v); *ARM* 22 153+170 (23/x). **Zimri-Lim 6**: *ARM* 7 133 (22/xii). **Zimri-Lim 7**: *ARM* 7 165 (4/vi); 169 (29/vi); **Zimri-Lim 8**: *ARM* 7 110 ([…]/ii); **Zimri-Lim 10**: *ARM* 6 51 (-/viii); *ARM* 21 2 (5/xi); *ARM* 18 64 (7/xii). **Zimri-Lim 12**: *ARM* 31 199 (4/ii); *ARM* 21 4 (18/v-bis); *FM* 4 44 (5/ix). **Zimri-Lim 13**: *ARM* 24 32 (8/i).
- **Undatable**: *ARM* 21 10; *ARM* 23 592.

Orthography

The orthography of the Sutean's name (for the common etymology "Southerner", cf. Streck 2002: 165) is found in several variations, the main one affecting the second dental consonant: two occurrences render it with DI instead of the more frequent Tv- signs (TE, TI, TU). This variant was understood by B. Groneberg for *ARM* 2 124 (= *ARM* 26/2 311) as an idiosyncrasy of the letter's scribe (Groneberg 1980: 212: "Schreiber verwechselt im Brief häufiger t und d"), but the publication by D. Charpin of a second example in a text dated from Yahdun-Lim (Charpin 2001) excludes this hypothesis. Given the limited number of examples, it is however highly improbable that this writing reflects an alternate spelling *Sudum, and it is safer to assume the use of the DI sign to render a [Tv] sound – according to a rather unusual orthography, attested only in a few contemporary documents.[4]

- *su-tu-um*: *ARM* 5 23: 7, 8, 15; *ARM* 26/2 497: 9'.
- *su-ti-im*: T.240: 6.
- *su-ti*$_4$(DI)-*im*: D. Charpin, *Mél. Veenhof* 8: 2'.
- *su-ti-i-im*: *ARM* 21 2: 3; *ARM* 31 199: 6.
- *su-te-im*: T.179: 5.
- *su-te-em*: *ARM* 5 23: 22.
- sú-te$_{10}$(DI)-*e-em*: *ARM* 26/2 311: 16.
- *su-ti-a-am-ma*: A.1289+: iv 15'.
- *su-tu-ú*: A.2801: 17; *ARM* 3 12: 10; *ARM* 6 15: 18; 44: 5; 51: 37, 57: 10, 4', 16', 58: 10, 11, *ARM* 7 110: 4'; 133: 3; 165: 15; 169: 13; 210: 21', 225: 19; *ARM* 13 106: 16; *ARM* 18 61: 22; 64: 7; *ARM* 21 4: 3; 10: 3; *ARM* 22 150: 19, 59, 7'; 327: 21, 35; *ARM* 23 241: 11; 243: 8; 446: 29'; 448: 21, 34, 60; 449: 20, 33, 59; 450: 21, 37; 451: 21, 33; *ARM* 26/1 225: 9; *ARM* 26/2 482: 5; 483: 3, 26; *ARM* 27 161: 33, 42; F. M.6159: 8'; M.6654+ (= *ARM* 30, p. 253): vi 9; M.6731 (= *ARM* 30, p. 251): 14; M.11009+: 5, 13;
- *su-tu-ú*.meš: *FM* 8 31: 24.
- *su-tu-ú*^{ki}: A.3297+: 20.
- *su-ti-i*: A.2279: 5; *ARM* 1 100: 6; *ARM* 6 44: 7'; *ARM* 6 51: 35, 39; 57: 5; *ARM* 7 19: 3; *ARM* 8 9: E1, E4, E6', EV4, T1, T4; *ARM* 9 244: 11; *ARM* 13 44: 8, *ARM* 13 106: 5, 9; *ARM* 14 78: 8, 3', 5', 6'; 79: 17; *ARM* 21 336: 5, 7; 407: i 19'; *ARM* 22 27: i 23; *ARM* 23 341: 2; *ARM* 26/2 482: 3'; 483: 11; *ARM* 28 8: 6'; D. Charpin, *MARI* 3 68: 2, *FM* 4 44: 2; M.6654+ (= *ARM* 30, p. 253): iv 15'; M.11009+: 4, 16; T.178.
- *su-ti-i*.meš: *ARM* 6 58: 5; *ARM* 14 79: 7; *FM* 8 31: 28, 35.
- *su-ti-i*^{ki}: A.3297+: 18, 23; *ARM* 22 153+170: 36.
- *sú-ti*: A.2993+: 39.
- *sú-ti-i*: *ARM* 24 32: 16; *FM* 10 68: 4.
- *su-te-i*: *ARM* 5 23: 20; T.259 : 5; T.260: 7.
- *su-te-tim*: *ARM* 26/2 483: 8.

Themes associated with Suteans in the Mari archives, according to the HIGEOMES indexes

Drawing from the indexation capacities of ARCHIBAB, project HIGEOMES has developed a series of tools, with the aim of establishing what could be defined as an 'enriched thesaurus' of toponyms. Beyond factual indications such as text reference, line, writing, etc., in documents, every piece of information is entered in the database, as extensively as possible, for each toponym, by each text where it occurs. The collected place names are thus inserted in a

4 See ex. gr., for DI=ti$_4$, *ARM* 23 219: 9; *ARM* 26/2 303: 18', 26', 31'; and 324: 6. For an example of DI=te$_{10}$ in contemporary documents from Upper Mesopotamia, see the Leilan letter PIHANS 117 106: 19 (which contains a lot of scribal peculiarities).

network of information which then allows for a geo-historical analysis, and can be computed, together with archaeological data linked to actual sites, with the help of web-semantics. As far as textual material is concerned, a hierarchical catalogue of parameters has been established for the following categories (table 1):

- a broad **geographical zone** (see map 1), roughly corresponding to the main ethno-political divisions of the Amorite age. Unsurprisingly, Suteans mostly belong to zones A and Y (the Syro-Jordanian desert); as well as Z (ethnics), even if several examples show them roaming along the Middle Euphrates (zone B) and in the Jezirah (notably zones E and F).
- a more detailed **geopolitical categorization**, based on ancient terms used to describe political entities and their subdivisions (kingdoms, districts), as well as cultural entities like the Zalmaqum or the Šubartum. This category proved not to be pertinent for Suteans.
- the **status** of the indexed toponym, be it natural (mountains, rivers) or political ('land', 'house', etc.). This category is also irrelevant for the Sutean nomads.
- the nature of its **administration** (king, leaders, sheikhs, administrators, but also elders and assemblies). As far as the Suteans are concerned, it is worth noting that only exceptionally is a king attested or can the name of a Sutean leader be identified as "king" (see below, p. 219). We don't know what rank exactly the other leaders had. One administrative text (*ARM* 24 32) documents the existence of Sutean sheikhs (*sugāgum*), which can also be inferred from *FM* 10 68, even if it is not stated explicitly: this text records deliveries of sheep, part of which are identified as 'from the *sugāgūtum* of PN', so it is plausible, if not certain, that such was also the case for the Suteans mentioned at the previous line.[5]
- the nature of its **population**. In most cases, Suteans mentioned in the Mari archives are men. The few occurrences of women related to Suteans refer to the distribution of female prisoners: in two cases Yaminite captives entrusted to Sutean men (*ARM* 21 407 and 22 27), and only one mention of Sutean women captured by the Mari troops (*ARM* 26/2 483). Suteans mentioned in the texts were soldiers or messengers, even if frequent mentions of their flocks also indicate their pastoral activities (below, p. 222). In four cases (*ARM* 5 23; 13 106; 23 92; and M.6159), they also appear as thieves, or raiders, launching attacks on flocks and people (below, p. 218).
  For this category, a reverse search can also prove useful, in order to see which towns or places are mentioned as having Suteans among their inhabitants, or at least are connected to Sutean activities (cf. table 2 and below p. 219ff.). These appear quite varied, and cover the whole geographical horizon of the Mari archives, from Western Syria to the Persian Gulf, including the Jezirah. Of special interest are three mentions of ethnonyms (Almutum, Mehalisayu, Yahmamum) which appear to define the term Sutum, and represent tribal subdivisions of the Suteans (see below, p. 218-219).
- all information regarding a toponym's **environment**, natural or anthropogenous. As nomad pastoralists, the Suteans are mostly associated with sheep, either as flocks, or as contributions to the Mari administration (especially taxes). Note however that in one case (*ARM* 21 4), an equid is mentioned, and that *ARM* 3 12 even records Suteans settled (*šaknū*) on 3 leagues of field (3 *bēri* a.šà) along the Euphrates, near Terqa; this need not imply that they were engaged in agriculture, but rather they had temporarily settled on the land (cf. below, p. 220).

5 Note, however, that the mention of Suteans here is not certain, since the middle sign is broken; these might as well be Suhean sheikhs (lú *s*[*ú-hi*]-*i* instead of lú *s*[*ú-ti*]-*i*); see the comment by Marti 2008: 108.

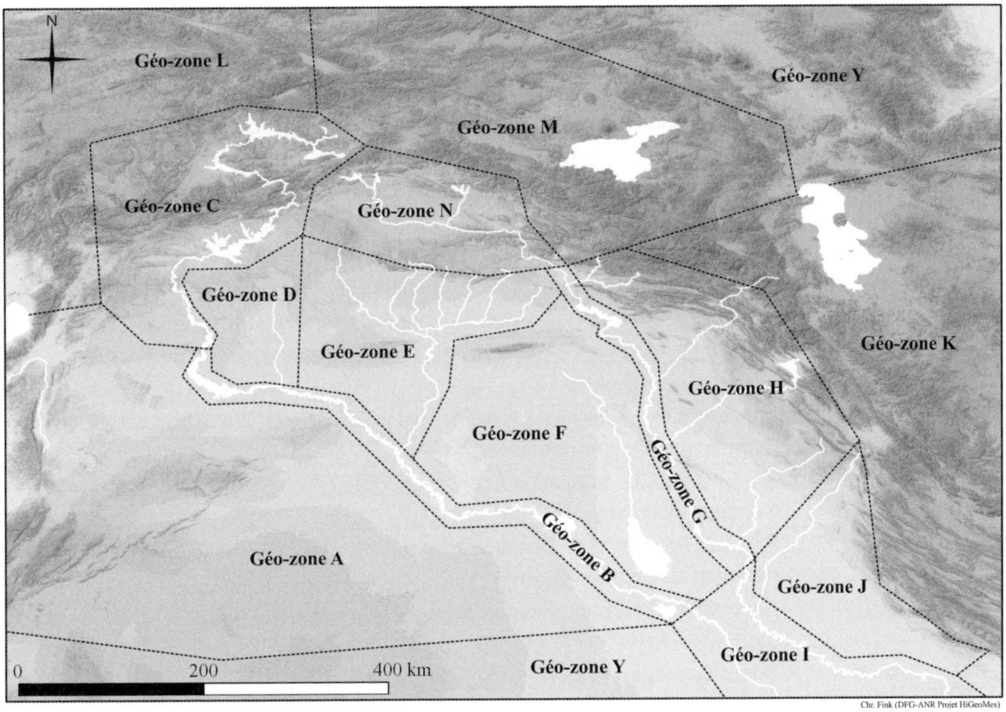

Map 1: "Geozones" of the HIGEOMES project

- **raw and manufactured goods** mentioned in relation to the toponym. These picture the Suteans engaged in a regular web of relations with the Mari palace administration, mostly in the form of sheep, delivered as tax, or textiles and silver, received as gifts (cf. below, p. 222 f.).
- several **other** indications, such as the belonging of the toponym to an itinerary, or the context where this toponym is mentioned. For Suteans, one notable mention is that of 'faraway Suteans' (see below, p. 219), and their presence along several routes in Upper and Lower Mesopotamia.

Table 1: Themes associated with Suteans according to the HIGEOMES-index[6]

A.1289+	Zone: Z , Zone: B, People: soldiers.
A.2279	Zone: Z, Environment: animal pen, Environment: ovines.
A.2801	Zone: Z, Zone: F, Zone: B, People: Nomads, Administration: chief, Route: Kurda—Mari—The King.
A.2993+	Zone: A, Zone: Y, Zone: Z, Historical context: war between Mari and Ešnunna.
A.3297+	Zone: A, Zone: Y, Zone: Z, People: men.
ARM 1 100	Zone: A, Zone: Y, Zone: Z, People: soldiers, Historical context: attempted raid against Yabliya.
ARM 3 12	Zone: B, Zone: Z, Environment: fields, Environment: water stream.

6 As indicated above § 1.3, in the HIGEOMES database, ethnics are indexed as "Zone: Z."

ARM 5 23	Zone: Z, Zone: A, People: men, People: thieves, Environment: ovines.
ARM 6 15	Zone: A, Zone: Y, Zone: Z, People: men, Commodity: raw material, Commodity: barley (še'um), Commodity: raw material, Commodity: varied plants.
ARM 6 44	Zone: Z, Zone: B, Zone: A, People: Sutum, clan: Mehalisayu, Commodity: raw material, Commodity: silver.
ARM 6 51	Zone: Z, Zone: I, Zone: B, People: messengers, Route: Babilim—Mari—The King.
ARM 6 57	Zone: Z, People: men, People: soldiers, Environment: ovines.
ARM 6 58	Zone: Z, People: men, People: soldiers, Environment: ovines.
ARM 7 19	Zone: Z, People: men, Commodity: manufactured goods, Commodity: oil.
ARM 7 110	Zone: Z, People: messengers (originate from), Commodity: gift (recipient), Commodity: silver.
ARM 7 133	Zone: Z, Administration: chefs, People: men, People: messengers, Commodity: gift (recipient), Commodity: silver.
ARM 7 165	Zone: Z, People: messengers (originate from), Commodity: gift (recipient), Commodity: silver.
ARM 7 169	Zone: Z, People: men, People: messengers (originate from), Commodity: gift (recipient), Commodity: silver.
ARM 7 210	Zone: Z, People: men, People: messengers (originate from), Commodity: gift (recipient), Commodity: silver.
ARM 7 225	Zone: Z, People: men, Environment: ovines.
ARM 8 9	Zone: Z, People: men, People: slaves.
ARM 9 244	Zone: Z, People: Sutum, clan: Mehalisayu, People: Sutum, clan: Almutum, Environment: ovines.
ARM 13 106	Zone: Z, People: men, People: thieves.
ARM 13 44	Zone: Z, People: men, Commodity: gift (recipient), Commodity: clothes.
ARM 14 78	Zone: A, Zone: Y, Zone: Z, People: deportee (destination), Distance: far-way Suteans (lú *su-ti-i ru-qú-tim*).
ARM 14 79	Zone: A, Zone: Y, Zone: Z, Zone: B, People: merchants.
ARM 18 61	Zone: Z, People: men, Commodity: gift (recipient), Commodity: clothes.
ARM 18 64	Zone: Z, People: men, Commodity: gift (recipient), Commodity: clothes.
ARM 21 2	Zone: A, Zone: Y, Zone: Z, Environment: ovines.
ARM 21 4	Zone: A, Zone: Y, Zone: Z, Environment: equides, Commodity: gift (originate from).
ARM 21 10	Zone: Z, Environment: ovines.
ARM 21 336	Zone: Z, People: men, Commodity: gift (recipient), Commodity: clothes, People: men.
ARM 21 407	Zone: Z, People: women, People: booty (destination).
ARM 22 150	Zone: A, Zone: Y, Zone: Z, People: men, Commodity: gift (recipient), Commodity: clothes.
ARM 22 153	Zone: A, Zone: Y, Zone: Z, People: men, Commodity: gift (recipient), Commodity: gift (recipient), Commodity: clothes.
ARM 22 27	Zone: Z, People: women, People: booty (destination), People: men.
ARM 22 327	Zone: Z, People: men, Commodity: gift (recipient), Commodity: clothes.

ARM 23 241	Zone: Z, People: men, People: messengers (originate from), Commodity: gift (recipient), Commodity: silver.
ARM 23 243	People: men, Commodity: gift (recipient), Commodity: meat.
ARM 23 341	Zone: Z, Environment: ovines.
ARM 23 446	Zone: Z, People: men, Commodity: gift (recipient), Commodity: clothes.
ARM 23 448	Zone: Z, People: men, Commodity: gift (recipient), Commodity: clothes.
ARM 23 449	Zone: Z, People: men, Commodity: gift (recipient), Commodity: clothes.
ARM 23 450	Zone: Z, People: men, Commodity: gift (recipient), Commodity: clothes.
ARM 23 451	Zone: Z, People: men, Commodity: gift (recipient), Commodity: clothes.
ARM 23 592	Zone: Z, Zone: B, People: thieves.
ARM 23 67	Zone: Z, Commodity: raw material, Commodity: gems.
ARM 24 32	Zone: Z, Administration: sheikh, People: men, Commodity: taxes, Commodity: ovines.
ARM 26/1 225	Zone: Z, Zone: A, Historical context: wedding of Yasmah-Addu and Damhuraṣi.
ARM 26/2 311	Zone: A, Zone: Y, Zone: Z, People: messengers (originate from).
ARM 26/2 482	Zone: Z.
ARM 26/2 483	Zone: Z, People: women, People: booty (originate from), Environment: equids.
ARM 26/2 497	Zone: Z, Zone: I, Route: Babilim–Sippar.
ARM 27 161	Zone: Z, People: soldiers.
ARM 28 8	Zone: Z, People: men, People: messengers.
ARM 31 199	Zone: Z, People: men, Commodity: gift (recipient), Commodity: manufactured goods.
D. Charpin, *MARI* 3 68	Zone: Z, People: men, Commodity: manufactured goods, Commodity: oil.
D. Charpin, *Mél. Veenhof* 8	Zone: A, Zone: Y, Zone: Z, Commodity: taxes, Commodity: ovines.
FM 10 68	Zone: Z, Commodity: taxes, Commodity: ovines.
FM 4 44	Zone: Z, People: men, Commodity: raw material, Commodity: silver, Commodity: manufactured goods, Commodity: clothes, Commodity: taxes.
FM 8 31	Zone: A, Zone: Y, Zone: Z, Zone: B, People: men
M.6159	Zone: Z, People: men, People: thieves, Environment: ovines.
M.6654+ (= *ARM* 30, p. 253)	Zone: Z, People: men, Commodity: gift (recipient), Commodity: clothes.
M.6731 (= *ARM* 30, p. 251)	Zone: A, Zone: Y, Zone: Z, People: men, Commodity: gift (recipient), Commodity: clothes.
M.11009+	Zone: A, Zone: Y, Zone: Z, Environment: ovines.
T.178	Zone: A, Zone: Y, Zone: Z, Commodity: taxes, Commodity: ovines.
T.179	Zone: A, Zone: Y, Zone: Z, Commodity: taxes, Commodity: ovines.
T.240	Zone: A, Zone: Y, Zone: Z, Commodity: taxes, Commodity: ovines.
T.259	Zone: A, Zone: Y, Zone: Z, Commodity: taxes, Commodity: ovines.
T.260	Zone: A, Zone: Y, Zone: Z, Commodity: taxes, Commodity: ovines.

Table 2 – Toponyms or ethnonyms associated with Sutean people according to the HIGEOMES-index

Akkad cf. p. 224	*ARM* 26/2 482	Zone: G, Zone: I, Zone: J, Environment: crossing (*nêberum*), Population: soldiers (passing by), Population: Sutum
Almutum cf. p. 218 f.	*ARM* 9 244: *ARM* 14 78	Zone: Z, Population: Sutum, clan: Almutum, Environment: ovines Zone: Z, Population: deportee (destination), Population: Sutum, clan: Almutum, Distance: far-away country
Ayabu cf. p. 219	*ARM* 23 592	Zone: B, Population: Sutum, Population: booty (originate from), Commodity: booty (originate from)
Babilim cf. p. 224	*ARM* 26/2 497	Zone: I, Population: Nomads, Population: Sutum, Population: garrison, Population: palace servants, Route: Babilim–Sippar
Bisir cf. p. 220 f.	*ARM* 6 44	Zone: A, Population: Sutum, clan: Mehalisayu, Commodity: raw material, Commodity: silver
Harbe cf. p. 219	*ARM* 23 592	Zone: B, Population: Sutum, Population: booty (originate from), Commodity: booty (originate from)
Hit cf. p. 219	*ARM* 14 79	Zone: B, Status: land, Population: Nomads, Population: Sutum, Population: slaves
Kurda	A.2801	Zone: F, Administration: king, Population: Sutum, Population: soldiers (originate from), Route: Kurda–Mari–The king
Mehalisayu cf. p. 218 f.	*ARM* 9 244: *ARM* 6 44	Zone: Z, Population: Sutum, clan: Mehalisayu, Environment: ovines Zone: Z, Zone: B, Zone: A, Population: Sutum, clan: Mehalisayu, Commodity: raw material, Commodity: silver
Našala cf. p. 221	*ARM* 5 23	Zone: A, Population: Sutum, Population: men, Population: thieves, Environment: ovines
Purattum	*ARM* 3 12	Zone: B, Environment: fields, Population: Sutum
Qaṭna cf. p. 221	*ARM* 5 23	Zone: A, Status: land, Population: Sutum, Population: men, Population: thieves, Environment: ovines
Sippar cf. p. 224	*ARM* 26/2 497	Zone: I, Population: Nomads, Population: Sutum, Population: garrison, Population: palace servants, Route: Babilim–Sippar
Suhum cf. p. 219	A.1289+	Zone: B, Population: Sutum, Population: soldiers
Tadmer cf. p. 221	*ARM* 5 23	Zone: A, Population: Sutum, Population: men, Population: thieves, Environment: ovines
Terqa cf. p. 220 f.	*ARM* 3 12	Zone: B, Environment: fields, Environment: streams, Population: Sutum
Yahmamum cf. p. 218 f.	*ARM* 14 78: 9	Zone: Z, Population: deportee (destination), Population: Sutum, clan: Yahmamum, Distance: far-away country

Suteans in the Mari Period

For the study of the Suteans[7] all textual evidence comes from non-Suteans. The Mari archives are actually the richest source; the information they give is manifold, as shown above. They record the presence of Suteans in their realm, and incoming or outgoing goods. Letters and contracts sometimes give us insight into the opinion the inhabitants of Mesopotamia had about them. And this opinion was generally speaking not a very good one: Suteans are seen as outsiders, foreigners, a generally hostile and dangerous tribe of the Syrian steppe. Whoever wanted to cross the latter had to trust in their protection … and pay for it.

The notion of Suteans being slave merchants

In the eyes of the people of Mesopotamia, Suteans were slave merchants of bad reputation and slavery in Sutean lands was a very severe punishment. This can be seen (among other examples) from an adoption contract from Southern Mesopotamia where the threat of being sold to Suteans should be a guarantee in favor of the respect of the contract:[8]

> "In the future, if NPf1 says to NPf2, her mother: 'You are not my mother!' they take away the son of NPf1 and they sell her to the land of Sutum or to the Yahmuteans."

In the same way, when Zimri-Lim definitely wanted to get rid of some people he proposed selling them to the Suteans:

> "One should sell these people to the far-away Suteans, to the Yahmameans or to the Almuteans, or to Eš/the desert,[9] a place from where no information will come to their country." (*ARM* 14 78 = *LAPO* 18 929)

To be sold to Suteans was a kind of "worst-case scenario" for the people of Mesopotamia: it clearly meant they had to leave their usual environment forever, to enter a completely different civilisation without any hope of redemption. For this reason, except as punishment, it was forbidden to sell the people of Mesopotamia to Suteans and the kings of Mesopotamia granted to their population that they would prevent this as well as they could.[10]

This royal protection was an effective one and those who didn't respect it were supposed to be punished. As an example we can quote a case submitted to Zimri-Lim by the governor of Saggaratum, Yaqqim-Addu:

7 For the different sources and studies on Suteans in general see Kärger and Minx 2012. Joannès 1997 is the fullest presentation of the Mari evidence. See also Ziegler 2004. All these publications mention former publications, notably Kupper 1957.

8 The adoption contract Ch.-F. Jean, *Tell Sifr* 13 (actually written and discovered at Ur) was republished by Charpin 1980: 46, 208-209; see also *UET* 5 97 and for further remarks Stol 2004: 913.

9 The description "far-away Suteans" seems to introduce two tribe names (Yahmamum and Almutum) and one unknown place, written E-AB-KI. This was interpreted as a place-name Eš (*e-èš*^{ki}) and commented on by Durand 2000: 67. But more recently, Durand 2007 argued that it should be read ⸢ia⸣-ab^{ki} and interpreted as a word for "desert" whose etymology relates it to the Arab word yab or yabāb.

10 It was not only forbidden to sell slaves to the Suteans. An edict of Samsu-iluna stipulated that neither was it allowed to buy slaves from them. *AbB* 3 1: "Nobody shall buy a man or a woman, citizen of Idamaraṣ or from Arrapha from the Suteans. [A merchant] who buys a citizen of Idamaraṣ or of Arrapha from a Sutean will lose his money."

> "NP, the no[mad] (*hanûm*) g[ave] one slave to the Suteans in the country of Hit. He got 2 donkeys and 3 sheep as his pay[ment]. When he crossed here (the river) to [Hit], the police didn't notice [him], but on his way home the police seized the 2 donkeys [and the 3 sheep] and brought him to me. I made an investigation against NP, who has given a slave to the Suteans. He said: '[I have bought] the slave in another country, I have sold the slave to another country. Why [do you investigate about me]?'" (*ARM* 14 79 = *LAPO* 18 1056: 5-23)

Selling slaves to Suteans was forbidden. But the seller argued that he had sold a foreign slave bought in a foreign country, so he didn't deliver a member of the society of the Mari-kingdom to the Suteans and thus didn't commit a crime. The governor sent the man to Mari and the king had to deal with this case. Unfortunately, we don't know which decision the king finally made.

The notion of Suteans being hostile wrongdoers

Many letters involving Suteans mention attacks on different towns,[11] or caravans,[12] notably diplomatic missions crossing the Syrian steppe; other texts mention the plundering of animals,[13] or people kidnapped by Suteans (*ARM* 13 106). The governors of the kingdom of Mari took measures against them. One text even mentions the nightmares the king of Mari had because of the danger they represented to caravans.[14] All these documents show that the principal opinion people from Mesopotamia held about Suteans was a rather negative one.

The Sutean clans

The Mari archives give insight into the tribal identities of the Mesopotamian world. Two ethnic groups are particularly important for the people from Mari: the Sim'alite and Yaminite tribes, which were both subdivided into smaller entities. It is clear that the Suteans belonged to neither of these tribes. But, just as the Yaminites and the Sim'alites, they were divided

11 Cf. below, p. 219, for attacks against the Suhum region and p. 221 for the kingdom of Qaṭna.

12 See M.6159 for an attack against the Babylonian delegation on the way to Qaṭna and for the proposal of such an attack, cf. Ziegler 2004.

13 In the time of Yasmah-Addu: M.7583 (unpubl. letter of Hammanum, the governor of Yabliya; courtesy D. Charpin) and an unpublished letter of Tarim-šakim M.8338, probably written during his stay in the Suhum region. For the reign of Zimri-Lim see *ARM* 6 57 and 58 (= *LAPO* 17 746 and 747). The general situation of these letters remains unclear. Lion 2001: 154-155 suggests that Menihum, author of the first letter, was the governor of Našer. The Sutean raid would thus have concerned the region south of Mari. See also ARM 26/2 483 where the Suteans raided *i-na ma-di-na-tim* "in the juridical districts". *ARM* 13 106 (= *LAPO* 17 749) is a letter of the governor of Terqa, Kibri-Dagan, for juridical reaction to the actions of Suteans. The order of Zimri-Lim is first quoted: "Send anybody whose stolen (objects) are with the Suteans and his witnesses who could assist in the making of a legal decision." Kibri-Dagan then added: "Now I have just sent NP, a man of Terqa, who has been robbed of a servant, and two witnesses."

14 Yasmah-Addu, waiting for his spouse, the princess of Qaṭna, is having nightmares about the danger represented by Suteans on the route between Qaṭna and Mari: "The dream that I saw was frightening. I am afraid the Suteans will seize Dam-hurași and you, and that they will say: 'As long as you do not return our homes, we will not release them." (*ARM* 26/1 225: 7-12). For this proposal of the chronological setting of the text cf. Ziegler 2004: 102 and 2007: 150-151.

into different "sub-tribes" or "clans". We know of three Sutean tribal entities, the Almutum, Mehalisayu and Yahmamum.[15] The Yahmamum and Almutum were both known in the Mari archives as "remote Suteans" (*ARM* 14 78: 8-9. Cf. above p. 217 and fn. 9).

For the time being, it is nearly impossible to identify the different Sutean chiefs as heads of one or other of these tribes. Only the Sutean chief, Hammi-tilu, can be linked to the "sub-tribe" of the Mehalisayu as e.g. the letter *ARM* 6 44 = *LAPO* 18 1047 shows. Kings of the Suteans are barely mentioned.[16]

Geographical distribution of Suteans in the Mari documentation

The HIGEOMES-database is especially useful for the examination of places where Suteans occur. First of all, the homeland of the Suteans was in the Syrian steppe.[17] Normally, we are not informed about the places they settled or the annual routes along which they migrated with their animals, but we get information about the places where they came into contact with the inhabitants of Mesopotamia. And in the Mari archives the largest number of attestations concerns the Euphrates region south of Mari, the Suhum.

The Suhum region

The governors and officials mention attacks on regions or towns of the Suhum[18]. All these examples show that raids of Suteans against this region occurred frequently.
A governor of Mari boasted about the strength of the defence of the Upper Suhum against the Suteans (M.7583, unpubl., courtesy D. Charpin):

> "May the Suteans attack the Upper Suhum region! May my lord see how numerous are his armies. My lord must not be afraid!"

Some letters speak about attacks on the city of Hanat, where the Suteans took 3 prisoners (e.g. the unpublished M.9607). Other letters tell about the attack of, as the author says, 1000 Suteans on Yabliya, opposite modern Shishin (Charpin and Millet Albà 2009). A memorandum mentions a Sutean raid on Harbe and Ayyabu (*ARM* 23 592: 4-6).

Other texts mention Sutean activities in the region of Hit. This town, or more probably its surroundings, was a place where nomads from the kingdom of Mari could meet Suteans (*ARM* 14 79 = *LAPO* 18 1056 quoted above) and engage in business activities. The letter *ARM* 14 79: 6 calls the place where a nomad had sold a slave the "land of Hit", even though there was no independent country of that name. So, perhaps this alludes to the fact that the Suteans were not allowed to enter the towns and had to do their business transactions outside of the city walls.

15 Cf. for bibliography Kärger and Minx 2012: 366b. Joannès 1997 enumerated 5 Sutean sub-tribes but in fact there is actually no evidence that the Pu-la-Ila and the Bar-Halanum, two clans of the Suhum region, were Suteans.

16 Unpublished A.304 (courtesy D. Charpin). See also below fn. 19.

17 Cf. fn. 9 for a possible mention of the desert in relation to Suteans.

18 For a recent study on the historical geography of the Suhum in the Mari period, see Charpin and Millet Albà 2009.

The region north of Mari and the Jebel Bishri-region

Less frequent than references to Suteans in the Suhum south of Mari are mentions of these people upstream on the Euphrates. This lack of evidence is surprising, because we know that the Suteans used to settle in the Jebel Bishri but might also settle down in the adjacent Euphrates valley. The governor of Terqa reported this to Zimri-Lim:

> "The Suteans settle 3 leagues upstream of Terqa on the Euphrates and come regularly to see me and go back. No mishap occurs currently." (*ARM* 3 12 = *LAPO* 17 748)

This letter shows that the Sutean presence on the Euphrates border was rather exceptional and that the local authorities closely observed the situation. The governor reported about Sutean delegations coming regularly to Terqa, which shows that the authorities were suspicious even if there was no actual reason to complain about the Suteans. Their residence in the territory of the kingdom of Mari was certainly not free of charge. An administrative text from Mari shows that Šamiš, the Sutean chief, had to pay at least a certain amount of silver and 6 textiles for the *igisûm*-contribution.[19]

Sometimes problems occurred between the Suteans and the other occupants of this region. One of the reasons for their confrontations is described in a letter from the king of the Yaminite Uprapeans to Zimri-Lim of Mari. In this letter, Hammi-ištamar complained about the Suteans who occupied the wells and water-installations of the Uprapeans on the Euphrates. Zimri-Lim was asked to convince them to settle downstream (*FM* 3 31, cf. the commentary of Durand 2005: 102).

The settling of these Sutean nomads in the Jebel Bishri was less problematic. The Sutean presence in this area, for instance, is known thanks to a letter from the governor of the Mari district to the king about a fortuitous discovery of silver:

> "The Suteans of the tribe of Mehalisayu have discovered a hidden treasure in the Bisir (mountains). It is reported to be a lot of silver. Some say it is up to [n] talents, others say it is up to 2 talents, even others say it is up to 10 minas. There is no agreement on the issue." (*ARM* 6 44 = *LAPO* 18 1047)

The circumstances of the discovery are not indicated. Charpin 2010: 243 proposed that the discovery could have been made while a grave was being dug. For funerary practices in the Bronze Age Bishri region cf. Fujii and Adachi 2010.

Uprapeans who wanted to avenge Sutean aggressions could attack them in their home regions close to the Jebel Bishri. An episode of the war between Uprapeans and Suteans is known from a letter in which an Uprapean raid on Sutean flocks is reported and the itinerary of the Uprapean nomads is given:[20]

19 *FM* 4 44: 2. The payment of Šamiš the Sutean is mentioned after that of the queen Dam-huraṣi and before that of Šibtu, followed by the governor of Mari and other high officials. It seems clear that this indicates the high rank of the "man of the Suteans", – probably a king (cf. above p. 212 and 219). The tax was collected as *igisûm*-contribution from the district of Mari and the Upper district, i.e. the district of Terqa the 5-ix-ZL 12. The presence of Šamiš among the people obliged to pay is exceptional and seems to indicate an exceptional sojourn in the region depending on Mari or Terqa – perhaps as just described in the letter from the governor of Terqa *ARM* 3 12.

20 An extant translation of the unpublished letter A.2200 has been given by Durand 2010: 261 fn. 30.

> "The (Uprapean) nomads who left to raid the sheep of the Suteans (…) raided the sheep of the Suteans in Burtum-Bardat and in Qabaqab. They left and came back below the mountain Ba'alta-Guratim[21] and continued downstream on the riverside. (…) It is some 10,000 sheep that they have taken. (…)"

Qabaqab, one of the place-names mentioned in this letter, corresponds to modern Qebaqeb, southeast of the Jebel Bishri. A Roman camp examined by a Finnish archaeological survey[22] and also a deep rock-cut well dated by the Finnish archaeologists to the Roman period have been found there. Some 18 km from Qebaqeb other water distribution and storage installations (a channel and well) from Roman times were discovered by the same team. The region on the southeastern edge of the Bishri was thus a place where waterholes, wells and channels were common from ancient times onwards. The place-name Burtum-Bardat (litt. "Cold Well", cf. Durand 2009: 261 fn. 30) shows that such water-collecting structures already existed in this region in the 18th century B.C. and were used by Sutean nomads. The long-lived toponym shows that the place was used without major interruption throughout the centuries.

Other mentions of Sutean settlements in the Syrian Steppe

The Suteans didn't only settle in the Bishri region, but also in other parts of the Syrian steppe - although precise information is lacking. Sutean raids against the kingdom of Qaṭna (Mišrife) are known from the Mari archives. A governor wrote to the king of Mari Yasmah-Addu:

> "Gazizanum, Abi-sare, Hammi-talu, the Suteans and 2,000 Suteans had assembled for a meeting and then went out to raid the pastures of the land of Qaṭna."
> (*ARM* 5 23 = *LAPO* 17 745)

According to this letter, the Suteans had to make a special journey to the pastures of the land of Qaṭna; this could indicate that there was normally no territorial proximity between the herdsmen pasturing the people of Qaṭna's sheep and the Suteans. According to this same letter, 60 Suteans attacked Palmyra (Tadmer) and Nazala (Qariyatain), but failed.[23]

We have seen above that a Sutean encampment in the Bishri-region bore a place-name formed with the Akkadian word for "waterhole, well" *būrtum*. Another Sutean settlement in the Syrian steppe known from textual evidence is Bugre-Burtim "Cow's well". Ziegler 2004 made the proposal that this place might be the modern Ain al Arnab, not far west of Hit (Ziegler 2004: 105 fn. 38, and for this toponym in neo-Assyrian times Cavigneaux 2010). Caravans between Babylon and Mari used to rest there.

21 This otherwise unknown name kur *ba-ha-al-ta-gu-ra-tim* is interpreted as "Lady-of-the-Guratum" by Durand 2009: 261 fn. 30. It is probably the name of a part of Jebel Bishri

22 We are indebted to M. Lönnqvist who gave us this information after our lecture. For the archaeological soundings at Qebaqeb see M. Lönnqvist *et. al.* 2009 : 18-19 and the maps on p. 26 and 33. We have not yet been able to consult the full publication (Lönnqvist and Lönnqvist 2011).

23 *ARM* 5 23: 14-21: "And before that, 60 Suteans, another military expedition, had left to raid Tadmer and Našala, but they came back without any booty. The people of Tadmer had even killed one of the Suteans."

Economic activities of the Suteans

Pastoral activities of the Suteans

The main income of the Suteans was clearly based on their pastoral activities, although they could have obtained additional income from other sources.

Suteans had sheep which seem to have been their principal wealth in the eyes of Mesopotamian sedentary people. They conducted raids to get sheep[24] and were also promised them as rewards.[25] The number of animals in the Sutean herds was probably enormous. When numbers are given, they amount to thousands: 10,000 or 15,000 sheep are promised to the Suteans (A.3297+, Ziegler 2004), or stolen by the Suteans, or stolen from the Suteans (cf. above A.2200 p. 220 f.), numbers that show that the overall flock sizes were much more impressive.

We don't know for sure if the Suteans were sometimes employed by the authorities of the Mari-kingdom for the herding of animals. But we see from the texts that Suteans had the obligation to deliver animals to the central authority of Mari: mainly sheep[26], sometimes donkeys[27]. These payments were a kind of tribute to be paid to the king in return for permission to stay occasionally in the Euphrates valley. The size and the regularity of this "tax" is clearly shown by an oath in which the oathtaker had to swear that he had indeed made his deliveries as he was obliged to (A.2279 edited by Durand 1993).

Some economic texts from Mari tell us how this tax-collection was done: notably a small group of texts from the reign of Yahdun-Lim published by Charpin 2001 show that small numbers of sheep were regularly transferred by Suteans to the royal estate as *miksum*-tax. But it seems that these deliveries were not the only ones, as *ARM* 9 244, an administrative document, might indicate: 280 sheep are delivered to the administration of Mari as *miksum*-tax by different Sutean sub-tribes.

Escorts for caravans in the Syrian steppe

The Sutean presence in the steppe can be confirmed in many other ways. They knew this region very well, the waterholes and routes through the arid zones, knowledge that would have been useful to those who wanted to cross, but they were also potential attackers of caravans. In the event, the various Mesopotamian and Western Syrian kings preferred to pay the Suteans for escorting their embassies which travelled between capital cities (Ziegler 2004).

Suteans were used as escorts for diplomatic missions between the cities of Mari and Qaṭna where they were escorts for ambassadors, and indeed even armies. Yet more frequently we see the Suteans in Mari texts being used as escorts for embassies from Babylon to Mari,[28] but also as far as Aleppo or Karkemish. They even could find a path between Babylon and Aleppo without passing by the territory of Mari.

24 Cf. *ARM* 6 57 and 58 (= *LAPO* 17 746 and 747) and also *ARM* 26/2 483: after a Sutean raid, Buqaqum stated that he succeeded in bringing back the stolen sheep. See also M.11009+ (= *LAPO* 18 1084) from the time of Yasmah-Addu.

25 A.3297+ is published by Ziegler 2004. Samsi-Addu promised 10,000 sheep for an attack on a delegation, or 15,000 if a royal gift worth this much was captured.

26 *ARM* 23 341 (24-xi-ZL 0): 10 sheep; *ARM* 21 2 (5-xi-ZL 10): 2 rams by Hammi-talu. *ARM* 21 10 (s.d.): 4 sheep by Rabi'um.

27 *ARM* 21 4 (18-vbis-ZL 12): 1 white donkey by Ballum.

28 Perhaps they were used as escorts as far as Kurda, cf. A.2801 (= *LAPO* 16 268)

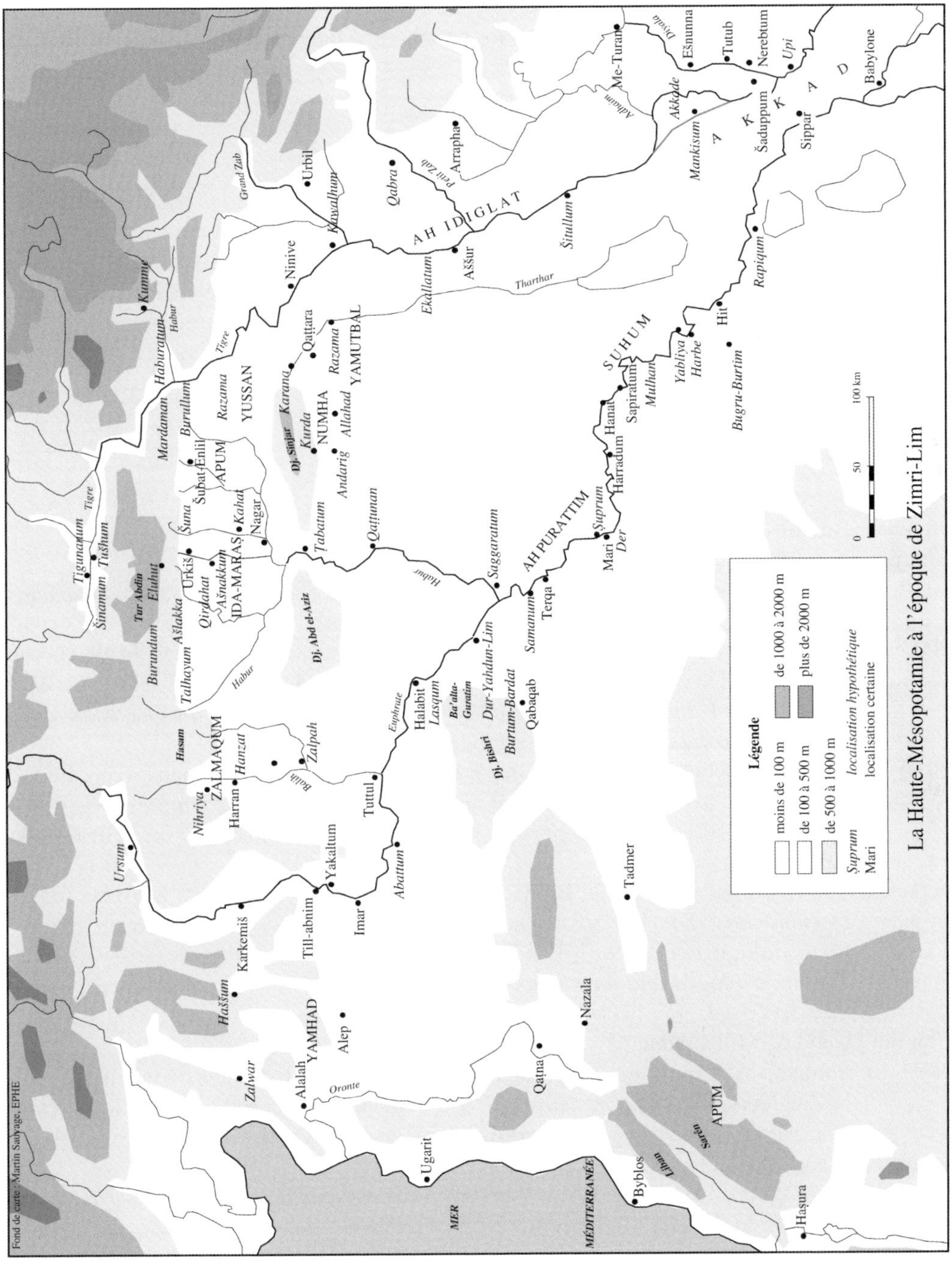

Map 2: Upper Mesopotamia and Syria in the time of Zimri-Lim (from Charpin and Ziegler 2003: 171).

On all these routes the Suteans could be guards, but also simple messengers, carrying letters quickly and reliably from one place to the other.[29] Many rewards for Suteans in Mari economic texts can be explained in this way.[30]

Suteans as mercenaries

Some letters show that Sutean troops could be engaged as mercenaries by Mesopotamian kings. In the war Hammurabi led against Larsa, Suteans seem to have served as mercenaries for the king of Larsa (*ARM* 27 161: 33-52). Sutean mercenaries are also attested on other occasions, such as *ARM* 26/2 482 which shows the problems a Sutean mercenary troop had after crossing the Tigris in Akkade and perpetrating some unsuccessful misdeeds. On these occasions the documentation of Mari gives indications that the Suteans moved further east than usual.

Bibliography

Cavigneaux, A.

2010 BuKrû-le-puits, *NABU* 2010/93: 111.

Charpin, D.

1980 *Archives familiales et propriété privée en Babylonie ancienne: étude des documents de «Tell Sifr»*, HEO 12, Genève/Paris.

2001 L'archivage des tablettes dans le palais de Mari: nouvelles données. In W. H. van Soldt, J. G. Dercksen, N. J. C. Kouwenberg and Th. J. H. Krispijn (eds.), *Veenhof Anniversary Volume. Studies Presented to Klaas R. Veenhof on the Occasion of his Sixty-fifth Birthday*, PIHANS 89, Leiden: 13-30.

2010 The Desert Routes around the Djebel Bishri and the Sutean Nomads According to the Mari Archives: In Ohnuma *et al., Formation of Tribal Communities: Integrated Research in the Middle Euphrates, Syria, Al-Rāfidān Special Issue:* Hiroshima: 239-245.

in press a The Historian and the Old Babylonian Archives. In H. Baker and M. Jursa (eds.), *Documentary Sources in Ancient Near Eastern and Greco-Roman History: Methodology and Practice*, London.

in press b *The Assyriologist and the Computer: The "Archibab" Project, Hebrew Bible and Ancient Israel.*

Charpin D. and A. Millet Albà

2009 Yabliya, Âl-kâpim et l'identification de Šišīn. In E. Cancik-Kirschbaum and N. Ziegler (eds.), *Entre les fleuves - I. Untersuchungen zur historischen Geographie Obermesopotamiens im 2. Jahrtausend v. Chr.*, BBVO 20, Gladbeck: 261-274.

Charpin, D. and N. Ziegler

2003 *Florilegium Marianum V. Mari et le Proche-Orient à l'époque amorrite: essai d'histoire politique,* Mémoires de *NABU* 6, Paris.

29 Hammurabi of Babylon used Suteans as tablet bearers, cf. *ARM* 6 51 (= LAPO 17 553) or *ARM* 28 8.

30 Cf. e.g. *ARM* 23 241: silver for (messengers): e.g. Napisum and Naṣirum two Suteans (28-xi). *ARM* 23 243: 8 (1 sheep for the Suteans).

Durand, J.-M.

1993 Enclos-*purrussâtum*, *NABU* 1993/55: 47.

2000 *Les Documents épistolaires du palais de Mari, tome III, LAPO* 18, Paris.

2005 *Florilegium Marianum VIII. Le Culte des pierres et les monuments commémoratifs en Syrie amorrite, Mémoires de NABU* 9, Paris.

2007 Le nom du désert en amorrite, *NABU* 2007/56: 71-72.

2010 The Banks of the Euphrates along the Bishri. In Ohnuma *et al.*, *Formation of Tribal Communities: Integrated Research in the Middle Euphrates, Syria*, *Al-Rāfidān Special Issue*, Hiroshima: 253-262.

Fujii, S. and T. Adachi

2010 Archaeological Investigations of Bronze Age Cairn Fields on the Northwestern Flank of Mt. Bishri. In Ohnuma *et al.*, *Formation of Tribal Communities: Integrated Research in the Middle Euphrates, Syria*, *Al-Rāfidān Special Issue*, Hiroshima: 61-77.

Groneberg, B.

1980 *Répertoire Géographique des Textes Cunéiformes 3. Die Orts- und Gewässernamen der altbabylonischen Zeit*, TAVO-Beih. B-7/3, Tübingen.

Joannès, F.

1997 Palmyre et les routes du désert au début du deuxième millénaire av. J.-C, *MARI* 8: 393-416.

Kärger, B. and S. Minx

2012 Sutäer, *RlA* 13: 365-369.

Kupper, J.-R.

1957 *Les Nomades en Mésopotamie au temps des rois de Mari*, Liège.

Lion, B.

2001 Les gouverneurs provinciaux du royaume de Mari à l'époque de Zimrî-Lîm. In J.-M. Durand and D. Charpin (eds.), *Mari, Ébla et les Hourrites: dix ans de travaux. Actes du colloque international (Paris, mai 1993).* Deuxième partie, Amurru 2, Paris: 141-210.

Lönnqvist, M., with the contribution of M. Törma, K. Könnqvist, J. Okkonen, M. Herles and M. Königsdorfer

2009 Archaeological Surveys of Jebel Bishri. The Preliminary Report of the Finnish Mission to Syria, 2005-2006, *Kaskal* 6: 1-42.

Lönnqvist, M. and K. Lönnqvist (eds.)

2011 *Jebel Bishri in Focus Remote sensing, archaeological surveying, mapping and GIS studies of Jebel Bishri in central Syria by the Finnish project SYGIS,* BAR S2230, Oxford.

Marti, L.

2008 *Florilegium Marianum X. Nomades et sédentaires à Mari: la perception de la taxe-sugâgûtum,* Mémoires de NABU 11, Paris.

Stolm, M.

2004 Wirtschaft und Gesellschaft in altbabylonischer Zeit. In D. Charpin, D.O. Edzard, M. Stol, *Mesopotamien. Die altbabylonische Zeit*, (= Attinger, P. – Sallaberger, W. – Wäfler, M. (eds.), *Annäherungen* 4), OBO 160/4, Fribourg/Göttingen: 643-975.

Streck, M.

2002 Zwischen Weide, Dorf und Stadt: sozio-ökonomische Strukturen des amurritischen Nomadismus am Mittleren Euphrat, *BaM* 33: 155-209.

Ziegler, N.

2004 Samsî-Addu et la combine sutéenne. In C. Nicolle (ed.), *Nomades et sédentaires dans le Proche-Orient ancien. Compte rendu de la XLVIe Rencontre Assyriologique Internationale (Paris, 10-13 juillet 2000)*, Amurru 3, Paris, 2004: 95-109.

2007 *Florilegium Marianum IX. Les Musiciens et la musique d'après les archives de Mari*, Mémoires de NABU 10, Paris.

FREDERICK MARIO FALES
University of Udine

Hamlets and Farmsteads in the Balīḫ River Valley

The Middle Assyrian and the Neo-Assyrian Evidence

On the basis of archaeological and textual evidence alike, the northwestern Jezirah –and specifically the sector of this steppeland around the Balīḫ river valley– proves to have been economically exploited both in Middle Assyrian (MA) times (15th - 12th centuries BC) and – following the "dark age" of abandonment and Aramean takeover– during the heyday of the Neo-Assyrian (NA) empire (8th - 7th centuries BC). The purpose of this contribution is thus to offer a brief overview of one and the same geographical area during these two periods, in search of the characteristics of the respective "rural landscapes", such as are drawn by the available documentation as well as by wider historical-ecological and socio-economic frameworks which have been traced in recent decades. This diachronic comparison has hitherto only been hinted at, despite the substantial growth of specialized literature on the subject of "land and people" on the Middle-Upper Euphrates river basin during pre-Classical Antiquity (Wilkinson and Tucker 1995; Wilkinson 1998; 2000).[1] I believe that it may prove particularly useful, in order to tackle some open questions concerning the historical and climatical bases for the respective settlement patterns, such as have in recent years been posed regarding the more eastward Ḫābūr (Reculeau 2011).

Portrait of a river valley

The Balīḫ (Fig. 1) is a small perennial tributary of the Euphrates flowing in a riverine plain which interrupts the surrounding rolling Tertiary limestone plains and gypsum hills of the northwestern Jezirah, with a record of settlement of very early date (Wilkinson 1998: 65; Wossink 2010: 185). The river, nowadays in a state of decided degradation, presents an average flow of 6-12 m^3/s, in sharp contrast with that of the Euphrates' 840 m^3/s and of the Ḫābūr's 45 m^3/s; however, this degradation could have begun already in Pre-Classical antiquity (Reculeau 2011: 20).

The river, forming "a narrow irrigated corridor between the Turkish border and the Euphrates" (Wilkinson 1998: 65),[2] runs for some 100 kms from the karstic springs of ʿAyn al-ʿArūs southward to the main river near ar-Raqqah;[3] its northernmost drainage basin is from periodi-

1 Cf. e.g. Wilkinson and Tucker 1995; Wilkinson 1998; 2000. General archaeological bibliography on the area may be found in Lehmann 2002; Anastasio *et al.* 2004; Tenu 2009. A wide-sweeping attempt at historical-economical analysis on northern Mesopotamia was made by Van Driel 2000, with numerous interesting insights.

2 Lyon (2000: 91) describes it instead as a "virtual oasis in the midst of the steppic zones to the east and west".

3 As recalled e.g. by E.I. Gordon (1967: 70), the geographical terms "Karst" and "karstic", derived from the name of the limestone district in the Dinaric Alps near the Adriatic coast of Yugoslavia and NE Italy (*Kras/Carso*), are applied generally to limestone regions in which most or all of the drainage is by underground channels. Since the calcium carbonate in the limestone is highly soluble in water, and especially in rain-water charged with carbon dioxide from the atmosphere, the water seeps through rock, forming caverns and channels beneath the surface. The logographic rendering for the Balīḫ in Sumero-Akkadian texts as [d]KASKAL-KUR, "underground water-course", reflects the ecological reality of the river source: cf. *ibid.*, 77ff.

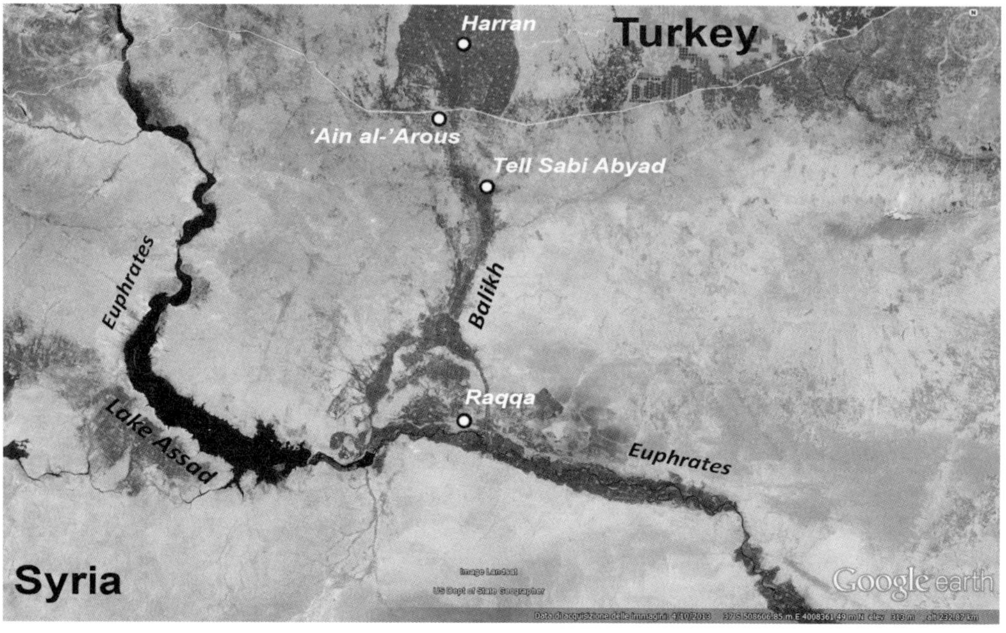

Fig. 1: The Balīḫ river valley, satellite view (source and date of photo: Google Earth, 4/10/2013; image © Image Landsat).

cal streams in the plain around Harrān, while other wadis lying E and W of the river valley also contribute to its water discharge. It flows through a broad plain, at times with "braided" courses, and even has some minute tributaries of the "Yazoo River" type.[4] Thus, assuming a ratio of 1-1.65 litres per second to irrigate 1 ha of land, it has been calculated that the base flow of the Balīḫ from ʿAyn al-ʿArūs could irrigate in Pre-Classical times between 3600 and 6000 ha, i.e. an area capable of supporting 2400-6000 people depending on land and water use (Wilkinson 1998: 81).

On the other hand, the Balīḫ should be considered within a more extensive drainage network extending as far north as Urfa in Turkey, specifically marked by the river Jullab and its valley, approx. 22 kms wide, between the limestone plateaux of the Top Dagh (W) and Tektek Dagh (E), with Harrān at its center. The overall Balīḫ catchment area from Urfa to Tuttul may be thus viewed as a significant component of the ecological and cultural boundary that at the same time connects and separates the Euphrates and Tigris regions, as well as the lowland and highland zones to the north and south. In this light, it may be recalled that a road along the Balīḫ river, linking the cities along the riverbank from Harrān to Tuttul was inserted in Old Babylonian itineraries from Southern Mesopotamia to Emar (Hallo 1964).

4 The first investigator of the area, M.E.L. Mallowan, described the river as having "a zigzag course ... with innumerable ramifications", thus wending it way "between steep and reedy banks to join the Euphrates at Raqqa" (Mallowan 1946: 112). As explained by Gordon (1967: 78, fn. 18) the term "Yazoo River" (derived from a tributary of the Mississippi) is applied to tributaries which are prevented from joining the main river because the latter has built up high natural levees of sediment. They therefore run parallel to the main river for quite a distance, eventually joining it downstream. However these alleged tributaries or ramifications of the Balīḫ might in part be understood as remnants of ancient canals. Marshland is attested here and there along the riverbanks, and is also known from ancient texts (Wiggermann 2000: 179).

Moreover, the river valley constituted the main discernible frontier of Middle Assyrian settlement westwards, although the Euphrates riverbank represented a more advanced expansion point, eventually attained by Tiglath-pileser I (Fales 2011, with prev. lit.).

The central (Syrian) Balīḫ river valley falls by and large in the "zone of marginal cultivation" established by T. Wilkinson in relation to its degree of natural moisture (Wilkinson 2000; Reculeau 2011: 15-17): in its northern sector, with rainfall around 300 mm per annum, rain-fed cultivation produces crops in most years, whereas to the south, closer to the Euphrates, the meager 200 mm rainfall results in crop failures roughly every second year (Wilkinson 1998: 65). In practice, the theoretical limit for rainfed agriculture occurs near the center of the Balīḫ valley, in the vicinity of Tell Zkero, 6 km south of Tell Hammam et-Turkman:[5] agricultural settlement could have depended to some extent on accessory irrigation in the central (and even more in the southern) sector of the Balīḫ, whereas dry-farming appears feasible and widely practiced in the northern part of the valley. Most specialists are of the opinion that localized, small-scale irrigation canals should have been in use and may have even had a certain historical impact throughout the valley: the problem rather regards whether, and to what extent, they were in use in the pre-Hellenistic period.[6] In sum, the basic scenario for agricultural purposes remains determined by climatic conditions –with little observable difference as regards the areas at a comparable latitude on the Ḫābūr, and on the Euphrates between the great bend and the Turkish foothills (Morandi Bonacossi 2000: 353).

Evidence for economic exploitation of the Balīḫ area in the Middle Assyrian period

Assyrian territorial expansion in the Jezira –and specifically around the axes of settlement along the Ḫābūr and the Balīḫ– began in the late 14th century, after the demise of Mittanian power, but attained its apex approx. a century later, under Tukulti-Ninurta I (1244-1208), thereafter entering into a phase of some decline –presumably marked by the earliest infiltrations of West Semitic nomadic groups dubbed traditionally as "Suteans" (Tenu 2009; Fales 2010: 69; Harmanşah 2012: 58-60). The textual evidence for this process of expansion of the MA state in the western sector, and its connected political and economic choices, has grown exponentially in the last generation due to archaeological discoveries in present-day Syria, and now represents a relatively fine "mesh" of partially interrelated data. The groups of administrative or similar texts from this period nowadays available are the following:

1. the 14th-century archives in Middle Babylonian ductus from Tall Munbāqa – *Ekalte*, (Mayer 2001; cf. *ibid.*, 16-17, for the dating), which are thereupon followed in time by the
2. letters and the administrative archives in Assyrian script from Tall Šēḫ Ḥamad – *Dūr-Katlimmu* on the Lower Ḫābūr. (Cancik-Kirschbaum 1996, for the letters; Rollig 2008 for the administrative texts)

5 Tell Zkero is marked as B(alikh)S(urvey) no. 152 in our Fig. 2; cf. also the map (only with numbers) in Wossink 2010: 186. The irrigation zone appears to have been "mainly downstream from Tell Zkero, but also penetrated northwards into the rainfed belt, extending perhaps as far north as Tell Sahlan" (Wiggermann 2000: 176-177).

6 Wossink 2010: 185, is very skeptical on this point, whereas Wilkinson 1998: 65-66 and Reculeau 2011: 79, take pains to quote the few hints in ancient written sources which seem to point to man-made irrigation structures. Most notable is a Mari letter on a conflict concerning water rights in the lowest part of the river valley, in Villard 1987.

The Dūr-Katlimmu letters and administrative texts – which constitute, for their very nature and their state of publication, a sort of central focus to any investigation on the MA presence/ impact in the Jezira –, are partially contemporaneous to the following corpora:

1. the mid-13th -century letters and administrative texts from Tell Ḫuwēra – Ḫarbe in the steppe between the Ḫābūr and the Balīḫ, and
2. the few MA tablets from Tell Faḫariyah on the Upper Ḫābūr, published in 1958;
3. the late 13th -late 12th century tablets from Tell Sabi Abyad on the Balīḫ (cf. below), and finally
4. the 13th-11th century texts from Tell Ṭaban – Ṭâbetu on the Lower Ḫābūr, both of official and of administrative nature.[7]

Many of these documents, also upon comparison with administrative materials from Assur and Kar-Tukulti-Ninurta in central Assyria, provide indirect documentation on the rural landscape in the Jezira in the MA period; thus this age "cannot be considered a dead angle of the history of Mesopotamian agriculture anymore, as it was in the late 1970's" (Reculeau 2011: 128). However, as regards the Balīḫ basin, the Tell Sabi Abyad material holds pride of place, especially with the support of some information from the Dur-Katlimmu letters. Thus this site will form here the main case-study for the 2nd millennium BC.

Tell Sabi Abyad (*Tall Ṣabî 'Abyaḍ*) lies in an intermediate position in the Balīḫ river valley between Tuttul/Tell Bi'a to the south and Harrān to the north (Fig. 2) (Wiggerman 2000: 172). The climate of the Sabi Abyad area is characterized by a precipitation of ca 250mm per annum, with virtually no rain in summer and a large variation in mean monthly/annual amounts. Mean annual temperature is around 20°, mean winter temperature about 8°, and the mean summer temperature reaches 28°. The location of Sabi Abyad is outside the main axes of the river valley, in an ill-defined zone stretching from around Tell Sahlan to the latitude of Tell Zkero (cf. §1). However, this image of Sabi Abyad lying isolated within steppeland fit for nomadic groups with their flocks (Wiggerman 2000: 172), may be to some extent corrected: first of all, a number of other agricultural sites lay within its catchment area (Duistermaat 2008: 28);[8] and secondly, it seems to have lain along an E-W road system which linked it to Dur-Katlimmu and thence to Assur (Faist 2001).

In MA times, after a first period of Mittanian occupation (Akkermans 2006: 205; Duistermaat 2008: 23), Sabi Abyad hosted a *dunnu*, a fortified agricultural farmstead allotted by the Assyrian king to Ili-ipadda, "Grand vizier", and also "king of Hanigalbat" –a title that presumably referred to an exceptional personal power, presumably both political and military, over the "difficult" zone of the Ṭūr ʿAbdīn range, where random Hurrian, and possibly also West Semitic, armed bands, still roamed (Fales 2011).

The "Grand Vizier"'s political and administrative authority – apparently with the support of a direct subordinate, the *sukkallu*,"Vizier"– stretched over a number of "districts", for which the term *pāḫutu* was used, with a "district governor" (*bēl pāḫete*) at the head of each (Jakob 2003: 14, 111ff). The settlements or nodes within each Assyrian enclave or "district" were dubbed according to their size and/or their more or less fortified nature, as *ālu*,

7 Cf. Maul 1992; 2005, for the official inscriptions; Shibata 2007 for a preliminary publication of the administrative texts. The archive of the *dunnu* of Giricano (Radner 2004) falls, instead, decidedly in the 11th century BC.

8 Cf. also Lyon 2000: 100: "Most of the sites with definite Middle Assyrian components are concentrated along the Balikh River between Tell Hammam et-Turkman and Tell Sahlan, i.e., north of the present-day 250 mm isohyet".

"city", *birtu*, "fort", *dunnu*, "(fortified) farmstead".[9] A number of sites endowed with fortifications of sorts thus systematically dotted the countryside of the Jezira.[10]

The Sabi Abyad *dunnu* should have been first assigned to Ili-ipadda's father, Aššur-iddin, who belonged to a secondary branch of the royal family. The most striking archaeological feature of the *dunnu* is its fortified nature, as shown by the massive central tower (21 x 23 meters), with its blind ground floor, which was in its turn surrounded by a massive mud-brick defensive wall, forming an enclosure of 60 x 60 m (=0,36 ha), and by the security measures conditioning the access to the structure. The site also shows a further residence for the *mašennu*, and quarters for servants and scribes (Akkermans 2006: 203). Textual evidence suggests the span of approx. a century (ca. 1225-1120 BC) of overall existence of the *dunnu*. After some years, a partial reoccupation took place with a renovation of parts of the structure, albeit on a much more modest scale, and finally its abandonment ensued (Wiggermann 2000; Duistermaat 2008).[11]

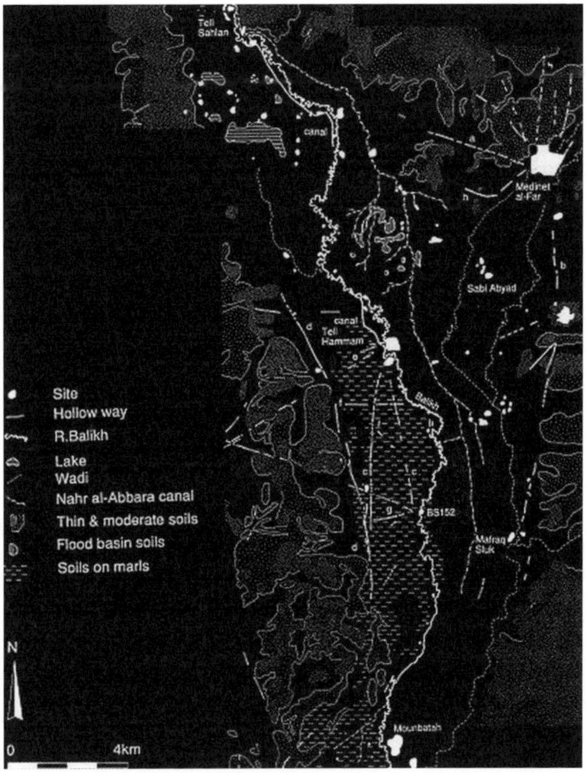

Fig. 2: Tell Sabi Abyad and the surrounding area (from Wilkinson 1998: 74, with color inversion and slight enlargement).

The texts from Tell Sabi Abyad have, unfortunately, still not been made available through full publication. Instead, only a selection of the documents have been presented here and there by the philologist F.M. Wiggermann.[12] Moreover, some of the main conclusions drawn

9 The *dunnu* appears to have been the terminological successor of the complex known as *dimtu*, "tower", i.e. a fortified farmstead with its territory, known from the Nuzi texts: cf. Zaccagnini 1979: 47-53; Koliński 2001; Wiggermann 2000: 172. The translation "Wehrgehöft" (Radner 2004: 69) is given underscoring the link of the word with the root *danānu*, "to be strong", while Herles' "befestigte Stadt"(2007: 415) exaggerates the dimensions and role of such settlements.

10 Cf. Lyon 2000, Duistermaat 2008: 27, for a total of 13 MA sites in the Balīḫ valley south of the Turkish border. It may be noted that Sabi Abyad text T98-115 (Wiggermann 2000: 25, 217) mentions five (adjacent?) sites, of which 3 bear the toponym-formative *dunnu*, while 2 are described as *ālus.*.

11 Specifically, after 1180, the site bears witness to vast changes in the layout and organization of the fortress, then to its devastation by a violent conflagration. Other buildings and stables, storage areas and other fixtures (e.g. wells and threshing-floors) have not been identified, and must have lain in the rural area outside the walls, thus accounting for the presence of some 900 dependents which Wiggermann deduces from the the administrative records.

12 Wiggermann 2000: 205-210, 217-22, presents resp. the transliterations and copies of 12 texts, but others are quoted piecemeal in the accompanying essay, 172ff.; Wiggermann 2008 presents 2 texts and 3 inscribed fragments on pots; Wiggerman 2006 and 2010 are concerned with one text each, with other mentions.

by this scholar have not withstood a historical-critical test;[13] what may be at present salvaged regarding the economic status of the *dunnu* is thus very general, and may be summarized as follows (Radner 2004: 70).

1. Farmland was an inseparable part of a *dunnu*-domain, which represented a royal endowment to the master (often present in the name of the domain itself) for purely agricultural purposes. Whether, how, and to what extent (parts of) the proceeds of the *dunnu* made their way to the central administration remains to be established.[14]
2. No other landowners were present in the settlement. The farmers were all employees of the master, some of them as *šiluḫlu*, i.e. of unfree status, who received rations for their work; among them were possibly foreign POWs of deportee status (Jakob 2003: 39).
3. The master had a residence elsewhere, and supervised the agricultural work by and large from afar, receiving its produce through specific shipments. Since, as "grand vizier", Ili-ipadda was often residing at Dur-Katlimmu or in the royal capital, his *dunnu* was administered by the *mašennu*, or "chief steward".[15]

On the positive side, the Sabi Abyad evidence in its entirety offers some interesting insights on the varieties forming the agricultural produce of the *dunnu*.[16] Barley proves from the texts to have been the staple crop, as in other low-moisture environments, but also wheat is attested from the botanical samplers retrieved in the *dunnu* –and might as such be an indicator for the use of supplementary irrigation locally (Wiggermann 2000: 178). A similar conclusion may be reached considering summer crops (such as cress and sesame), as well as garden produce.

Cress (*kudimmu*): the largest quantity of this herb used for seasoning known from Sabi Abyad is 100 *imēru* requested in a letter by Ili-ipadda to his *mašennu* for his urban residence because "there is no cress for the banquet (*ana šākulte*)".[17] Minute quantities occur instead in other texts among garden products, and in a further letter of request by the master besides onions and lentils (Wiggermann 2000: 197).

13 The evidence presented by Wiggermann (2000: 180-182) for the amount of arable land and the size of the property of the *dunnu*, with deductions regarding the seed-corn /yield ratio, stems basically from only one text of 17 lines, T 98-115 (pp. 205, 217). However, an in-depth analysis of the document by H. Reculeau (2011: 97-99) has now made Wiggermann's calculations invalid; cf. fn. 39, below.

14 On this point, the main flaw in Wiggermann's analysis of text T 98-115 would seem to hinge (cf. fn. 37, above): its reading-out was based on the assumption that a commonly attested procedure of MA times, the "dissolution of the grain-heap" (*pišerti karu'e*), underlay the account. This procedure –well known from Assur and elsewhere– was used to reckon the annual production of a specific area of land, and the surplus it generates, in the perspective of a supervising administration, viz. the Crown. However, Reculeau (2011: 97-99) has contested the existence of *pišerti karu'e* in the Tell Sabi Abyad document – because it is not named, and because the point of view of the text itself is that of the "inner" registrar, consistently with the nature of Tell Sabi Abyad as an oikos whose primary beneficiary is the master of the domain, and not the king. Thus the data of the text are merely random and unusable for generalizing purposes on area, produce and yields.

15 Three such individuals are known by name: letters from the *mašennus* to Ili-ipadda regarding domestic goods to be shipped make it likely that the latter's residence was in the region: Wiggerman 2000: 173.

16 Unfortunately, the material from Sabi Abyad (both textual and paleo-botanical) still awaits definite publication, and thus fails to provide an all-rounded picture such as is possible for the Dur-Katlimmu of the same age: cf. e.g. Röllig 2008, Fales 2010, Postgate 2012.

17 From the further description of this text given in Wiggermann 2008: 560, it seems that this "banquet" was held on a yearly ceremonial occasion in the Assyrian capital, possibly in connection with the *akītu*-festival, when Ili-ipadda was expected to pour scented oil on behalf of the king. This occasion would explain the exceptionally large quantity required.

Sesame (ŠE.GIŠ.Ì = *šamaššammū*) has been the object of a long debate, centered on the alternative possibility that linseed was meant, especially in the earliest texts, but now the presence of sesame is fully confirmed by archeobotany (Reculeau 2009, with review of previous literature). In particular, the seeds identified at Sabi Abyad are reliable archaeobotanical evidence for the cultivation of this summer crop requiring irrigation, and not salt tolerant (Bottema and Cappers 2000: 52; Wiggermann 2000: 197; Duistermaat 2008: 227).[18] The largest attested quantity is 20 *imēru*; normally it would be collected and filtered at the *dunnu* itself.[19]

Fruit and vegetables are present in the texts only in very small quantities: terebinth nuts (*buṭnātu*), lentils (*abšu*), onions, fennel (*taḫšu*), chick peas seem to be attested. Spices included coriander (*kisibirru*), cumin (*kamūnu*), black cumin (*sibibiānu*), possibly mint (*enāniu*) and some unidentified types are attested (Wiggermann 2000: 197). Garlic is also known from botanical samples (Duistermaat 2008: 227).

The overall diet of the personnel of the *dunnu* was thus, on the whole, based on bread and dishes made of grain, with the addition of some vegetables (onions, cress, lentils, fennel, and chick peas) and sesame oil, while fruit seems to have been scarce indeed, and honey was probably consumed only in exceptional cases. The texts do not contain any information on the consumption of meat or milk products (Wiggermann 2000: 198), but it is likely that butter or cheese formed a substantial part of the diet, especially due to the large amounts of sheep in the care of the settlement.

As for animal husbandry in general, ovicaprids (sheep and goat) are well attested in the written documentation, and dominant in the archaeozoological material, whereas suids and bovids are much less attested. Horses and mules were kept and stabled at the *dunnu*, but possibly employed only for strategic purposes –i.e. for Ili-ipadda's occasional military needs (Wiggermann 2000: 196, 199). The faunal remains also indicate that gazelles were hunted. The use of the meat, fat and skin of the wild boar is attested in the texts, while fallow deer is present among the faunal remains (Wiggermann 2000: 179). It seems that ostriches were known at the site, and doves and other birds were eaten (Duistermaat 2008: 250).

An interesting case regards alcoholic beverages: wine is not attested in the Sabi Abyad texts (Duistermaat 2008: 250), whereas beer seems to have been locally produced –not only in the Sabi Abyad *dunnu* but also in adjacent rural sites. A letter from the time of Aššur-iddin, sent from a regional official to the *mašennu* Mannu-ki-Adad, indicates that beer was a token of good social relations, such as the Assyrians were carrying out with the local Sutean pastoralists:

> Speak to Mannu-ki-Adad, thus says Mudammeq-Assur: What is this, that you do not execute my orders as I tell you? Why did you not give orders to your brewer to send a potter to Dunni-Aššur? Let a written assignment go out from you to your brewer in Saḫlalu that he should make available beer and *tariḫu*-vessels when the Suteans come to have dinner with me. Who else would I ask, who else would give it to me? Send out these tablets (with commands) promptly![20]

18 For the chronology of sesame cultivation, and its need for moist ground, cf. Reculeau 2009: 34-36.

19 One of the published texts runs as follows (Wiggermann 2000: 192, 209-210: T 93-7 = no. 12): "The oil-pressers must filter their sesame. If there is no sesame in your *dunnu*, go to the *dunnu* of the [Sub]araeans, and give out sesame to the oil-pressers there, so they can filter it. They should not be idle, but filter the sesame. It is not necessary that they press (olive) oil; filtered sesame(-oil) is requested to be supplied, and they must quickly filter it." (translation by the present author, with the verb *ḫalāṣu*, attested many times in this passage, rendered as "to filter [sesame oil]", rather than "to dehull [the sesame-plant]").

20 T93-3: see Wiggermann 2008: 561-562; also Wiggermann 2010: 27 (Dutch translation)

A generation later, Ili-ipadda executed a treaty with the "sheikhs" (GAL.MEŠ) of the Sutean Niḫsānu-tribe.[21] Two clauses of this text indicate that the Assyrians in the Balīḫ valley were regular producers of beer, while the pastoralists were not; thus the former could impose some measures of protectionism regarding this product on the latter:

§ 5. When a Sutean goes to an (Assyrian) settlement, he may not drink his beer in the tavern (there), (but) he must take the beer, and return to (his own) camp; (only then) may he drink the beer. Beer should be sold, he may not take it for free.

§ 6. When a Sutean wants to drink on credit then he should (first) issue some sort of deposit, and one may give him the beer. The innkeeper (/brewer) may charge an interest for the beer, but no other Sutean (at random) should attempt to recover it, one may burden the interest (only) on the Sutean who has been given the beer, when he encounters him".[22]

The Balīḫ area in the Neo-Assyrian period: the Ḫarrān census

After the *reconquista* of the Euphrates riverbank from the Arameans under Shalmaneser III (ca. 855 BC), the Balīḫ valley became once more a part of "inner Assyria", while presumably retaining its role as area of intersection for trade routes northbound to Anatolia, westward to the Transeuphratene and the Mediterranean coast, and southward to the Middle Euphrates and Babylonia. A number of provinces were drawn up in this territory by Tiglath-pileser III, with Harrān holding pride of place, as sometimes seat of the *turtānu* (Radner 2006: 54-55).[23] The site of Harrān (present-day Altınbaşak) is at present undergoing an unhurried excavation (Fig.3) (Yardımcı 2008), and –more widely– no significant combination of texts and archaeological finds for the NA phase has been hitherto discovered in one site within the area save for Sultantepe-Ḫuzirīna, which was however only subjected to a brief dig in the 1950s (Lloyd-Gokçe 1953).[24] This situation forces us to fall back on the NA written record, which –apart from random notations in a number of real estate sale deeds– finds its main focus in the texts of the so-called "Harran Census" (Fales 1995, Introduction, xxx-xxiv).

The texts of the "Harran Census" are a group of larger or smaller multi-columned tablets and fragments, which were considered relevant to a single archive in the first edition by C.H.W. Johns in 1901; later editions have fully confirmed this status (Fales 1973: 5ff.; Fales 1995: xxx). The texts are undated, with no unifying label; however their contents show an internal homogeneity from various points of view. Although they stem from the central archives of Nineveh, their geographical outlook (Fales 1973: 93 ff.) yields a NW Mesopotamian regional context (Fig. 4): that of the Harrān region and areas adjacent to it, essentially in the basin of the Balīḫ, together with zones immediately to the west (Sarugi, Ḫadattu) and the north (the region of Qipāni) (Lipiński 2000: 129, with previous bibl.). The surviv-

21 T04-37: mention in Wiggermann 2008: 561; copy in Wiggermann 2010: 29 (Dutch translation of §§5-6 ibid., 28).

22 Translation after Wiggermann 2010: 28.

23 The main sites in the Balīḫ valley, together with areas to the north, are comprised in the titulary of Bēlu-lû-balaṭ, eponym for the year 814 BC, on a stela from Assur ("commander-in-chief, great herald,administrator of temples, commander of vast troops, governor of Tabitu, Harrān, Ḫuzirīna, Dūru, Qipāni, Zallu and Balīḫu,": RIMA 3, 178: 2002).

24 For survey results in the Harran area, cf. Yardımcı 2004.

ing fragments also yield one toponym (Magrisu) to be sought eastwards, in the Ḫābūr triangle. No urban sites are directly described in the texts: the listed real estate pertains to the rural world of the village and the hamlet (*kapru*), which may at best be located (*ina*) *qanni*, "near" a larger town or a city, or may lie deep *ina madbar*, "in the steppe," its sole reference being a particular province.

The typology of the lists also presents a number of particular and unifying aspects. A common feature for most of the documents in this group is a division of the tablets through vertical rulings in columns and thereupon horizontal ones in sections. The contents of a typical section correspond in general to a rural holding, comprising land and people; these are in the main listed in the following order:

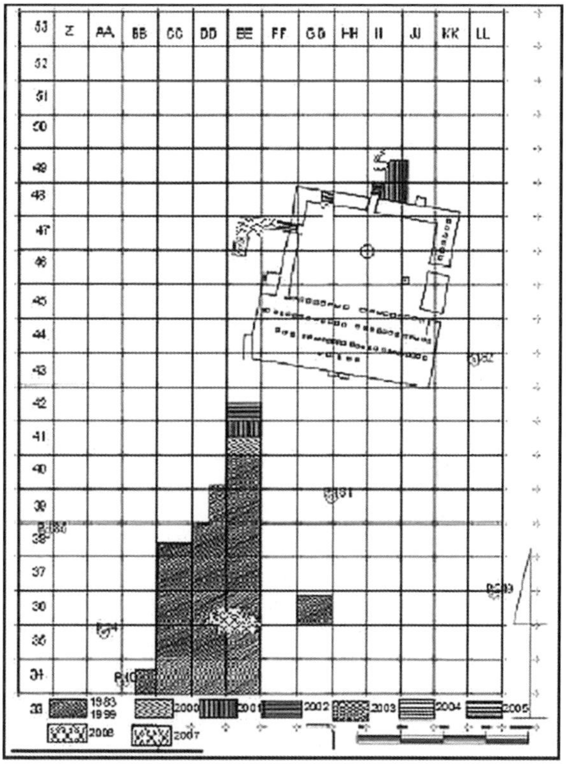

Fig. 3: Topographic plan of excavation areas at Harran, 1983-2007 (from Yardımcı 2008: 378).

1. lists of people in families:
   (a) father noted by personal name and professional designation;
   (b) other members defined in their relation to him, viz.
      - sons or brothers, counted or sometimes named, with indication of age;
      - wives and daughters, merely counted, the latter bearing indication of age;[25]
2. lists of agricultural holdings, which include
   (a) real estate, listed by type and quantity (cornland, in *emāru*; vineyards, in shoots of vine (by the thousands); orchards, usually not quantified; etc);
   (b) livestock (sheep and goats; oxen), by number;
3. localization of the people and the estate - in relation to
   (a) the surrounding or neighbouring village/hamlet or the "steppe," and thereupon to
   (b) the larger urban or regional district/province.

As a token sampler of the type of listing of the plots in the "Harran census", the following sections from the first, and most extended, text of the group may be charted as follows:

25 The classification of people by age/size categories effected here, represents, for all its crude appearance, one of the most far-reaching and most efficiently standardized results of the Assyrian administrative machine, which we find also in use in other cadastral documents, in sales documents of land and people, and in tallies of displaced persons (Fales 1995: xxix-xxx).

Text	No. of people (profession of *paterfamilias*)	Arable land (in *emārus*)	Vineyard (stalks)	Houses	Other agricultural fixtures	Live-stock	Geographical notations
SAA XI 201, I, 1-8	5 (gardener)	10	10000	2			In TOWN, in district of CITY (*Sarugu*)
Ibid., 12-24	8 (gardener)	6	15000	1			In TOWN, near CITY (*Til-Abnā*)
Ibid., 25-40	11 (gardener)	10	29000	2		2 oxen	In VILLAGE (PN), near CITY (*Dimmetu*), in REGION (*Qipāni*)
Ibid., I, 42- 50	7 (gardener)		6000				In VILLAGE (PN), near CITY (*Duma*), in REGION (*Qipāni*)
Ibid., II, 1-7	4 (gardener)		5000	1	1 orchard		In TOWN, in REGION (*Qipani*)
Ibid., 8-15	4 (gardener)		4000				In VILLAGE, in REGION (*Qipāni*)
Ibid., 16-26	8 (gardener)					61 horses, cattle	In TOWN (GN : springs) near CITY (*Haurina*)
Ibid., 27-32	4 (gardener)					57 goats	In TOWN, in district of CITY (*Ḫasame)*
Ibid., 33-37	4 (gardener)						In TOWN, near CITY (*Ḫarrān*)
Ibid., 38-45	3 (gardener)				1 grove of *šaššugu*		In TOWN, (GN : *spring, well*) in REGION (*Qipāni)*

A further common trait of the collection, which appears once every few sections, regards the name of individuals explicitly described as *owners* of the previously listed holdings. Since the estates were of quite different sizes and types, the tablets allotted different spaces to each. Thus we find up to two landowners named in some texts;[26] while in other texts, single landowners are mentioned, their titles indicating various degrees of power in the local NW Mesopotamian élite –and their obviously absentee status.[27]

Despite this wealth of data, the exact purpose or occasion of the series is unspecified, and thus it remains a matter of theoretical reconstruction, whether remnants of a general-purpose "inventory" or of a more "project-specific" tally are represented here. On internal grounds, this corpus seems scarcely plausible as a random portion of a general inventory or "open-ended" archive of land use and occupation regarding all Assyrian territory. Quite to the contrary, in fact, the goals or limits of the lists were probably as specific as what we find in the extant sample: i.e. the identity of a number of large landowners of NW Mesopotamia with the register of their individual holdings, broken down by plots and the relevant workgroups.[28] On the basis of the identification of one of the landowners, Gir-Dadi "from the town of Ti[l- ... ", (SAA XI: 206) with one Gir(i)-Dadi, known from a letter to Sargon as city lord of Til-turi, a date for the series during the late reign of

26 Specifically, in the preserved parts of SAA XI, nos. 203 and 207.

27 SAA XI, 202: II, 14'-15': 213: IV, 12-13: 203: III, 7-9, resp.

28 On the social status of the people attached to the land, cf. most recently Galil 2007; Fales 2009-10.

Sargon II, or perhaps under Sennacherib, has been suggested (Parpola 1974: 108). As for the occasion, it has been proposed that the "lands enumerated are those properties in the area in question, which were exempt from taxation" (Postgate 1974: 229 ff). Connecting these two ideas, the compilation of the "Harran Census" could be connected to the vast operation of tax-exemption undertaken by Sargon upon his ascent to the throne, by which the *kidinnūtu* that the cities of Assur and Harrān had enjoyed of old, and that had lapsed, was "restored".

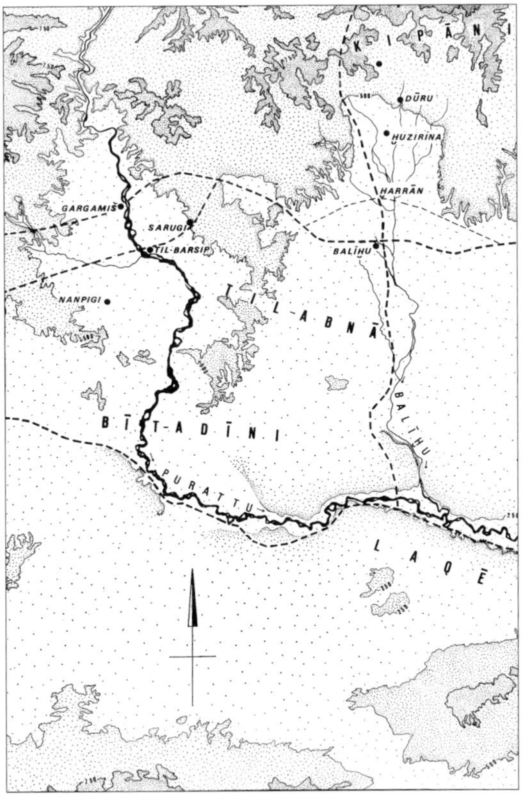

Fig. 4: The geographical setting of the "Harran census" texts (from Fales 1973: fig. 4).

Comparing the evidence

In any case –and to whatever extent the "Harran census" may be considered a faithful mirror of its time and place– it is a fact that this group of texts provides us with a "snapshot" of the upper Balīḫ valley (and adjacent sectors) which contrasts with the one from MA Sabi Abyad under many significant aspects. First of all, the settlement pattern: against a "landscape" of autonomous fortified *dunnus* dotting sparsely but staunchly a still somewhat hostile environment, peopled by Hurrian escapees or Sutean nomads, we have here a presumably denser mesh of non-walled rural farmsteads or hamlets, which –even when they lay far off in the steppe– had administrative links (and physical connections along safe roads) to the main cities around and to the nearby regional/provincial centers of the Empire.

Secondly, the social factor: whereas Ili-ipadda's *dunnu* would seem to have supported a vast population of male and female dependents with different work specialties –as probably was also the case for his neighbors– the landowners from Harrān and environs merely required small families of servile personnel (again, formed by many deportees) to run their holdings, which were centered on one or two main forms of produce or types of animal breeding. And finally, the crops themselves: while cereals are present in both periods (and probably were always mainly barley with some wheat), a fundamental difference is represented by the relatively extensive vineyards of the Harrān region and northwards which were widespread from one holding to the next.[29]

29 Cf. Fales 1973: 128, for the suggestion that the (presumably free-standing) shoots of vine were planted in quadrilateral layouts, so as to ease the count of the plants.

This difference in crops, of course, opens up the issue of a significant variation in yearly rainfall, which could have set decidedly apart the northernmost reaches of the valley from the area some 40 kms. to the south. In practice, whereas Sabi Abyad might have required supplementary irrigation to water its barley crops (also for its prized beer production), the hilly zone of Qipāni probably enjoyed sufficient precipitation to support its vine shoots by the thousands. As for general climatic conditions over longer periods, the two corpora in mutual comparison are largely silent: from both sides, we gain a picture of an area of limited agricultural possibilities and relatively sparse population –whether as "outer frontier" in MA times, or as "inner marginal area" within the later Empire– which would have been strongly affected even by minute variations in temperature over more than yearly periods.

This difference in crops, of course, opens up the issue of a significant variation in yearly rainfall, which could have set decidedly apart the northernmost reaches of the valley from the area some 40 kms. to the south. In practice, whereas Sabi Abyad might have required supplementary irrigation to water its barley crops (also for its prized beer production), the hilly zone of Qipāni probably enjoyed sufficient precipitation to support its vine shoots by the thousands. As for general climatic conditions over longer periods, the two corpora in mutual comparison are largely silent: from both sides, we gain a picture of an area of limited agricultural possibilitifes and relatively sparse population –whether as "outer frontier" in MA times, or as "inner marginal area" within the later Empire– which would have been strongly affected even by minute variations in temperature over more than yearly periods.[30]

Conclusions

In this light, the long-range conclusions drawn by Wilkinson's survey of the Balīḫ valley, relevant to a dispersal "from nucleated centers into smaller rural settlements in the plain", starting from the LBA, and with greater emphasis in the IA, down "through Hellenistic into the Late Antique/Early Islamic periods" (Wilkinson 1998), may be retained; specifically, both the MA and NA scenarios depict rural holdings in the hands of absentee landlords, tied to the higher levels of state and provincial organization. Further, as noted by Lyon, survey data from the Balih valley depict a general decline in population during the 2nd millennium, both in the total number of settlements occupied and in the aggregate settled area. By the first centuries of the 1st millennium the low continuity trend continues, but overall the number of settlements increases; these sites, however, are generally smaller and occupy almost one-third less aggregate area than did the pre-MA period sites (Lyon 2000). This last picture, to be sure, does not contrast with the shift from dispersed but well-manned MA *dunnus* to a thicker and more organized network of minute family-level *kaprus* in the NA period, as seen above through the texts.

30 In sum, the river valley fits particularly well the definition of "high risk - low yield environment" put forth by Van Driel 2000: 269.

Bibliography

Akkermans, P.M.M.G.

2006 The fortress of Ili-pada. Middle Assyrian architecture at Tell Sabi Abyad, Syria. In P. Butterlin *et al.* (eds.), *Les espaces syro-mésopotamiens. Hommage à J.-Cl. Margueron*, Turnhout: 201-211.

Anastasio, S., M. Lebeau and M. Savage (eds.)

2004 *APUM: Atlas of Preclassical Upper Mesopotamia*. Turnhout.

Bottema, S. and Cappers, R.T.J.

2000 Palynological and Archaeobotanical Evidence from Bronze Age Northern Mesopotamia. In Jas 2000: 37-70.

Cancik-Kirschbaum, E.

1996 *Die mittelassyrischen Briefe aus Tall Šēḫ Ḥamad*, Berlin.

Duistermaat, K.

2008 *The Pots and Potters of Assyri*a, Turnhout.

Faist, B.

2001 *Der Fernhandel des assyrischen Reiches zwischen den 14. und. 11. Jh. v.Chr, Münster.*

Fales, F.M.

1973 *Censimenti e catasti di epoca neo-assira*, Roma.

2009-2010 On Assyrian "Lower-stratum" Families, *SAAB* 18: 163-186.

2010 Production and Consumption at Dur-Katlimmu: A Survey of the Evidence. In Kühne 2010: 67-86.

2011 Transition: The Assyrians at the Euphrates between the 13th and 12th century BC. In K. Strobel (ed.), *Empires after the Empire*, Vicchio (FI): 9-59.

2012 "Ḫanigalbat" in the Early Neo-Assyrian Royal Inscriptions: A Retrospective View. In G. Galil *et al.* (eds.), *The Ancient Near East in the 12th –10th Centuries BCE: Culture and History*, Münster: 99-119.

Fales, F.M. and J.N. Postgate

1995 *Imperial Administrative Records, Part II: Provincial and Military Administration* [=SAA XI], Helsinki.

Galil, G.

2007 *The Lower Stratum Families in the Neo-Assyrian Period*, Leiden.

Gordon, E.I.

1967 The Meaning of the Ideogram ᵈKASKAL.KUR = "Underground Water-Course" and Its Significance for Bronze Age Historical Geography, *JCS* 21: 70-88.

Güterbock, H.G.

1958 The Cuneiform tablets. In C. McEwan *et al.* (eds.), *Soundings at Tell Fakhariyah*, Chicago: 80-90.

Hallo, W.W.

1964 The Road to Emar, *JCS* 18: 57-88.

Harmanşah, Ö.

2012 Beyond Aššur: New Cities and the Assyrian Politics of Landscape, *BASOR* 365: 53-77.

Herles, M.

2007 Assyrische Präsenz an Euphrat und Balīḫ, *UF* 39: 413-449.

Jakob, S.
2003 *Mittelassyrische Verwaltung- und Sozialstruktur: Untersuchungen*, Leiden.
2009 *Die mittelassyrischen Texte aus Tell Chuēra in Nordost-Syrien*, Wiesbaden.
Jas, R.M. (ed.)
2000 *Rainfall and Agriculture in Northern Mesopotamia,* Leiden.
Koliński, R.
2001 *Mesopotamian* dimātu *of the Second Millennium BC*, Oxford.
Kühne, H. (ed.)
2010 *Dur-Katlimmu 2008 and Beyond*, Wiesbaden.
Lehmann, G.
2002 *Bibliographie der archäologischen Fundstellen und Surveys in Syrien und Libanon*, Rähden/Westf.
Lloyd, S. and N. Gokçe
1953 Sultantepe: Anglo-Turkish Joint Excavations, 1952, *AnSt 3*: 27-47.
Lyon, J.D.
2000 Middle Assyrian Expansion and Settlement Development in the Syrian Jezira: The View from the Balikh Valley. In Jas 2000: 89-113.
Mallowan, M.E.L.
1946 Excavations in the Balīḫ Valley, 1938, *Iraq* 8: 111-159.
Maul, S.M.
1992 *Die Inschriften von Tell Bderi*, Berlin.
2005 *Die Inschriften von Tall Ṭaban (Grabungskampagnen 1997-1999). Die Könige von Ṭabetu und das Land Māri in mittelassyrischer Zeit,* Tokyo.
Mayer, W.
2001 *Tall Munbāqa – Ekalte – II. Die Texte*, Saarbrücken.
Morandi Bonacossi, D.
2000 The Syrian Jezireh in the Late Assyrian Period: A View from the Countryside. In G. Bunnens (ed.), *Essays on Syria in the Iron Age,* Leuven – Paris – Sterling (Va.): 349-396.
Parpola, S.
1974 A Note on the Neo-Assyrian Census Lists, ZA 64: 96-115.
Postgate, J.N.
1974 Some Remarks on Conditions in the Assyrian Countryside, *JESHO* 17: 225-243.
2012 Assyrian Percentages? Calculating the Birth-rate at Dur-katlimmu. In G..B. Lanfranchi *et al.*(eds.), *Leggo! Studies... F.M. Fales*, Wiesbaden: 677-685.
Radner, K.
2004 *Das mittelassyrische Tontafelarchiv von Giricano / Dunnu-ša-Uzibi,* Turnhout.

2006 "Provinz. C. Assyrien", *RlA* XI/1-2: 42–68.
Reculeau, H.
2009 Le point sur la "plante à huile" : réflexions sur la culture du sésame en Syrie-Mésopotamie à l'âge du Bronze, *JMC* 13: 13-37.
2010 The Lower Ḫābūr before the Assyrians. In Kühne 2010: 187-216.
2011 *Climate, Environment and Agriculture in Assyria in the 2nd Half of the 2nd Millennium BCE*, Wiesbaden.

Röllig, W.
2008 *Land- und Viehwirtschaft am Unteren Habur im mittelassyrische Zeit*, Wiesbaden
Shibata, D.
2007 Middle Assyrian Administrative and Legal Texts from the 2005 Excavation at Tell Taban: A Preliminary Report. In H. Numoto (ed.), *Excavations at Tell Ṭaban, Hassake, Syria*, Tokyo: 169-180 (="*Al Rafidan*" 28: 63-74).
Tenu, A.
2009 *L'expansion médio-assyrienne: approche archéologique*, London.
Van Driel, G..
2000 The Mesopotamian North: Land Use, An Attempt. In Jas 2000: 265-299.
Villard, P.
1987 Un conflit d'autorité à propos des eaux du Balih, *MARI* 5: 591-596.
Wiggermann, F.A.M.
2000 Agriculture in the Northern Balikh Valley: The Case of Middle Assyrian Tell Sabi Abyad. In Jas 2000: 171-231.
2006 The Seal of Ilī-padâ, Grand Vizier of the Middle Assyrian Empire. In P. Taylor (ed.), *The Iconography of Cylinder Seals,* London – Turin: 92-99.
2008 Cuneiform texts from Tell Sabi Abyad Related to Pottery. In Duistermaat 2008: Appendix E, 559-564.
2010 Wein, Weib und Gesang in een midden-assyrische nederzetting aan de Balikh, *Phoenix* 56: 18-60.
Wilkinson, T.J.
1998 Water and Human Settlement in the Balikh Valley, Syria: Investigations from 1992-1995, *JFA* 25: 63-87.
2000 Settlement and Land Use in the Zone of Uncertainty in Upper Mesopotamia. In Jas 2000: 3-3.
Wilkinson, T.J and D.J. Tucker
1995 *Settlement Development in the North Jazira, Iraq*, Warminster.
Wossink, A.
2010 Climate, history, and demography : a case-study from the Balikh Valley, Syria. In A. Alabashi *et al.* (eds.), *Regards croisés sur l'étude archéologique des paysages anciens*, Lyon: 181-192.
Yardımcı, N.
2004 *Harran Ovası Yüzey Araştırması/ Archaeological Survey in the Harran Plain, I-II.* Istanbul.
2008 2007 Yılı Harran Kazıları, in 30. *Kazı Sonuçları Toplantısı* 4. Cilt, Ankara 2008: 371-384.
Zaccagnini, C.
1979 *The Rural Landscape of the Land of Arraphe*, Rome.

OLIVIER ROUAULT and MARIA GRAZIA MASETTI-ROUAULT
University Lyon 2, CNRS, UMR 5133; EPHE-Sorbonne, CNRS, UMR 8167

From One Valley to Another

Bir el-Haddad, a Neo-Assyrian Trading Post?

Bir el-Haddad, a settlement in the Syrian Jazireh steppe between the valleys of the Lower Khabur and of the Lower Middle Euphrates (Fig. 1), has been explored by a team associated with the Syro-French Missions in Terqa and in Tell Masaikh only during two short campaigns, in Fall 2008, when the site was re-discovered[1], and in 2010, when a first study has been carried out by O. Rouault (Masetti-Rouault and Rouault 2012; Rouault in press).

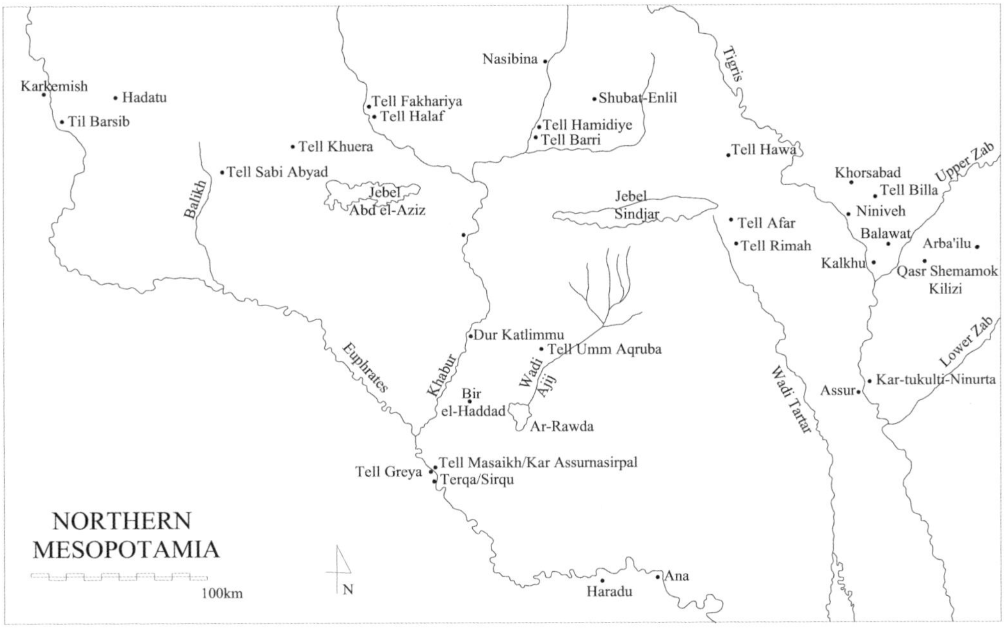

Fig.1: General map of Northern Mesopotamia

The site is in the steppe, as the crow flies, 45 km north/north-east of Tell Masaikh – Kar Assurnasirpal, on the Jezirah bank of the Middle Euphrates, in front of Terqa/Ashara. It is a little bit more than halfway between Masaikh and Tell Sheikh Hamed – Dur-Katlimmu, to the south of it, and around 15 km east of the River Khabur (Fig. 2). It is situated in the centre of a *faida* depression, marked by the presence of a well, hence the toponymy records, still

1 Bir Haddad corresponds to the site Ğilīb al-Adad that was mapped during the ʿAǧīǧ-survey by Bernbeck and Pfälzner but not surveyed because it was situated beyond the limits of the survey permission (Bernbeck 1993: 15, 128 footnote 167; cf. also Kühne 2010: 123 Fig. 05-06 [site no. 134]).

Studia Chaburensia 4 (2014), pp. 243–256.

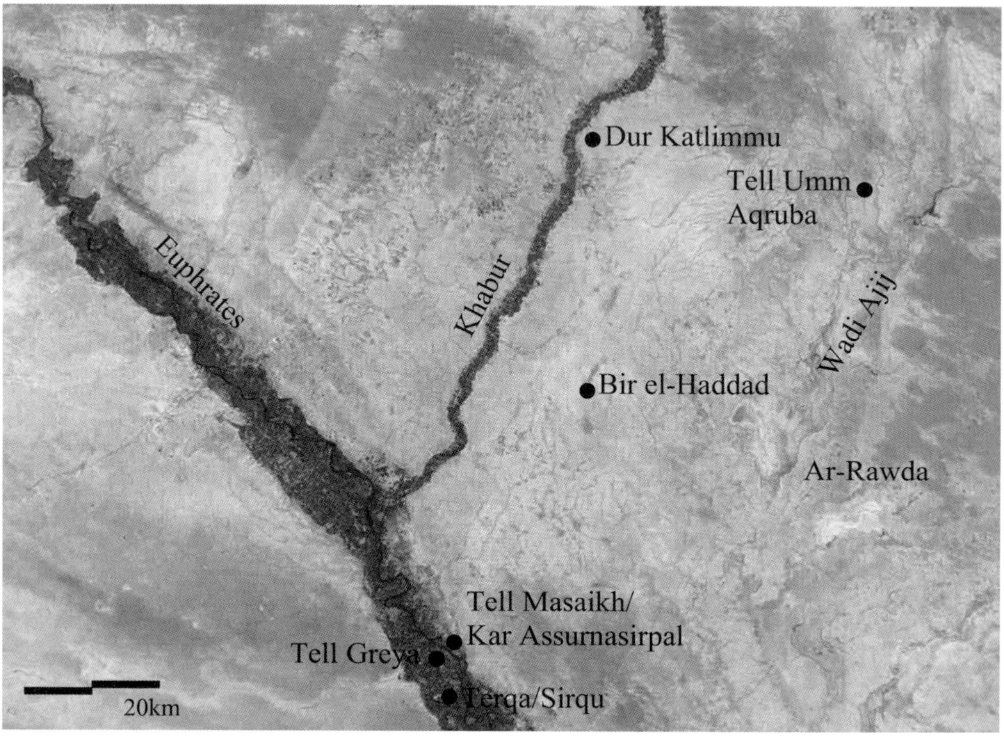

Fig. 2: Satellite image of the region of Bir el-Haddad

used today by pastoralists, shepherds and local hunters. When we took it into consideration, Bir el-Haddad was already identified as a Late Roman period settlement. Considering its geographical position, when we discovered on its surface Iron II ceramic material associated with extensive structures, we quickly realized its importance.

Bir el-Haddad, with its large surface of more than 15 hectares (Fig. 3) and without apparent limits, is crossed by new tracks connecting gas extraction installations in this part of the steppe. As it has been already seriously disturbed and eroded by modern activities, we got a permit to study it as a salvage operation, thanks to the quick agreement of the Syrian archeological authorities, the help of the Deir-ez-Zor Direction of Antiquities and to the financial support of the French Centre National de la Recherche Scientifique (CNRS). After the first surface survey, when we found Iron II-III sherds, – along with evidence of Roman-Parthian and Early Islamic occupations – we carried out a magnetic survey[2] and some soundings during the 2010 campaign, which, while limited, have revealed the complexity of the site. It has been, and continued to be, impossible to develop a study of the natural environment and of the characteristics of the site itself, as well as of its economy and production activities, in order to explain the success of the settlement, at least for a period. As a consequence, what we present here are just the macroscopic aspects of the Iron II-III settlement, which we considered mainly in the context of the communication net of the region, in the larger system of the Assyrian Empire.

2 The geo-magnetic survey has been carried out by a team composed by Dr. C. Brinon and dr. F.-X. Simon, under the direction of Prof. A. Tabbagh (Paris VI – Pierre and Marie Curie University, Paris, UMR 7619 Sisyphe).

Fig. 3:View of a part of the site with houses, and the faida in the back

A settlement in the steppe, between Kar-Assurnasirpal and Dur-Katlimmu

Kühne, Bernbeck and Morandi Bonaccossi's works have largely explored the strategy of the south-western Jezirah steppe occupation from an Assyrian perspective (Kühne 2010, 2011a; Bernbeck 1993; Morandi-Bonaccossi 1996, 2000), compared to the situation in the river valleys. On the other hand, the Tell Sheikh Hamed team has highlighted the presence of a chain of settlements situated on an East-West itinerary, following the wadi el-Agig, connecting the Khabur valley to the Tigris since prehistoric times, and, since the Late Bronze Age, the western provinces of the Middle Assyrian "empire" to their capital (Bernbeck 1993).

Bir el-Haddad however is not part of the system of occupations belonging to this net, being much further south. It is situated in an area where no canals are documented (Erzenzinger and Kühne 1991; Kühne 1990, 2012a, 2012b), and where no dry agriculture is possible, even if the faida situation provides some soil moisture and if the well existed already in antiquity (Fig. 4). In fact, there are no reasons to think that a climatic change in that period could have modified to such an extent the conditions to make a permanent settlement in the steppe, and dry agriculture, seem like a good idea, worth the human and economic investment it would need (Neumann and Parpola 1987; cf. Van Driel 2000; Wilkinson and Barbanes 2000). Therefore, the reasons to create and develop such a large sedentary settlement almost in the middle of nowhere should be looked for in the need to connect two places, Tell Masaikh and Tell Sheikh Hamed, both heavily marked by Assyrian presence, tracing an itinerary passing through the site itself. More precisely, the Bir el-Haddad establishment could be connected with the development of the Neo-Assyrian "colony" of Tell Masaikh, founded in the Iron II period, while the Assyrian Dur-Katlimmu had already a much older story, since Middle Assyrian period (Kühne 1995, 2011b).

Fig. 4: Photo of the Bir el-Haddad well with camels, at the southern end of the *faida*

Excavations in Tell Masaikh, since 2002, have led to the identification of this large site as the Neo-Assyrian colony of Kar-Assurnasirpal, "Assurnasirpal's trading post" founded in the first half of the 9th century at a crossing point on the Euphrates (Masetti-Rouault 2010, 2013; Masetti-Rouault and Salmon 2010). During the reign of Adad-nirari III, at the beginning of the 8th century, the settlement was reconstructed, probably by Nergal-eresh, the Rasappa province governor charged with the reorganization of this region (Grayson 1993; Postgate 1995; Radner 2002: 4-7). The construction, in the same period, of a very long canal built on the left bank (Masetti-Rouault 2008), linking the hub point of Hindanu, 120 km downstream – the ancient region of Mari - to the Lower Khabur area, possibly reveals a special interest of the Assyrian state, or maybe of Assyrian economic elites, in controlling the growing exchange network between Northern Mesopotamia and Babylonia, in particular with the South-Western markets on the Arabian Coast (Masetti-Rouault 2010, 2013).

The Iron II canal associated with Kar-Assurnasirpal is now called "Nahr Dawrin" (Fig. 5), as its layout was later followed by a canal of the Early Islamic period, known in Arab documentation with this name (Berthier 2001: 32-69; Geyer and Monchambert 2003: 199-217). It takes its waters from the Lower Khabur, just before its confluence in the Euphrates. Beyond its use for irrigation of the East bank, this canal could have been exploited also for navigation and transport of goods upstream, in the times of the year when the Euphrates is too difficult to sail. After the Nahr Dawrin, cargo and people could continue on the Khabur, or in one of the canals depending on it, up to Dur-Katlimmu, eventually taking there the road eastward to Assyria, or northbound, to the *harrān šarri*, the royal road (Kessler 1997). Now, Bir el-Haddad's Iron II development opens the way to reconstruct another possible connection: A shorter road crossing the steppe, which, avoiding the area around the mouth of the Khabur, would have joined the colony on the Euphrates directly with the cluster of Assyrian centers in the Lower Khabur region, and Dur-Katlimmu in particular. While it reshaped in some way the ancient unity of these regions during Middle Bronze II-III Khana period, and later, of

Fig. 5:Tell Masaikh and the Nahr Dawrin

the Aramean Upper and Lower Laqê country - with Terqa/Sirqu as capital (Masetti-Rouault 2001; Rouault 2009) -, the creation of this itinerary needed a change in the perception of the steppe itself and in its exploitation methods during Iron I-II periods, managed not only by semi-nomad pastoralists, but by societies and administrations under Assyrian rule.

A settlement in the steppe, an Assyrian creation?

Assyrians did not like the steppe in general. Even if their king list kept the memory of ancient ancestors who lived in the tents – actually, the remote Amorites fathers of Samsi-Addu -, the more common Assyrian landscape consisted of the plains and mountains of the Northern Mesopotamia Tigris valley, around and beyond Nineveh, made greener by canals (Dalley 2001-2; Bagg 2000; Altaweel 2008; Kühne 2012a). Assyrian kings considered for a long time the Jezirah steppes, where barbarians roamed, to be wild and unsafe places, *ubi sunt leones,* good only for safaris – lion, wild bull and ostrich hunts - a Far West frontier where they could prove their courage and manhood. The foundation of the Bir el-Haddad site highlights, for the Iron II period, a different perspective, a practical knowledge of the natural environment and its possibilities, and clear confidence and trust in the durable success of the system it belonged to.

That the Iron II-III sedentary occupation of Bir el-Haddad, was not, or not only, connected with new rural and agrarian politics, or a change in the economic or cultural habits of local societies, can be seen through the structure of the settlement itself, as we have been able to define it up to now (Rouault in press; Masetti-Rouault and Rouault 2012). The site, of at least 15 hectares, covers the Northern and Eastern slopes of the *faida*, the dividing line between the cultivable area and the residential one being marked by the well. Even today water is found here 10 meters below the surface, and some drinking structures for sheep and camels have been built around the well. Around it, we have identified no ancient structures, only traces of camps and tents, some with constructed fireplaces, but the dates of these occupations are possibly quite recent.

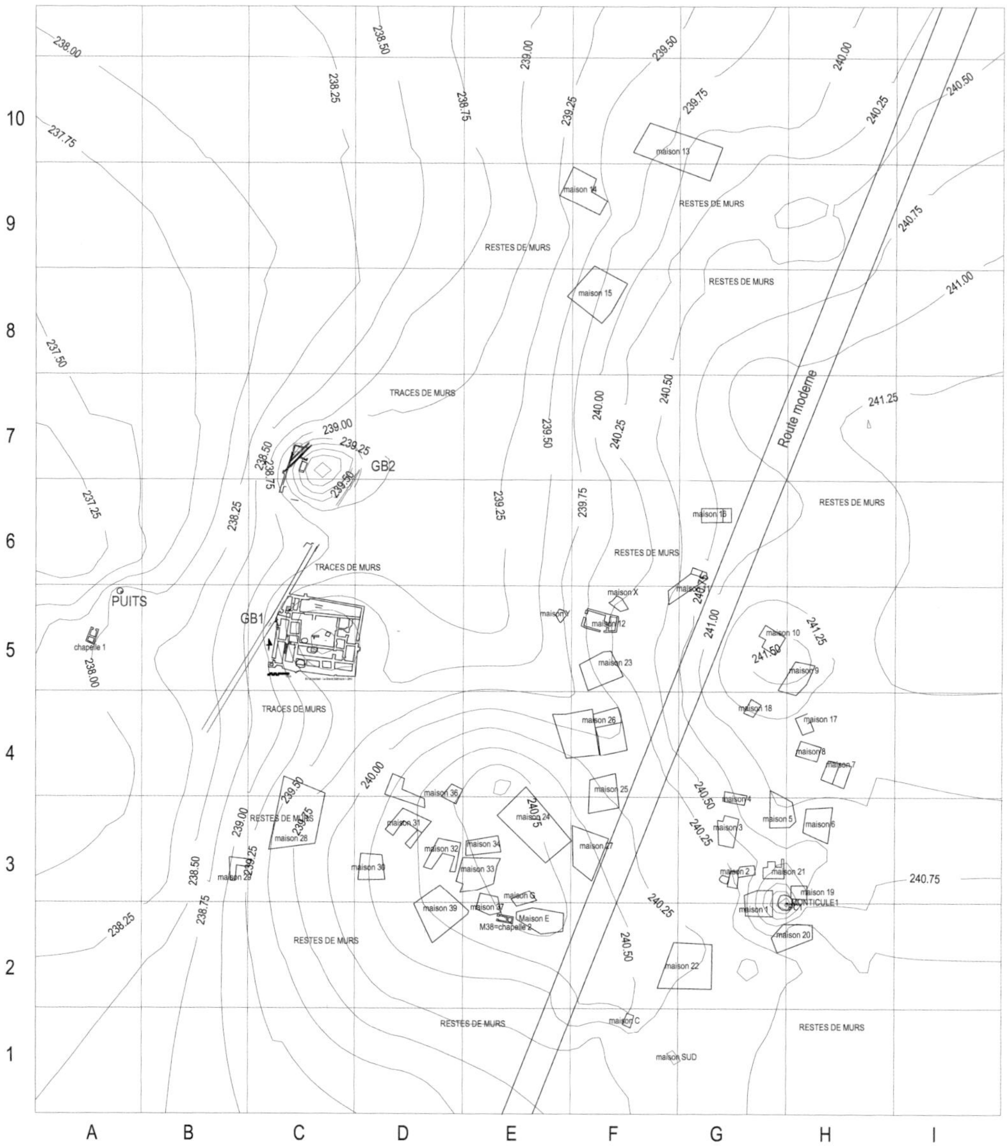

Fig. 6: General plan of Bir el-Haddad site (2010).

A settlement in the steppe: the plan (Figs. 6-10)

Beyond this empty and free space that surrounds the well, at a distance of around 70 meters, on a slight elevation of the ground of around 2 m over the rest of the settlement, we have distinguished the outline of two large mudbrick buildings, oriented North-West/South-East. These structures form a line, beyond which the domestic occupations are distributed eastward. With their stone foundations visible on the rocky surface of the Pleistocene terrace, the houses appear have been built in a quite regulated and structured way, oriented and aligned along some passages or streets, forming definite blocks and neighborhoods.

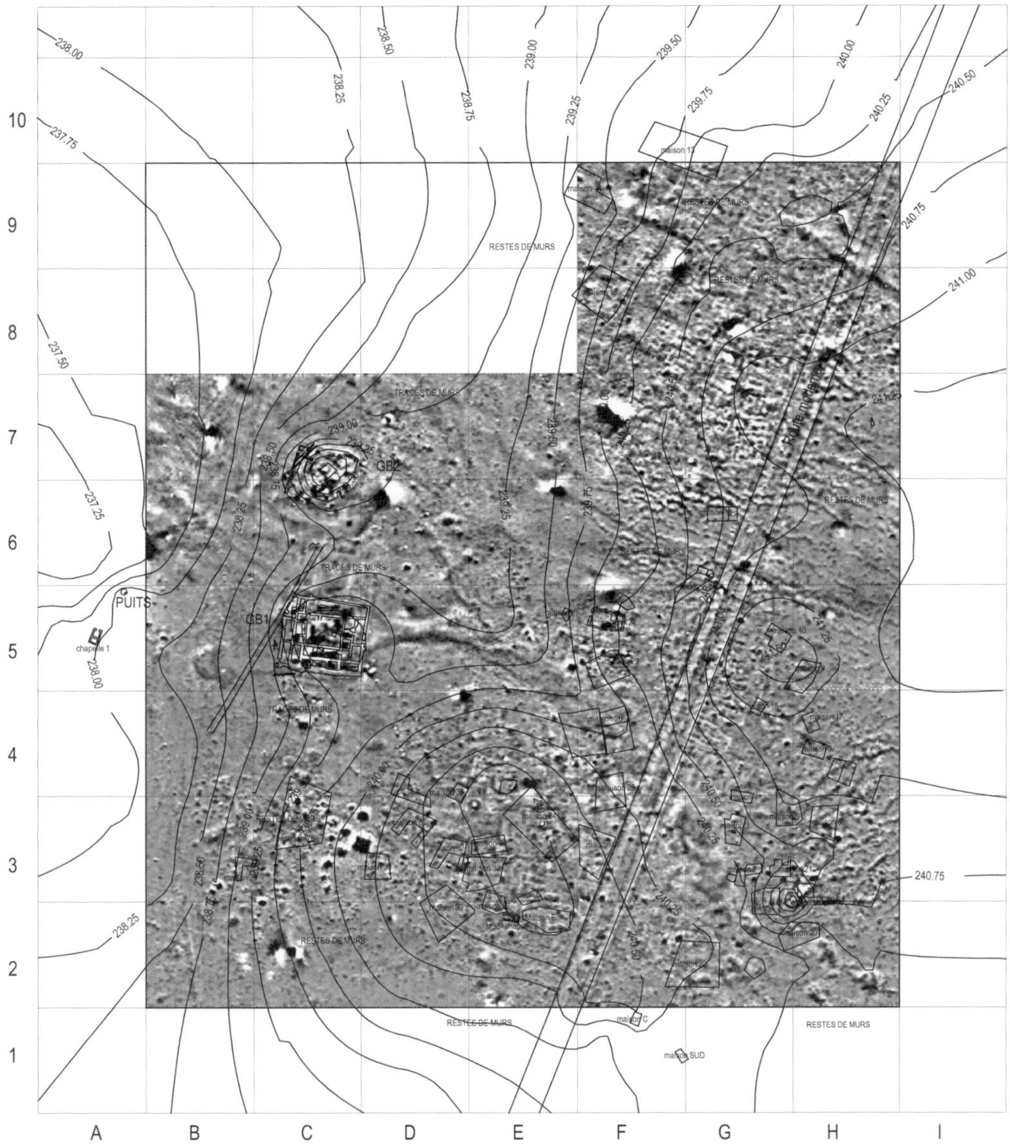

Fig. 7: General plan with the results of the magnetic survey (2010).

The two mudbrick buildings – called respectively to the South "Grand Bâtiment One", and to the North "Grand Bâtiment Two" –, clearly separated from the rest of the settlement, form a kind of citadel. Even if they are badly eroded, and partially destroyed by illegal excavations, we were able to identify the visible limits and the articulations of the two structures, confirmed by the results of a magnetic survey. Grand Bâtiment One has a surface of at least 36 m by 40 m, and Grand Bâtiment Two, 30 m by 30 m. While they are both certainly autonomous, they share, or they are linked by, some structures. The buildings seem to be connected also by a large mudbrick wall, around one hundred meters in length. This wall should be the limit, and the protection, of the well

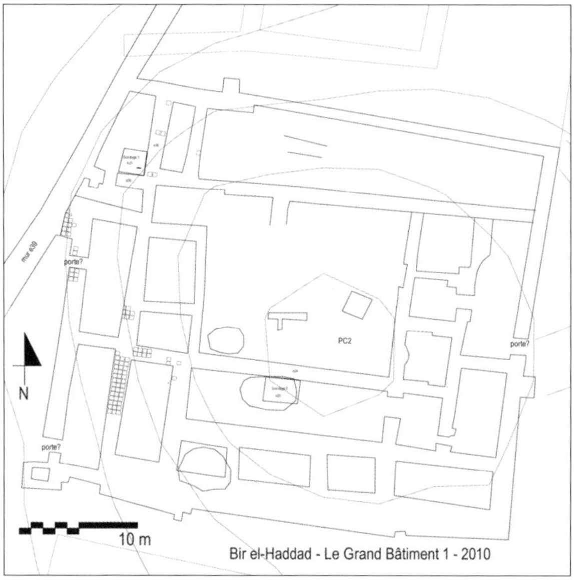

Fig. 8: Sketch plan of "Grand Bâtiment One"

area itself, isolating it from the residential context. The plan of the mudbrick buildings are clearly similar to the Iron II domestic, and also elite houses, found in all of Northern Mesopotamia, as well as in Assyrian settlements, but they also reminded us, some way, of the Middle-Assyrian *dunnu* (Akkermans and Schwartz 2003: 348-351). A double series of spaces, rooms, some of which could have been workshops, is organized around a central open courtyard. Both buildings seem to have two openings, one toward the well, in the West, the other toward the houses, and beyond into the steppe. What could be hollow ways are visible in the image generated by the geo-magnetic survey, arriving to their entrance, and then leading to the well.

Concerning the private houses, we could study only a small portion of them, around fifty, sometimes organized in blocks. With their varying dimensions – we found some houses of more than 600 square meters, and others with around 80 square meters –, these domestic structures show a regular classic and coherent plan, with one to three spaces associated with a central area, possibly open, where a private well, or a cistern, was occasionally situated. A staircase is sometimes visible on a side of the building. Closer to the well, a small temple, with a simple in antis plan, has been maybe identified.

A settlement in the steppe: the stratigraphy

Limited soundings have been carried out in the two mudbrick buildings. Their stratigraphic record is simple, indicating at least two phases of construction. In Grand Bâtiment One, we found two levels of earth or mudbrick floors (Fig. 11), with the most recent often covered by ashes – possibly traces of workshop activities, if not evidence of the final destruction of the building. On the more ancient floor, a fragment of a cuneiform tablet, a Neo-Assyrian administrative document with a list of names, has confirmed for us the chronology of the construction, but also the official, non-domestic function of the building itself. Grand BâtimentTwo has still some white plaster preserved on its walls. It shows the same sequence of levels, but it may have been used for a longer period than its twin. On the more ancient floor, we have found a docket (Fig. 12), with an Aramean inscription, and a sealing (Fig. 13) with classic Neo-Assyrian iconography.

In the residential area, soundings in some of the houses showed that often the smallest of them did not have real foundations, the walls being built on the rocky terrace. The big mansions show a certain skill in their organization, and, moreover, a rather complex stratigraphy (Fig. 14). However, almost no clear archaeological material has been collected in these domestic contexts. So, it is for the moment impossible to date them, and to be sure they belong

Fig. 9: Photo of the remains of "Grand Bâtiment One" after a heavy rain. "Grand Bâtiment Two" is in the back.

Fig. 10: The "plan" of a house, visible on the surface

to the same period as the Grands Bâtiments of the "Neo-Assyrian" level. The fact that the plan of the residential neighborhoods seems to fit well with the structure of the settlement defined by the Assyrian buildings supports the idea of the foundation of this settlement in the context of the same plan, to be a station in the itinerary linking the Euphrates and the Khabur banks. But we cannot exclude the possibility that some of these houses date to a much later period, built maybe during the Roman-Parthian period, or even during the Early Islamic one.

Fig. 11: Sounding in "Grand Bâtiment One"

Fig.12: Docket with Aramean inscription, in situ

Some conclusions

Waiting for other, more complete researches to be conducted in Bir el Haddad, we have no doubt that the Iron II settlement we have identified – be it limited to the two Grands Bâtiments with an official function, or integrating also the residential occupation – is the result of a coherent project, managed by the elites of Kar-Assurnasirpal, and of Dur-Katlimmu, maybe Nergal-ersh himself, which included the creation of the "Nahr Dawrin" canal, in or-

Fig. 13: Neo-Assyrian Sealing

der to assure and to monitor the exchanges between the Hindanu area, the Babylonian and Arabian markets, and Northern Mesopotamia, the Khabur cluster and Assyria itself.

In Bir el-Haddad, the Grands Bâtiments were built to host personnel and services to control not only local society, but also travelers and goods - maybe also to protect them, as in a kind of caravanserails. People living in the associated settlement would eventually exploit the richness produced by this circulation and this market business, which could sustain a large population better than the dry agriculture economy, integrated with pastoralism. But this high living standards, and the complex urbanism and social organization associated with it, were only an ephemeral reality, the balance of which depended on the proper functioning of the entire system, the "Assyrian" system, not only political, but also economic and cultural. Living in the steppe continued to be a rather risky business, for a long time.

Fig. 14: Cleaning and sounding in one of the houses

Bibliography

Akkermans, P.M.M.G. and G.M. Schwartz

2003 *The Archaeology of Syria. From Complex Hunter-Gatherers to Early Urban Societies (ca 16,000-300 BC),* Cambridge.

Altaweel, M.

2008 *The Imperial Landscape of Ashur: Settlement and Land Use in the Assyrian Heartland*, Heidelberger Studien zum Alten Orient 11, Heidelberg.

Bagg, A. M.

2000 Assyrische Wasserbauten: Landwirtschaftliche Wasserbauten im Kernland Assyriens zwischen der 2. Hälfte des 2. und der 1. Jahrtausends v. Chr., *Baghdader Forschungen* 24, Mainz am Rhein.

Bernbeck, R.

1993 *Steppe als Kulturlandschaft. Das 'Agig-Gebiet Ostsyriens vom Neolithikum bis zur islamischen Zeit,* Berliner Beiträge zum Vorderen Orient Ausgrabungen I, Berlin.

Berthier, S.

2001 Le peuplement rural et les aménagements hydroagricoles dans la Moyenne vallée de l'Euphrate entre la fin du VIIe et le XIVe siècle. Etude archéologique. In S. Berthier (ed.), *Peuplement rural et aménagements hydroagricoles dans la Moyenne vallée de l'Euphrate. Fin du VIIe -XIXe siècle*, Damas: 29-264.

Dalley, S.

2001-2 Water Management in Assyria from the Ninth to the Seventh Century BC", *Aram* 13-14 (2001-2): 443-460.

Ergenzinger, P.J. and H. Kühne

1991 Ein regional Bewässerungssystem am Habur. In H. Kühne (ed.), *Die rezente Umwelt von Tall Šēh Hamad und Daten zur Umweltrekonstruktion der assyrischen Stadt Dūr-katlimmu,* Berichte der Ausgrabung Tall Šēh Hamad 1, Berlin: 163-190.

Geyer, B. and J.-Y. Monchambert

2003 *La basse vallée de l'Euphrate syrien du Néolithique à l'avènement de l'Islam,* Beirut.

Grayson, A.K.

1993 Assyrian Officials and Power in the ninth and eighth Centuries, *State Archive of Assyrian Bulletin* VII/1: 19-52.

Kessler, K.

1997 "Royal Roads" and other Questions of the Neo-Assyrian Communication System. In S. Parpola and R.M. Whiting (eds.), *Assyria 1995. Proceedings of the 10th Anniversary Symposium of the Neo-Assyrian Text Corpus Project,* Helsinki September 7-11, 1995", Helsinki : 129-136.

Kühne, H.

1990 Ein Bewässerungssystem des Ersten Jahrtausends v.Chr. am Unteren Hābūr. In B. Geyer (ed.), *Techniques et Pratiques Hydro-Agricoles Traditionelles en Domaine Irrigué,* Paris: 193-215.

1995 The Assyrians in the Middle Euphrates and the Habur. In M. Liverani (ed.), *Neo-Assyrian Geography,* Rome: 69-85.

2010 The Rural Hinterland of Dūr-Katlimmu. In H. Kühne (ed.), *Dūr-Katlimmu 2008 and Beyond,* Studia Chaburiensia vol. 1, Wiesbaden: 115-128.

2011a Dur-Katlimmu und die Steppe. In J. Renger (ed.) *Assur- Gott, Stadt und Land, 5. Internationales Colloquium der Deutschen Orient-Gesellschaft, 18.-21. Febrar 2004 in Berlin*, Wiesbaden: 101-114.

2011b Urbanism in the Assyrian Homeland. In B. Düring, A. Wossink, and P.M.M. Akkermans (eds.), *Correlates of Complexity. Essays in Archaeology and Assyriology Dedicated to Diederik J.W. Meijer, in Honour of his 65th Birthday*, PIHANS 116, Leiden: 143-152.

2012a Water for Assyria. In D. Matthews and J. Curtis (eds), *Proceedings of the 7th International Congress for the Archaeology of the Ancient Near East, (12 April – 16 April 2010 in London,)*, vol. I. Mega –Cities and Mega-Sites. The Archaeology of Consumption and Disposal, Landscape, Transport and Communication, Wiesbaden: 559-572.

2012b Wasser für Dur-Katlimmu – Wasser für Assur. In F. Klimscha, R. Eichmann, Ch. Schuler and H. Fahlbusch (eds.), *Wasserwirtschaftliche Innovationen im archäologischer Kontext von den prähistorischen Anfängen bis zu den Metropolen der Antike*, Mahden/Westf: 203-209.

Masetti-Rouault, M.G.

2001 *Cultures locales du Moyen-Euphrate. Modèles et événements, IIe-Ier mill. av. J.-C.*, Subartu 8,Turnhout.

2008 Living in the valley: State, Irrigation and Colonization in the Middle Euphrates Valley. In H. Kühne, R. M. Czichon and F.J. Krepper (eds.), *Proceedings of the 4th International Congress of the Archaeology of the Ancient Near East, 29 March-3 April 2004*, Freie Universität Berlin, Vol. I, The Reconstruction of the Environment: Natural Resources and Human Interrelations through Time. Art History : Visual Communication, Wiesbaden: 129-141.

2010 Rural Economy and Steppe Management in an Assyrian Colony in the West: a View from Tell Masaikh, Lower Middle Euphrates, Syria. In H. Kühne (ed.) *Dūr-katlimmu 2008 and Beyond,* Studia Chaburensia 1, Wiesbaden: 129-150.

2013 Interpreting the Changes in the Plan of the Assyrian Palace in Tell Masaikh – Kar Assurnasirpal (Lower Middle Euphrates Valley, Syria). In D. Kertai and P.A. Miglus (eds.), *New Research on Late Assyrian, Conference at Heidelberg January 22nd 2011,* Heidelberger Studien zum Alten Orient 15, Heidelberg: *31-40.*

Masetti-Rouault, M.G. and S. Salmon

2010 The Neo-Assyrian Colony of Tell Masaikh, in the Syrian Lower Middle Euphrates Valley : a Report about Last Excavations. In P. Matthiae, F. Pinnock, L. Nigro and N.Marchetti (eds), *Proceedings of the 6th International Congress on the Archaeology of the Ancient Near East, University La Sapienza, Rome, 5th-10th May 2008,* vol. 2: Excavations, Surveys and Restorations in the Ancient Near East, Wiesbaden: 385-396.

Masetti-Rouault, M.G. and O. Rouault

2012 Tutte le strade portano a Roma (ed a Assur): dernières nouvelles de l'empire néo-assyrien dans le Bas Moyen-Euphrate syrien. In G.B. Lanfranchi, D. Morandi Bonaccossi, C. Pappi and S. Ponchia (eds.), *Leggo ! Studies presented to Frederick Mario Fales,* Wiesbaden: 459-475.

Morandi Bonacossi, D.
1996 Landscapes of Power. The political organization of space in the lower Khabur valley in the Neo-Assyrian Period, *State Archives of Assyria Bulletin* 10/2: 15-49.
2000 The Syrian Jazireh in the Late Assyrian Period; a View from the Countryside. In G. Bunnens (ed.), *Essays on Syria in the Iron Age, Ancient Near Eastern Studies Supplement* 7, Leuven: 349-396.

Neumann, J. and S. Parpola
1987 Climate change and the eleventh-tenth-century eclipse of Assyria and Babylonia, *Journal of Near Eastern Studies* 46: 161- 182.

Postgate, J.N.
1995 The Home Provinces. In M. Liverani (ed.), *Neo-Assyrian Geography*, Rome: 1-17.

Radner, K.
2002 *Die neuassyrischen Texte aus Tall Šeh Hamad, Berichte der Ausgrabung Tall Šeh Hamad* 6, Berlin.

Rouault, O.
2009 Assyrians, Aramaeans and Babylonians: The Syrian Lower Middle Euphrates Valley at the End of the Bronze Age, *Syria* 86: 133-139.
In press Rapport préliminaire de la campagne de prospections et sondages à Bir el-Haddad (vallée du Khabour, Syrie), *Akh Purattim* 4.

van Driel, G.
2000 The Mesopotamian North: Land Use, an Attempt. In R.M. Jas (ed.), *Rainfall and Agriculture in Northern Mesopotamia, Proceedings of the third MOS Symposium, (Leiden 1999),* Istanbul: 265-299.

Wilkinson, T. J. and E. Barbanes
2000 Settlement Patterns in the Syrian Jazira During the Iron Age. In G. Bunnens (ed.), *Essays on Syria in the Iron Age, Ancient Near Eastern Studies Supplement* 7, Leuven: 397-422.

DOROTHÉE SACK, MARTIN GUSSONE and DIETMAR KURAPKAT
TU Berlin, FG Historische Bauforschung, German Archaeological Institute Berlin

A Vivid City in the 'Syrian Desert'

The case of Resafa-Sergiupolis/Rusafat Hisham

Its origins as a military post alone make the city of Resafa a good example to show how human interventions can render possible spatial development even in a hostile environment with unfavorable climatic conditions. Despite its location in the desert-like landscape of this stretch of the Roman eastern limes, Resafa grew to a vivid pilgrim's and merchant's centre, existing for nearly one millennium.[1]

Resafa is situated in the steppe approximately 25 km south of the Euphrates at the edge of the 'Syrian Desert'.[2] By virtue of its natural conditions this site by no means would have been a good choice for the establishment of a city. The average annual precipitation is about 150 mm with a minimum of 100 mm (Wolfart 1966: 28; Wirth 1971: 88-93; current statistic: Berking *et al.* 2010: 819-820), and there is no freshwater available, and wells only provide brackish groundwater due to gypsum soil conditions (Wolfart 1966: 35; Wirth 1971: 150, 442). The summer months are hot and dry without any rain; all rainfall occurs in the cold and wet winter months, whose runoff had to be collected and stored to provide drinking water (Brinker 1991: 135-136; Garbrecht 1991: 239-241; Beckers *et al.* 2012). Even if the amount of rainfall was not the only unfavorable condition in North Syria (Widell 2007: 47-70), the ability to safeguard the provision with drinking water was of crucial importance for the success of a settlement.

According to the hagiographic account here at Resafa, the high-ranking Roman officer Sergios, companion of Bacchus, suffered martyrdom, probably about 312 AD (Fowden 1999: 8-17). The execution is said to have been carried out in front of the Roman fortress called 'Rosapha/R(h)esapha' belonging to the north-eastern part of a line of fortifications on the Roman eastern frontier (*Limes Arabicus*) (Konrad, M. 1992: 313-383, 2001).

The fortified city and its settled surroundings were well integrated into the natural environment, its position at the edge of Wadi as-Saila, marking a geological fissure, was by no means accidental, but had been carefully chosen to take advantage of specific features of this site for water harvesting of the seasonal rainwater runoff (Beckers *et al.* 2012). These qualities of the site had been enhanced by hydraulic engineering, which was permanently maintained and expanded by its inhabitants. A recent satellite image combined with a super elevated digital terrain model (DTM) displays clearly the layout of the late antique city walls, and the general layout of the structures in its surroundings, particularly the southern *extra muros* area, which was mainly settled during the Umayyad period (Sack *et al.* 2004: 207-213) (Fig. 1).

1 The spelling of 'Resafa' follows the usage, established by the Resafa-Project (DAI), Mackensen 1984: 1, fn. 1. This general overview of Resafa's development had been supplemented at the conference by a 2nd paper going more into details, especially in respect of the water-collection and its management, which had been topic of an interdisciplinary joint PhD thesis-project between archaeology (Ch. Konrad) and physical geography (B. Beckers) supported by excellent-cluster TOPOI, s. Schütt and Meyer 2011: 12-16.

2 'Syrian Desert' used as geographical, not as climatic term, s. Grant 1937: 6-16; Lewis 1987: 3-24

Studia Chaburensia 4 (2014), pp. 257–274.

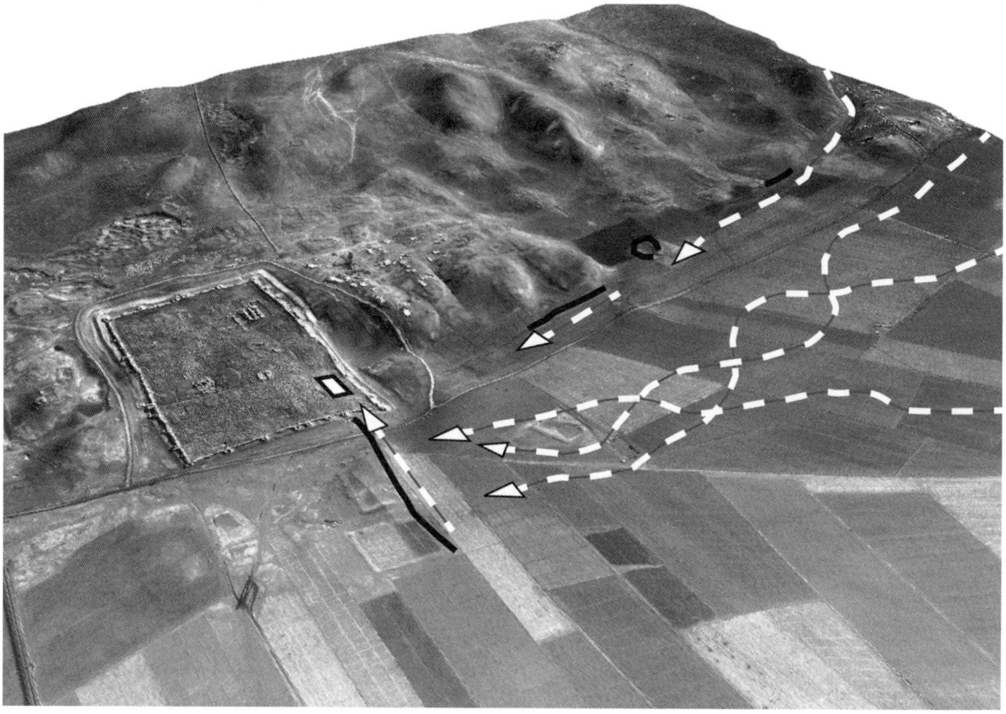

Fig. 1: Resafa, digital terrain model (superelevated) combined with satellite image (IKONOS 2009) shows the rainfall-runoff (dashed arrows). The water is directed to the Great Cisterns (white rectangle with black contour) by ramparts (black lines: archaeological testified structures), Beckers 2009.

On behalf of the German Archaeological Institute since 2006 a new project has been established, entitled "Resafa-Sergiupolis/Rusafat Hisham. City of Pilgrimage and Caliphal Residence" (Sack *et al.* 2010: 102-129). The final aim of the project is to generate an overview of the settlements complete development, thereby also summarizing the work of the past 60 years in Resafa, supplemented by additional studies (Kollwitz 1959: 45-70; Ulbert 2008: 19-30).

The research for this overview on the history of the settlement is still ongoing, nevertheless it already produced some new evidence for interaction processes between the natural environment and the human presence at Resafa. In the following the current state of work is presented in example through reconstructions of selected moments in the city's lifetime.[3]

Resafa, early settlement and Roman period

The origins of settlement at this place are still uncertain. Widely agreed upon is the assumption of Resafas position at a junction of caravan tracks leading from the Mediterranean coast (Tyre/Tyrus) and the Red Sea (Aqaba) via Damascus in Syria to Şanlıurfa (Edessa) in North Mesopotamia with a branch coming from Homs (Emesa) (Grant 1937: 33-44). The other sites

3 The final results of this research will be published in the Resafa-series of the German Archaeological Institute (DAI), therefore the explanation for the plans are given here in an unavoidable reduced extent

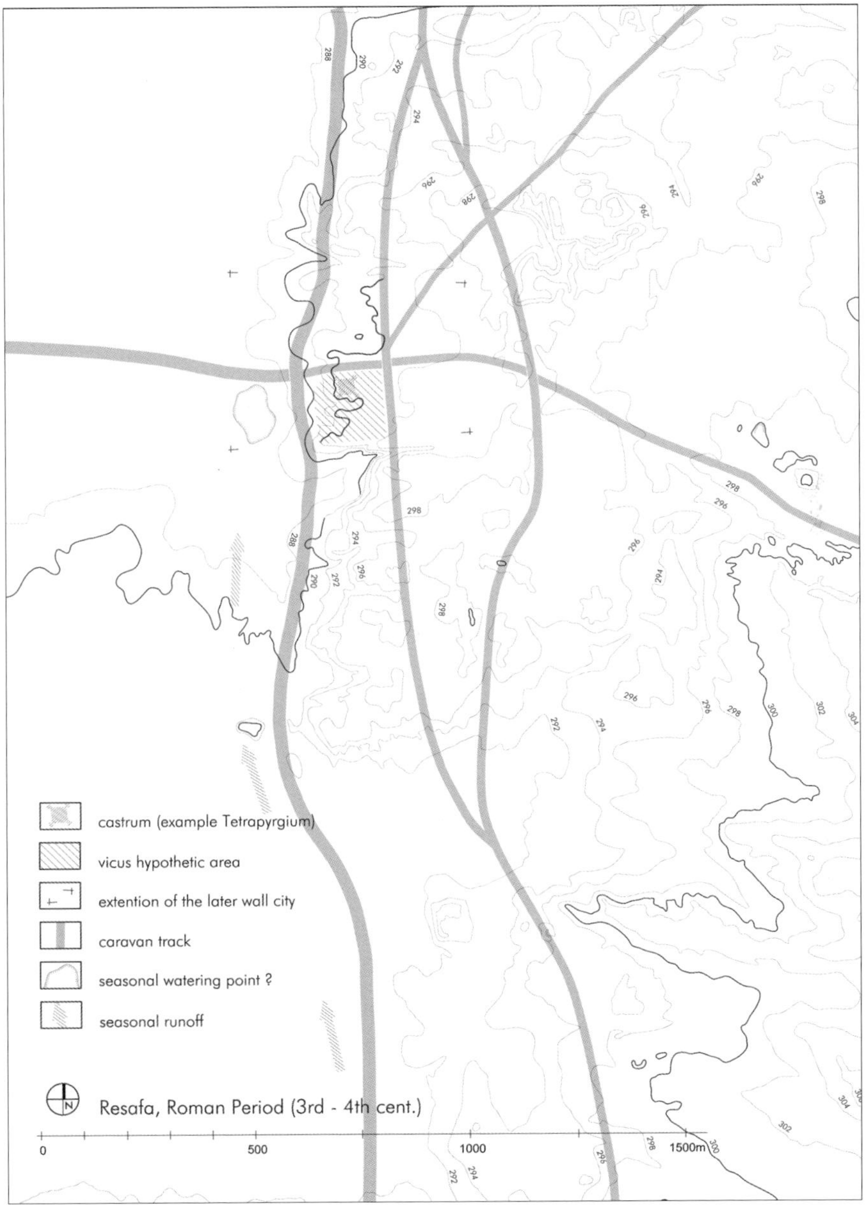

Fig. 2: Resafa, Archaeological Map, Roman Period (3rd-4th cent.), 2013.

of Roman forts along the border are provided with fresh water by springs, which do not exist at Resafa. With the exception of wells with brackish water there is no permanent water supply, but we have to expect the initial existence of a watering point, fed by seasonal runoff from the main *wadi* and its tributaries. This site is thought to have been a gathering place for caravans and nomads with their flock alike since ancient times, and later appeared to be an acceptable location for a fortified outpost (Bauzou 1989, according to Fowden 1999: 67-77) (Fig. 2).

Traces of settlement date back to the first century AD, but the Roman *castrum* and its *vicus* are buried by later occupation layers. The archaeological strata sum up to a total thickness of approx. 4.5 m, making research on this early occupation phases difficult (Konrad, M. 1992: 313-402, 2001: 4, 14-15).

However, the examination of the neighboring fortresses of *Tetrapyrgium*, also known as Qasr as-Saila in the North and al-Khallul/Cholle to the South generated insights of comparable Roman strongholds (Konrad, M. 2001), which allow for an assumption of the size of the former Roman site at Resafa. Fresh water was probably supplied by small cisterns within the castrum and larger cisterns in the main *wadi* (Garbrecht 1991: 240).

The place where Sergios was executed and buried outside the roman fortress of Resafa, soon became the site of constantly increasing pilgrimage, induced by reported miracles (Fowden 1999: 17-29). The exact place of execution is an issue of scholarly debate since the first studies on Resafa. The only written source, which mentions explicitly this incident, the *passio,* tells us, that it took place outside of the *castrum* (Fowden 1999: 10-11). But in which direction and how far outside remain unclear. Now recent excavations outside of the south-eastern extension of the Tetraconch Church ('Zentralbau') brought to light remains of a building structure and ceramic sherds, which prove, that this area (surely situated outside the *castrum* at the time of Sergios' martyrdom) was the place of activities a long time before the Tetraconch Church was built in the first half of the 6th century (Schuhmann 2011: 40). We know nothing of which kind these activities were and whether or not they stood in relation to the cult of S. Sergios. But the fact, that some time later this spot was chosen for the in many regards unusual Tetraconch Church could be explained by the assumption, that it should commemorate the place of Sergios' martyrdom and the believer's original aim of pilgrimage[4] (Brands 2002: 175).

Resafa in Late Antiquity

'Long time' later Christians from the nearby settlement of Sura/al-Suriyya had tried to get hold of the relics, but were held up by the watchful garrison; as compensation or by repentance the thieves built a first modest memorial at the place of execution of Sergios (Fowden 1999: 10-11). The relics of S. Sergios were transferred to a church, built within the *castrum,* and kept there from this time on, in order to safeguard the mortal remains of the saint. This church was presumably the shrine built by bishop Alexander of Hierapolis shortly before 431 AD, as we know also from the *passio* with supplementary details in the records of the Council at Ephesus (Fowden 1999: 11-12, 21, 26-27; archaeological record s. Konrad, M. 1992: 345-350).

Probably at the same time the *vicus* adjoining the *castrum* developed and was surrounded by a wall due to the increasing number of pilgrims and prosperity. This first defense is said to have been constructed from mud bricks (Karnapp 1976: 4; Brands 2002: 212-235). The extension of this wall is unknown and until now we have no archaeological evidence of this early fortification. Presumably its construction was comparable to the first building phase made of mud bricks at Sura, where a second, more solid construction of gypsum had not substituted the earlier structure, but was erected adjacent to the old fortified *vicus* (Konrad, M. 2001: 5-15).

4 Assumption of the area of the al-Mundhir-Building as site of Sergios' martyrdom s. Konrad, M. in: Konrad, M. and Ulbert in press.

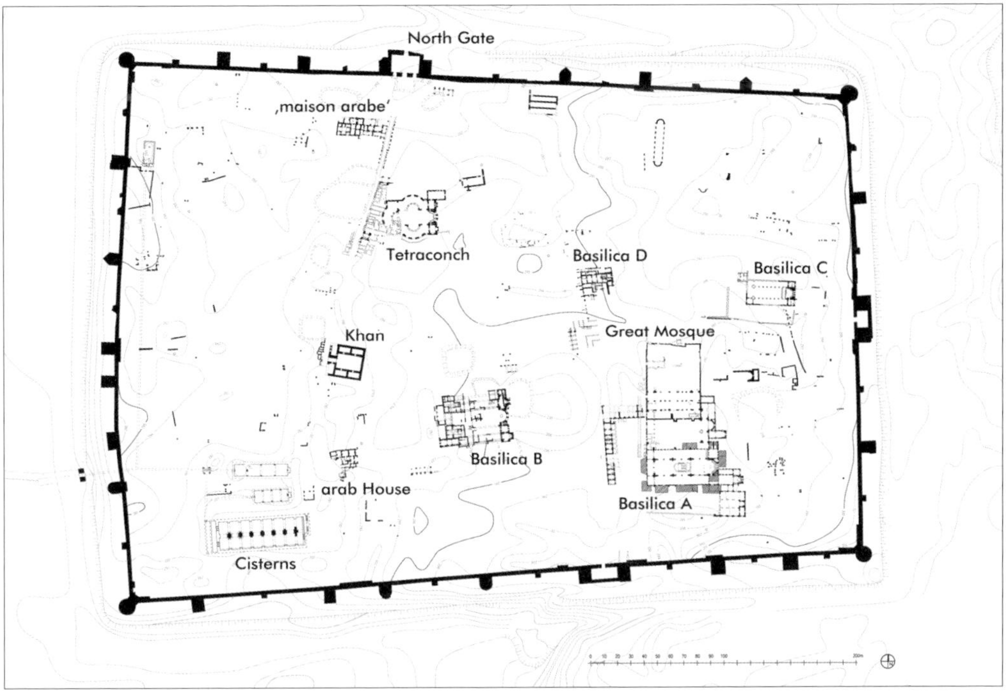

Fig. 3: Resafa, intra muros - new city plan, M. Gussone and G. Hell with N. Erbe and I. Salman 2010.

The final large pilgrim shrine, Basilica A, its courtyard to the North and a predecessor building of the neighboring so-called Four-Pillar-Building, perhaps a *baptisterium*, now can be assumed to have been built towards the end of the 5th century (Brands 2002: 9-77). At the same time, the construction of the new city wall made from gypsum-stone, surrounding a much larger area, begun (Current research on the city wall: Hof 2009, 2010, in press; Sack *et al.* 2010: 129-132). The building activities on the new enclosure started at two different points. The first parts of the new city wall consisted on the one hand of a stretch in the North-East, which had been erected as a sample for the planned construction and therefore had been built with particular care; on the other hand of a small part of the western wall, covering the water inflow to the cisterns inside the city (Hof 2010: 233-248, in press). The original, ambitious buildings scheme is characterized by a round tower at the corner followed by alternating small and larger towers with different shapes linked by curtain walls with two guard's wall walks above one another. The stretch of the city wall on the west side, displays clearly the importance of the cross point, where the water entered the city. In alignment with this water gate lies one of the large cisterns, supposedly conceived and built at the same time with the city wall, thus indicating the existence of a general layout and city planning scheme to coordinate and safeguard these elementary construction units of the city's infrastructure (Brinker 1991: 125-131; Westphalen and Knötzele 2004: 345-348) (Fig. 3).

Epigraphical finds in Resafa are very limited, so it was of great importance, when a lintel with an inscription regarding a ‚new church' was found, during the excavation of the Great Mosque (Sack 1996: 4), providing – for the first time – an exact date for the construction of a monument in the city. Originally it was installed over the main entry of the Basilica B and

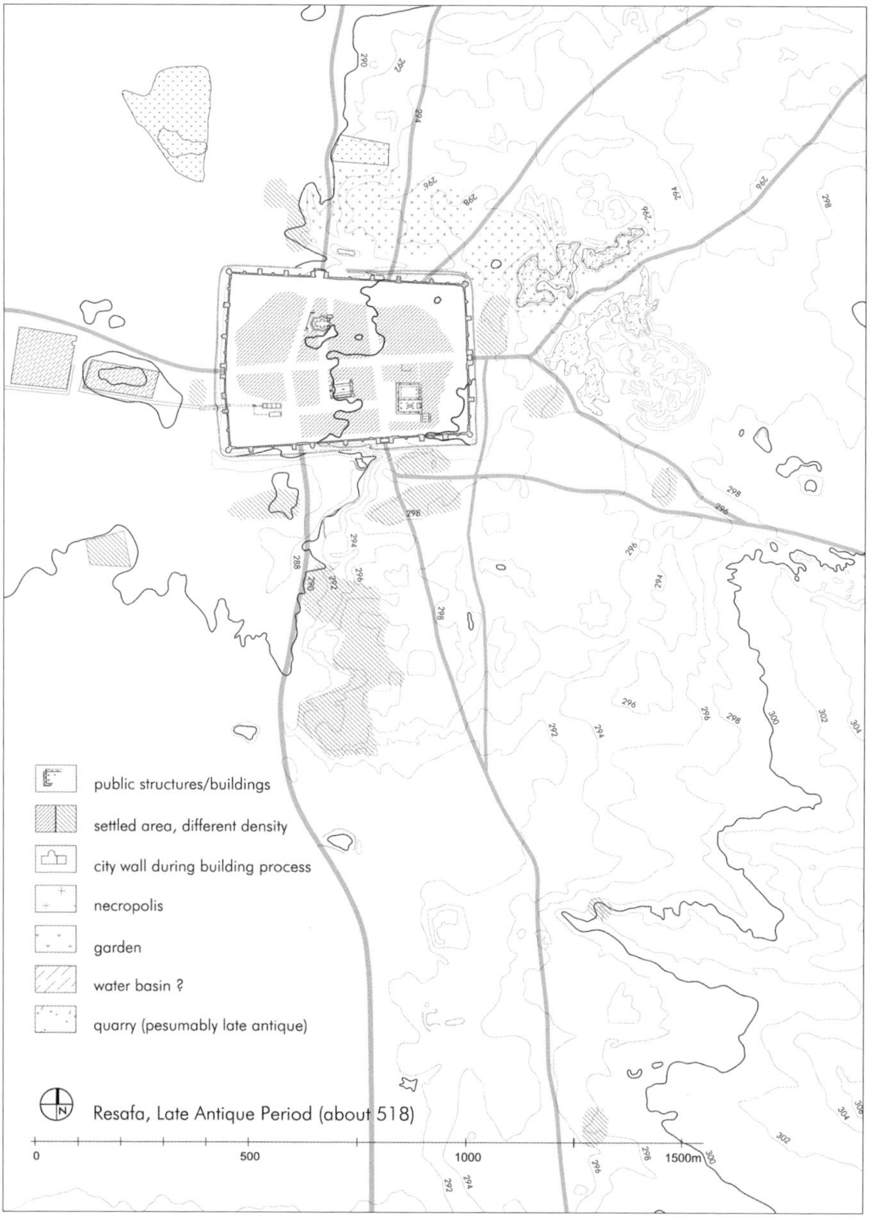

Fig. 4: Resafa, Archaeological Map, Late Antique Period (about 518), 2013.

indicates the spring of 518 AD as date for the beginning of its erection (Gatier and Ulbert 1991: 169-182). This inscription not only reveals the date of the construction of Basilica B, but also refers to the complexity of the building activities which took place at this special location (Brands 2002: 114-117). According to the inscription the ‚new church‘ (Basilica B) succeeded an older one on the same spot, which, as excavations have shown was constructed of mud brick (archaeological evidence see Konrad, M. 1992: 313-383). The elder church accommodated the relics of S. Sergios until another 'venerable church' (identified with Basilica A) had been built to house the holy shrine from then on. Based on this argument Basilica A had been identified as the oldest existing monument still standing.[5]

In this period (around 518 AD), also the Tetraconch Church had been built (Brands 2002: 176-179) and a second large cistern was constructed (Brinker 1991: 130-132). At the very end of this construction phase the city wall was completed. This construction phase included some sections of the wall left open as temporary entries to the city as well as the North Gate, the main entry to the city and the most representative one, even though it was left unfinished in its decoration (Karnapp 1976: 37-44. 51-52) (Fig. 4).[6]

So in this period, dating about 518 AD, the first completion of the centre of pilgrimage took place, but also traces of settlement in the surroundings of the city have been found: building activities near to the gates, especially to the South and East of the city, as well as small settled areas in the southern surroundings are to be noticed.

As motivation for undertakings outside the enclosure we have to take in account the scenic features providing a beautiful view from the escarpment towards the *wadi*; the distribution of find sites with late antique sherds in the southern surroundings indicate also an influence of the track of the military road for the location of settlements (Mackensen 1984: 1: fig. 2, 72-84). Further on a first necropolis had developed North and North-East to the city (Konrad, M. in: Konrad, M. and Ulbert in press).

The second phase of completing the centre of pilgrimage ranges in the time mid of the 6th cent., when the Basilica C (Ulbert in: Konrad, M. and Ulbert in press; Sack *et al.* 2013: 54-55) and Basilica D (Westphalen 2000: 325-365) were constructed, what seems to have been necessary to serve the still increasing number of inhabitants, pilgrims and travelers. The unabated growth demanded also for the enlargement of the cisterns, so the double cistern – a real 'cathedral' of water-collection – had been built as well as the cistern in the North-West (Brinker 1991: 126, 132-133).[7]Another sign for this development is the erection of a *khan,* situated next to the main east-west axis (Karnapp 1978: 136-150). Furthermore the Basilica A, the centre of the pilgrim's cult with the relics of S. Sergios, had been enlarged by a double-storey-building, which possibly hosted the bishop (Ulbert 1986: 125; Brands 2002: 67). Yet the locations of the main accommodations for the pilgrims remain unclear (Severin 1997: 329-339).

Finally the city wall was reinforced by domes and vaults made from stone, replacing the wooden ceilings in the towers. This measure was thought in regard of fire as a weapon, but had also negative side effects, due to the weight and resulting thrusts of the supplemented constructions (Hof in press). In Resafa *extra muros*, the al-Mundhir-Building, interpreted as reception hall as well as a church, had been built by the Ghassanids, erected on the site of an already existing necropolis (Konrad, M. and Brands in separate contributions with divergent conclusions in: Konrad, M. and Ulbert in press; see Sack *et al.* 2013: 75-80; cf. Shahid 2009 arguing for a secular use of the structure; interpretation as church s. Brands 1998: 211-235; usage as church as well as reception hall s. Fowden 2000: 303-324).

5 The up to then stated later time of origin of Basilica A (also known as Basilica of the Holy Cross, dated to the second half of 6th cent., s. Ulbert 1986: 147-154) proved indefensible, extensive discussion s. Brands 2002: 48-56. Recent revision of the building s. Kurapkat 2011: 36.

6 Discussion of chronology s. Brands 2002: 195-201; recent research and new insights on the dating s. Hof 2010: 235-247, in press.

7 Outdated chronology s. Brinker 1991: 144-146; sequence of cisterns presented here according to the more convincing argument of Westphalen and Knötzele 2004: 345-348.

Until now the period of the late 6th/7th century has not been in the focus of research, which is due to the relatively sparse information available. There is only limited archaeological evidence (e. g. Mackensen 1984: 62-63) and the written sources are silent regarding building activities in this period. We estimate this period as mainly determined by stagnation, economic decline and decay caused by pandemics,[8] earthquakes and military conflicts with Persian forces. Traces of these events can be identified everywhere in the city, corresponding to the general situation in Greater Syria (Foss 1997: 189-269; Wamsley 2007: 34-47; recently: Haldon 2010: 1-11). Evidently Basilica B was destroyed (Sack 1996: 25-26, 39; Brands 2002: 117), as well as the Tetraconch Church, where a provisional church had been installed in the main aisle (Kollwitz 1963: 330-331; Brands 2002: 122, 127-128). But also Basilica A suffered from these earthquakes and its wide arcades – the characteristic feature of this still high rising and marvelous monument – had to be supported by intermediate arcades, reusing columns from the demolished Basilica B (Ulbert 1986: 19-22,118-127).

Resafa in Early Islamic period

So the city was partly in ruins, when the Umayyad caliph Hisham b. Abd al-Malik (724-743 AD/105-124 H), presumably also an admirer of S. Sergios, decided to reside in Resafa and built his residence in the surroundings of the walled city (Sack 1996: 1-6, 155-156; Kellner-Heinkele 1996: 133-154; Haase 1995). The case of Resafa is of special importance for Early Islamic archaeology, as it is one rare example of a caliphal residence, which has not been covered by later remains (Fig. 5).

A far larger extent of the Umayyad settlement to areas North and South of the walled city as estimated before has been traced by a recent survey focusing on a detailed differentiation of the Islamic finds (Gussone and Müller-Wiener 2012a: 569-584, 2012b: 27, 2011: 27). Nevertheless the centre of Hisham's residence is thought to be situated mainly to the south of the city (Sack *et al.* 2004: 207-213). Substantial progress regarding the documentation of building remains was accomplished by staking out surface features and subsequent measurements, revealing a variety of ground plans unknown as yet (Sack *et al.* 2004: 221; Siegel 2010: 37). At several find sites a differentiation between presumably Umayyad phases and later occupation was achieved (Gussone 2010: 36; Gussone *et al.* 2009: 30), by the combined interpretation of prospection results.[9]The results of archaeological soundings at 15 find sites *extra muros* are providing more detailed insights regarding the architecture and its construction, as well as the usage and chronology of the excavated buildings (Sack *et al.* 2013: 97-104 with further references). The analysis is still going on.

According to the current state of research, the main buildings of the residence consisted of an palace near to the south-east corner of the walled city (FP 106), surrounded by buildings for members of the court, and a palace with a large irrigated garden-area, situated in the South (FP 220).[10]

8 The plague reached Syria in 542 and the pandemic returned continuously during the following century, afflicting also the most remote places. Its impact was debilitating for the whole Middle East. Horden 2005: 153-156; Meier 2005: 91-96.

9 Interpretation initially based on geophysical and geodetic methods (Sack *et al.* 2004) was later improved by the integration of the above mentioned measurement of surface features.

10 M. Mackensen identified (1977) 310 find sites (*Fundplätze*, FP), which are the reference for all works there, cf. Sack *et al.* 2004: 207-209; Sack 2010: 34; Konrad, Ch. 2011, 31.

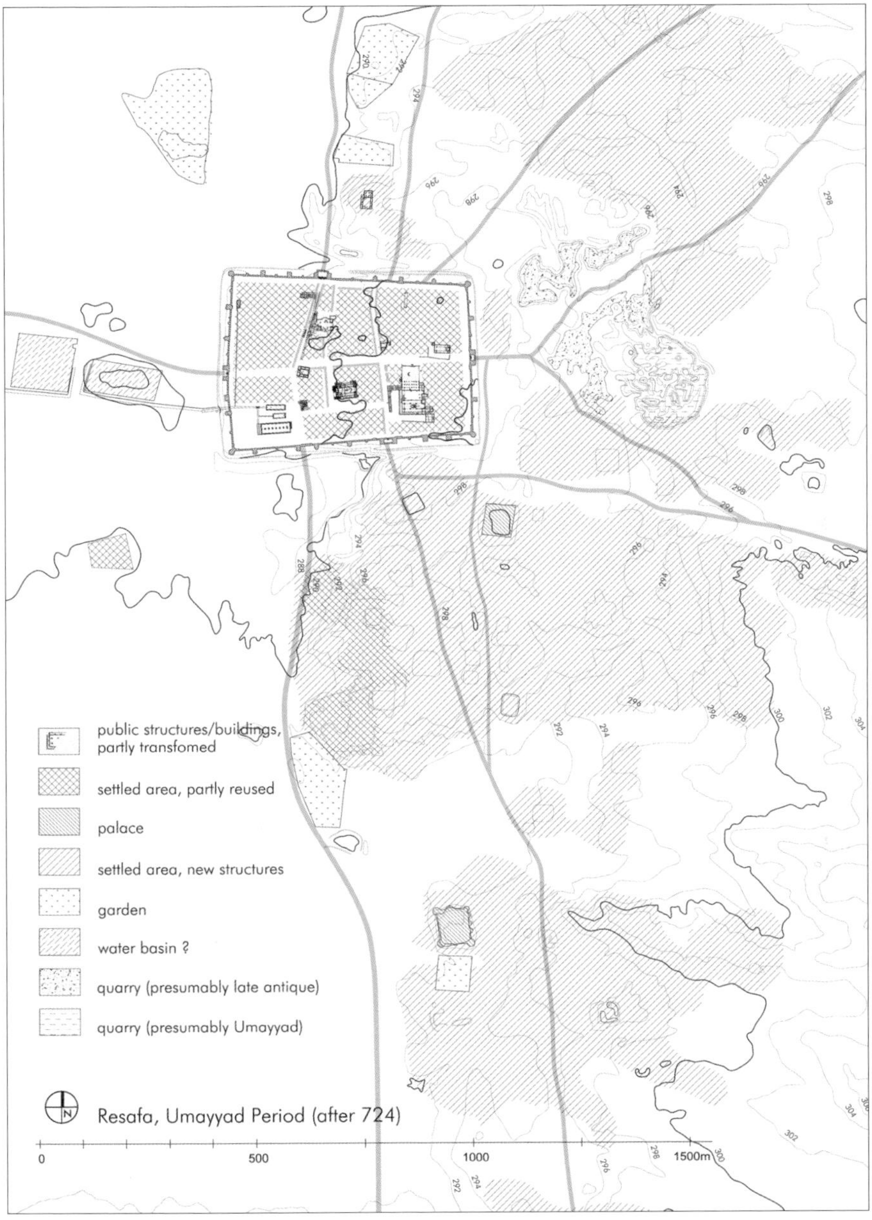

Fig. 5: Resafa, Archaeological Map, Umayyad Period (after 724), 2013.

Even though Hisham himself settled *extra muros* of the city, the reanimation of the still existing town surrounding the pilgrim's centre, which formally was used by the Christian community and later, was settled also by Muslim inhabitants, took place under his rule. He built the Great Mosque, north to Basilica A, close to the relics shrine of S. Sergios, integrating in this new building already destroyed parts of the North court of Basilica A. This building project is equally to be seen as a manifestation of his rule as well as an expression of his devotion to the saint (Sack 1996: 155-156). In this conjunction we do not see the superiority of one of these aspects, but a correlation of political and religious spheres, which fostered Hisham's position in relation to the still influential Christian tribes, who worshipped Sergios (Fowden 1999: 174-183). Hisham's improvement of Resafas urban infrastructure served presumably likewise 'mixed' purposes as maintenance in functional respect as well as a prestigious, symbolical intervention. He reconstructed the water distribution system as we know from the sources, by restoring the cisterns, presumably mainly the large cistern, which displays multiple phases of repair. Hisham's investments in the system of water harvesting and irrigation were even more extensive in the surroundings of the walled city as recent research has revealed.[11]Ramparts to direct the flow of water were maintained or newly constructed, probably together with water basins for temporary storage; irrigated garden areas were established or revitalized (Beckers and Konrad, Ch. 2010: 38; Berking *et al.* 2010: 815-836; Beckers *et al.* 2012: 149-153) (Fig. 1).

All this measures were both, of practical use as of highly symbolic meaning, representing the caliph's rule over cultivated pasture and his command about the capabilities to establish these luxurious installations in a semi-arid environment (Müller-Wiener 2012: 44-57; for Islamic gardens in general cf. Leisten 1993: 47-67).

The activities to reinforce and reconstruct the destroyed parts of the city wall appear similarly – in consideration of the unfortified settlement and palace buildings south of the city – first and foremost as an act of representation, not as a mere defensive measure. To evaluate the urbanistic situation of Resafa in the Umayyad period we have to take in account the increasing commerce[12] and re-flourishing life in the city during Hisham's reign: Bazaar-areas have been added in the street leading towards the North Gate (Kollwitz 1959: 50-52) and at the fringe of the Western Court of the Basilica A (Ulbert 1992: 403-416) and on the west side of the Great Mosque (Westphalen 2000: 326-336).

These projects went together with the reanimation of urban life, as to be seen at the streets, which were reduced in width by installing shops, a development comparable with those noticed in Palmyra, Latakia and Damascus (Wirth 2002: 29-48; cf. Raymond 1994: 3-18; Walmsley 2007: 76-90). Building activities have been recognized in the ruins of several churches previously destroyed by earthquakes (Basilica B) (Kollwitz 1959: 48; cf. Kurapkat 2009: 27), Basilica D (Westphalen 2000: 349-358) and the Tetraconch Church (Kollwitz 1963: 330; cf. Salman and Schuhmann 2009: 37). They were now used as habitation areas. Considering alone the expansion and densification of settled areas (Gussone and Müller-Wiener 2012a: 573-574) – in addition to our knowledge of Resafas political importance as residence of caliph Hisham – we have to estimate the Umayyad period as a phase of economic recuperation and urban growth for the city.

This situation changed again after the death of Hisham in 743 AD/125 H, who was buried here at his favorite haunt. Finds concerning settlements in the surroundings of Resafa in the early Abbasid period are rare. Presumably the abandonment was caused by decline at the end of Umayyad rule. As few diagnostic finds indicate, after the turn to the 9th century, only small parts of the territory were settled further on, whilst large areas had been abandoned (Gussone and Müller-Wiener 2012a: 569-584, 2012b: 27).

11 Beckers and Ch. Konrad 2010: 38; Berking *et al.* 2010: 815-836; Beckers *et al.* 2012: 149-153.

12 Current research: 'regionalism and interaction: ceramic tradition as indicator of socio-political and economic transformation in Early Islamic Syria and Iraq' (by M. Müller-Wiener).

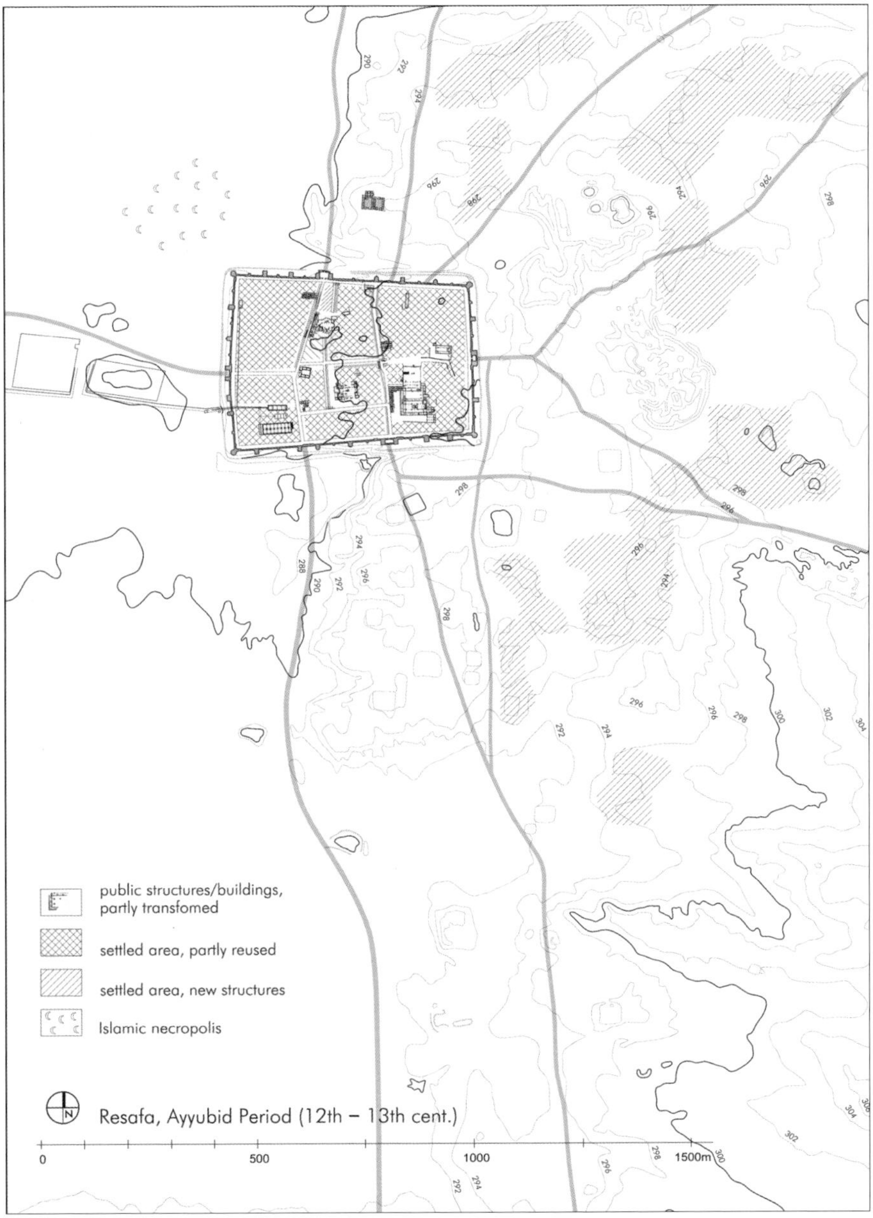

Fig. 6: Resafa, Archaeological Map, Ayyubid Period (12th-13th cent.), 2013.

Despite this peripheral abatement, inside the walled city new consolidation work was accomplished in the early Abbasid period as destructions by earthquakes went on. Basilica A as centre of pilgrimage was restored, its main apse re-plastered. Its style resembles the decorations of certain palaces in Samarra, thus we may assume that the work dates to the second third of the 9th century (Meinecke 1991: 235; cf. Ulbert 1986: 131, dating the stucco ca. 880 AD). Destructions, probably as a result of political conflicts, are to be noticed at the Great

Mosque, which was presumably reconstructed at the beginning of the 10th century with a flat roof (Sack 1996: 22, 44-45, 52, 158).

Nevertheless, Resafa continued to be a Christian pilgrimage city as the shrine of S. Sergios still attracted visitors (Ulbert 1986: 150-154; Ulbert 1990: 1-6; Fowden 1999: 183-184). According to the sources, urban life continued in Resafa in the Abbasid period (Kellner-Heinkele 1996: 133-154). During the 9th-11th centuries we expect at least *intra muros* an urban settlement of a certain size. There was a Christian community in Resafa, as well as tribes in its wider vicinity, which still held sympathies for the Umayyads (Qaddūmī 1996: 123-125). Supposedly activities were commenced on a rather low level, although some visitors described Resafa as a flourishing city (Kellner-Heinkele 1996: 133-135, 143; Sack 1996: 52).

Resafa in the Ayyubid period and later

By contrast in Ayyubid times a distinctive recovery of urban life can be traced. The excavators in the 1950s did not find an undisturbed late antique city, the archaeological findings indicated instead the whole area *intra muros* beeing covered by multiple building structures of 'medieval' origin (Kollwitz 1959: 48; ceramic finds s. Legner 1964; Saliby 1990). This impression was approved by the archaeological excavations since 1976, which all unearthed rich findings testifying a vivid urban community in the 12th and 13th centuries (e. g. Logar 1992, 1996; Knötzele 2006; Korn 2004; cf. Ulbert 2008 summarising). Also in the surroundings of the walled city a notable revival of settlement activities in this period has been detected by recent surveys. Concentrated in the East and North-East of the city, the large amount of diagnostic small finds dating to the 12th and 13th centuries and an agglomeration of building structures demonstrate a strong increase of habitation in these areas in the Ayyubid period (Gussone and Müller-Wiener 2012a: 41-42; 2012b: 27) (Fig. 6).

The same impression is given by the written sources, which depict a town with difficult environmental conditions, but nevertheless beautiful and with impressive buildings. Furthermore, the Arabic geographers report about markets with vigorous trade, far reaching commercial connections and prosperous merchants (Kellner-Heinkele 1996: 144-146). Even so, regarding the living conditions in Resafa the statements of the chronicles are more critical. Without having running water or perennial springs at disposal, but only salty water in deep wells, the inhabitants of Resafa had to meet their need of water from cisterns. They ran dry in poor years, so the inhabitants - if wealthy enough - had to send servants to the Euphrates to fetch drinking water (Kellner-Heinkele 1996: 146-150). Obviously the maintenance of Resafa's water supply had been crucial for survival in these unfavorable environmental conditions in every period. This is attested by repair measures at the water distributor and the inflow to the great cisterns dating to the middle of the 13th century (Logar 1991: 147-149, 161), as well as by graffiti in the greatest cistern, which had been explained as dedicatory inscriptions of repairs (Westphalen and Knötzele 2004: 347).

This period ended with the abandoning of the city after the setback of the Mongol invasion in Syria in the last third of 13th century. Some buildings may have been deliberately destroyed, as in Basilica A layers of ashes were found, but the city is said to have been spared by the invaders (Ulbert 1986: 151-152; Sack 1996: 52; Kellner-Heinkele 1996: 151, 153). The inhabitants were forced to move to Hama or Salamiya when the Euphrates became the frontier between the Mamluk and the Mongol empires due to economic reasons (Kellner-Heinkele 1996: 153; Ilisch 1996: 132; Sack 1996: 160). Afterwards the city, its infrastructure and its buildings were not maintained any longer and fell into decay. It is likely, that the adjoining caravan tracks had been used further on. Moreover, there are traces of secondary, small settlements in the surroundings as well as inside the city, but without datable remains (Gussone and Müller-Wiener 2011: 27). Information about Resafa in the following centuries is sparse, with the exception of reports about a visit of English merchants from Aleppo at the end of the 17th century (Chapot 1903: 280-282).

The surroundings of Resafa are partly settled since the 1960s and a small settlement to the South of the city wall evolved since then. Modern building activities and intensive development of agricultural production since the establishment of the Tabqa Euphrates dam project (1968–75) posed a severe threat to the cultural heritage and its archaeological traces (Karnapp 1976: 4; Mackensen 1984: VIII). To protect the heritage of Resafa and its surroundings an archaeological zone has been designated in 1994 by the Syrian ministry of cultural affairs (Ulbert 2008: 20).

The study of the modern settlement and its inhabitants have been integrated in our scientific program, as the history of the village and the family living there as well as the recent transformation of the landscape, are also part of the history of Resafa (Gussone and Mollenhauer 2009: 28).

Further topics of the ongoing Resafa-Project under the archaeological license of German Archaeological Institute (DAI) in cooperation with Direction Générale des Antiquités et des Musées de la Syrie (DGAMS) are the planning, preparation and execution of consolidation measures for Basilica A, which from the very beginning is suffering from constructive problems (Horn and Klessing 2010: 41). Further preliminary studies are concerning the Tetraconch Church (also known as 'Zentralbau'), already subject to excavation activities of Johannes Kollwitz in the 1950s-1960s. For this ruin the DGAMS requested scientific research leading towards a partial anastylosis, to visualize the magnitude of its architecture (Salman and Schuhmann 2009: 37).

Last but not least the process of installing a site management in Resafa has been a special task to explain this outstanding archaeological heritage to the local population and the visitors and safeguard the site for the future (Mollenhauer and Khoury 2009).

Conclusion

The former pilgrimage city and caliphal residence appeared during our campaigns 2006-2011 not any longer as lying in desolate solitude, as it was, when it had been discovered by travelers at the end of the 17th century up to the middle of 20th century, before systematic excavations began. We perceived an outpost of the nearby, prosperous Euphrates valley and a much frequented sightseeing attraction on the touristic highway between Aleppo, the Euphrates and Palmyra.

So 'Settlement Dynamics and Human-Landscape Interaction in the Steppes and Deserts of Syria' are not historically contained affairs at Resafa, but vital processes of sustainable progress. At the moment we cannot continue our work on the site, due to the current turmoil, and our profound sympathies belong to the people living there.

Bibliography

Bartl, K. and A. Moaz, (eds.)

2008 *Residences, Castles, Settlements. Transformation Processes from Late Antiquity to Early Islam in Bilad al-Sham.* OrA 24. Rahden/ Westfalen.

Bauzou, T.

1989 *A finibus Syriae : recherches sur les routes des frontières orientales de l'Empire Romain* (PhD. University Paris).

Beckers, B., J. Berking and B. Schütt

2012 The Elaborated Ancient Water Supply System of Resafa. Risk and Uncertainty of Water Harvesting in the Syrian Desert Steppe. In W. Bebermeier, R. Hebenstreit, E. Kaiser and J. Krause (eds.), *Landscape Archaeology. Proceedings of the International Conference Held in Berlin, 6th – 8th June 2012*, Topoi Journal for Ancient Studies. Special Journal 3: 149-153.

Beckers, B. and Ch. Konrad

2010 Resafa – Rusafat Hisham, Syrien. Archäologie und Prospektionen. Palastanlagen, Paläoumwelt und Wasserwirtschaftssystem, *JAHRBUCH MSD* 2008-10: 38.

Berking, J., B. Beckers, and B. Schütt

2010 Runoff in two semi-arid watersheds in a geoarchaeological context – a case study of Naga, Sudan, and Resafa, Syria, *Geoarchaeology* 25.6: 815-836.

Brands, G.

1998 Der sogenannte Audienzsaal des al-Mundir in Resafa, *DaM* 10: 211-235.

2002 *Die Bauornamentik von Resafa - Sergiupolis: Studien zur spätantiken Architektur und Bauausstattung in Syrien und Nordmesopotamien*, Resafa 6. Mainz.

Brinker, W.

1991 Zur Wasserversorgung von Resafa-Sergiupolis, *DaM* 5: 119-146.

Calvet, Y. and B. Geyer,

1992 *Barrages antiques de Syrie.* Lyon.

Chapot, V.

1903 Resapha-Sergiopolis, *BCH* 27.1: 280-291.

Foss, C.

1997 Syria in Transition, A.D. 550-750: An Archaeological Approach, *DOP* 51: 189-269.

Fowden, E.-K.

1999 *The Barbarian Plain, Saint Sergius between Rome an Iran.* Berkeley and Los Angeles.

2000 An Arab building at al-Ruṣāfa-Sergiopolis, *DaM* 12: 303-324.

Garbrecht, G.

1991 Der Staudamm von Resafa-Sergiupolis. In G. Garbrecht (ed.), *Historische Talsperren* 2. Stuttgart: 237-248.

Gatier, P.-L. and T. Ulbert

1991 Eine Türsturzinschrift aus Resafa-Sergiupolis, *DaM* 5: 169-182.

Grant, C. P.

1937 *The Syrian Desert: Caravans, Travel and Exploration.* London.

Gussone, M.

2010 Resafa – Rusafat Hisham, Syrien. Archäologie und Prospektionen. Digitaler Fundplatzkatalog am Beispiel von Fundplatz 109, *JAHRBUCH MSD* 2008-10: 36.

Gussone, M. and A. Mollenhauer

2009 Resafa, Syrien. Archäologische Karte – Dorf und Familie. Entwicklung eines Dorfes im Umfeld einer archäologischen Grabung, *JAHRBUCH MSD* 2007-09: 28.

Gussone, M. and M. Müller-Wiener

2011 Resafa - Rusafat Hisham, Syrien. Archäologie und Prospektionen. Am Rande der Militärstraße – Die Nachbegehung der Fundplätze südlich der Wadi-Bucht, *JAHRBUCH MSD* 2009-11: 27.

2012a Resafa - Rusafat Hisham, Syria. ‚Long-term survival' of an Umayyad residence – First results of the extended surface survey. In R. Matthews, J. Curtis *et al.* (eds.), *Proceedings of the 7th International Congress of the Archaeology of the Ancient Near East, London 2010*, vol. 2. Wiesbaden: 569-584.

2012b Resafa - Rusafat Hisham, Syrien. Archäologie und Prospektionen – Die Nachbegehung der Fundplätze im Osten und Nordosten der Stadt, *JAHRBUCH MSD* 2010-12: 27.

Gussone, M., D. Sack and U. Siegel

2009 Resafa - Rusafat Hisham, Syrien. Archäologie und Prospektionen – Auswertung der Prospektionsdaten im Hinblick auf die Erschließung und Gebäudeorientierung, *JAHRBUCH MSD* 2007-09: 30.

Haase, C.-P.

1995 Encyclopaedia of Islam 2.8 (1995), s.v. Ruṣafa, 630–631.

Haldon, J. F.

2010 *Money, power and politics in early Islamic Syria: a review of current debates.* Farnham, Surrey.

Hof, C.

2009 Masonry Techniques of the Early Sixth Century City Wall of Resafa, Syria. In K.-E. Kurrer, W. Lorenz and V. Wetzk (eds.), *Third International Congress on Construction History. Cottbus 2009.* Cottbus: 813-820.

2010 Die Stadtmauer von Resafa – Spuren früher Planänderung und deren Datierungsrelevanz. In Koldewey-Gesellschaft, *Bericht über die 45. Tagung*, Regensburg 2008. Dresden: 233–248.

In press The city wall of Resafa/Sergiupolis (Syria): The transition of a late Roman representative and permeable town wall over a sturdy fortification to an introverted early Islamic ḥaram enclosure. In Proceedings of the conference: "Fokus Fortifikation" zur Erforschung antiker Befestigungen, Athens 2012.

Horden, P.

2005 Mediterranean Plague in the Age of Justinian. In M. Maas (ed.), *The Cambridge Companion to the Age of Justinian.* Cambridge and New York: 134-160.

Horn, T. and J. M. Klessing

2010 Resafa, Syrien. Konsolidierungs- und Restaurierungsmaßnahmen – Basilika A, Konservierungsmaßnahmen, *JAHRBUCH MSD* 2008-10: 41.

Ilisch, L.

1996 Die islamischen Fundmünzen. In Sack 1996, *Die Große Moschee von Resafa – Ruṣāfat Hišām,* Resafa 4. Mainz.: 111-132.

Jahrbuch, MSD [Jahrbuch MSD 2005–07 – Jahrbuch MSD 2011-13]

2005–2013 Short reports of the Resafa-Project. In D. Sack *et al.* (eds.), Masterstudium Denkmalpflege der TU Berlin, *Jahrbuch* MSD 2005–07 (Heft 3, 2007): 16-23. 66; *Jahrbuch* MSD 2006–08 (Heft 4, 2008): 30. 33-41. 77-80; *Jahrbuch* MSD 2007–09 (Heft 5, 2009): 24-38; *Jahrbuch* MSD 2008–10 (Heft 6, 2010): 28-43. 92-93; *Jahrbuch* MSD 2009–11 (Heft 7, 2011): 22-43; *Jahrbuch* MSD 2010–12 (Heft 8, 2012): 22-32; *Jahrbuch* MSD 2011-13 (Heft 9, 2013): 28-34.

Karnapp, W.

1976 *Die Stadtmauer von Resafa in Syrien,* DAA 11. Berlin.

1978 Der Khan in der syrischen Ruinenstadt Resafa, *AA* 93: 136-150.

Kellner-Heinkele, B.
1996 Ruṣāfa in den arabischen Quellen. In Sack 1996, *Die Große Moschee von Resafa – Ruṣāfat Hišām,* Resafa 4. Mainz: 133-154.
Knötzele, P.
2006 Resafa: Die Gefäßkeramik der Stadtgrabung 1997-1999. In F. Bloch, V. Daiber and P. Knötzele (eds.), *Studien zur spätantiken und islamischen Keramik. Hirbat al-Minya - Baalbek - Resafa,* OrA 18. Rahden/ Westfalen: 167-208.
Kollwitz, J.
1959 Die Grabungen in Resafa. In E. Boehringer (ed.): *Neue deutsche Grabungen im Mittelmeergebiet und im vorderen Orient.* Berlin: 45-70.
1963 Die Grabungen in Resafa Frühjahr 1959 und Herbst 1961, *AA*: 328-360.
Konrad, Ch.
2011 Resafa - Rusafat Hisham, Syrien. Archäologie und Prospektionen. Die Qusur FP 106 und FP 220, zwei Paläste in der Residenz des umaiyadischen Kalifen Hisham b. Abd al-Malik, *JAHRBUCH MSD* 2009-11: 31.
Konrad, M.
1992 Flavische und spätantike Bebauung unter der Basilika B von Resafa-Sergiupolis, *DaM* 6: 313-402.
2001 *Der spätrömische Limes in Syrien: Archäologische Untersuchungen an den Grenzkastellen von Sura, Tetrapyrgium, Cholle und in Resafa,* Resafa 5. Mainz.
Konrad, M. and T. Ulbert, (eds.)
In press *Forschungen in Resafa-Sergiupolis. I. al-Mundir-Bau und Nordnekropole. II. Basilika C.* Resafa 7.
Korn, L.
2004 Resafa: Fundmünzen der Stadtgrabungen 1997-1999, *DaM* 14: 197-206.
Kurapkat, D.
2009 Resafa, Syrien. Archäologische Karte – Basilika B. Zeitschichtenplan des Areals der Basilika B mit Vorgängerbauten und späterer Überbauung, *JAHRBUCH MSD* 2007-09: 27.
2011 Resafa, Syrien. Archäologische Karte – Basilika A-Komplex. Zeitschichtenplan der Basilika A, ihrer Annexbauten und der Großen Moschee, *JAHRBUCH MSD* 2009-11: 36.
Legner, A.
1964 Die Grabungen in Resafa, IV. Islamische Keramik in Resafa, *AAS* 14: 98-108.
Leisten, T.
1993 Die Gärten des Islam. In H. Forkl, J. Kalter, T. Leisten and M. Pavaloi (eds.), *Die Gärten des Islam.* Stuttgart: 47-76.
Lewis, N.
1987 *Nomads and settlers in Syria and Jordan, 1800-1980.* Cambridge and New York.
Logar, N.
1991 Katalog der Keramikfunde aus dem Wasserverteiler, *DaM* 5: 147-168.
1992 Die Kleinfunde aus dem Westhofbereich der Großen Basilika von Resafa, *DaM* 6: 417-478.
1996 Die Keramik und andere Kleinfunde. In Sack 1996, *Die Große Moschee von Resafa – Ruṣāfat Hišām,* Resafa 4. Mainz: 77-110.

Mackensen, M.
1984 *Eine befestigte spätantike Anlage vor den Stadtmauern von Resafa. Ausgrabungen und spätantike Kleinfunde eines Surveys im Umland von Resafa-Sergiupolis,* Resafa 1. Mainz.

Meier, M.
2005 »Hinzu kam auch noch die Pest ...« Die sogenannte Justinianische Pest und ihre Folgen. In M. Meier (ed.), *Pest. Die Geschichte eines Menschheitstraumas.* Stuttgart: 86-107.

Meinecke, M.
1991 Early Abbasid Stucco Decorations in Bilād al-Shām. In M. A. Bakhit and R. Schick (eds.), *Bilād al-Shām during the Abbasīd Period* (132 A.H./750 A.D.-451 A.H./1059 A.D.). Amman: 226-267.

Mollenhauer, A. and Y. Khoury,
2009 Resafa-Sergiupolis/Rusafat Hisham. Pilgrimage City and Caliphal Residence - The Process of Developing a Visitor's Concept. In *Conference Proceedings: Traditions and Transformations: Tourism, Heritage and Cultural Change in the Middle East and North Africa Region.* Amman, Jordan, 2009. Leeds: CD-ROM 1-16.

Müller-Wiener, M.
2012 *Die Kunst der islamischen Welt.* Stuttgart.
1996 *Book of gifts and rarities: selections compiled in the fifteenth century from an eleventh-century manuscript on gifts and treasures = Kitab al-hadāyā wa al-tuḥaf.* Harvard Middle Eastern monographs 29. Cambridge, Mass.

Raymond, A.
1994 Islamic City, Arab City: Orientalist Myths and Recent Views, *BJMES* 21.1: 3-18.

Sack, D.
1996 *Die Große Moschee von Resafa – Ruṣāfat Hišām,* Resafa 4. Mainz.
2008 Resafa-Sergiupolis/Rusafat Hisham – neue Forschungsansätze. In Bartl, K. and A. Moaz, (eds.), *Residences, Castles, Settlements. Transformation Processes from Late Antiquity to Early Islam in Bilad al-Sham.* OrA 24. Rahden/ Westfalen: 31-44.
2010 Resafa, Syrien. Archäologie und Prospektionen – Kalif und Hofstaat, *JAHRBUCH MSD* 2008-10: 34.

Sack, D., H. Becker, M. Stephani and F. Chouker
2004 Resafa-Umland, Archäologische Geländebegehungen, geophysikalische Untersuchungen und digitale Geländemodelle zur Prospektion in Resafa – Rusafat Hisham. Bericht über die Kampagnen 1997 – 2001, *DaM* 14: 207-232.

Sack, D., M. Gussone and A. Mollenhauer (eds.)
2013 *Resafa-Sergiupolis/Rusafat Hisham. Forschungen 1975–2007. Reader – Resafa-Sergiupolis/Ruṣāfat Hišām. Dalil ad-dirāsāt wa'l-abḥāṯ almunǧaza ḫilāl al-fitra* 1975–2007.

Sack, D., M. Sarhan and M. Gussone
2010 Resafa-Sergiupolis/Ruṣāfat Hišām, Syrien. Pilgerstadt und Kalifenresidenz. Neue Ansätze, Ergebnisse und Perspektiven, *ZOrA* 3: 102-129.

Saliby, N.
1990 Une maison arabe a Resafa. In P. Matthiae, M. van Loon and H. Weiss (eds.), *Resurrecting the Past. A Joint Tribute to Adnan Bounni.* Istanbul: 277-296.

Salman, I. and A. Schuhmann
2009 Resafa, Syrien. Konsolidierungs- und Restaurierungsmaßnahmen. ‚Zentralbau' – Bauarchäologische Untersuchung und Planung einer Teil-Anastilosis, *JAHRBUCH MSD* 2007-09: 37.

Schuhmann, A.
2011 Resafa-Sergiupolis/Syrien. Der ‚Zentralbau'. Archäologische Untersuchungen im Bereich des südlichen Anbaus, *JAHRBUCH MSD* 2009-11: 40.

Schütt, B. and M. Meyer
2011 A-I Central Places and Their Environment – Preliminary Results from the Research Group (2011), *eTopoi, Journal for Ancient Studies. Special Volume* 1.

Severin, H.-G.
1997 Pilgerwesen und Herbergen. In E. Dassmann and J. Engemann (eds.), *Akten des XII. Internationalen Kongresses für Christliche Archäologie, Bonn* 1991, JbAC. Ergbd. 20.1. Münster: 329-339.

Shahid, I.
2009 The Ghassanid Structure outside Sergiopolis/Rusafa: Church or Praetorium? In J. D. Alchermes with H. C. Evans and T. K. Thomas (eds.), *Anathēmata heortika: studies in honor of Thomas F. Mathews*. Mainz: 283-287.

Siegel, U.
2010 Resafa - Rusafat Hisham, Syrien. Archäologie und Prospektionen. Auswertung der Oberflächenbefunde am Fundplatz 109, *JAHRBUCH MSD* 2008-10: 37.

Ulbert, T.
1986 *Die Basilika des Heiligen Kreuzes in Resafa-Sergiupolis,* Resafa 2. Mainz.
1990 *Der kreuzfahrerzeitliche Silberschatz aus Resafa-Sergiupolis*, Resafa 3. Mainz.
1992 Beobachtungen im Westhofbereich der Großen Basilika von Resafa, *DaM* 6: 403-416.
2008 50 Jahre Forschungen in Resafa/Sergiupolis: Struktur und Kontinuität. In K. Bartl and A. Moaz (eds.), *Residences, Castles, Settlements. Transformation Processes from Late Antiquity to Early Islam in Bilad al-Sham.* OrA 24. Rahden/ Westfalen: 15-30.

Walmsley, A.
2007 *Early Islamic Syria: An Archaeological Appraisal*. London.

Westphalen, St.
2000 Resafa. Bericht über die Ausgrabungen 1997 bis 1999, DaM 12: 325-365.

Westphalen, St. and P. Knötzele
2004 Water Supply of Resafa, Syria – Remarks on the Chronology of the Big Cisterns. In H.-D. Bienert and J. Häser (eds.), *Men of Dikes and Canals: The Archaeology of Water in the Middle East*, OrA 13. Rahden/ Westfalen: 345-348.

Widell, M.
2007 Historical Evidence for Climate Instability and Environmental Catastrophes in Northern Syria and the Jazira: the Chronicle of Michael the Syrian, *Environment and History* 13.1: 47-70.

Wirth, E.
1971 *Syrien. Eine geographische Landeskunde.* Darmstadt.
2002 *Die orientalische Stadt im islamischen Vorderasien und Nordafrika.* Mainz.

Wolfart, R.
1966 *Zur Geologie und Hydrogeologie von Syrien*. Geologisches Jahrbuch. Beiheft 68. Hannover.

ELISABETTA RANDAZZO
Independent Researcher

La Steppe Syrienne à la période omeyyade

L'étude des modes d'occupation du sol dans les terres arides de la période omeyyade est un aspect très intéressant pour la compréhension de la politique d'occupation et l'utilisation des territoires de nouvelle acquisition de la part de la dynastie islamique. En effet, les formes d'occupation du territoire d'une culture doivent être lues comme un référentiel d'une pluralité d'informations, non seulement sur elles-mêmes, mais en particulier sur la société, l'économie et les formes de pouvoir dans lequel elles ont été produites.

Nous savons que la conquête du Bilad al Sham a commencé durant le règne d'Abū Bakr, en particulier la Palmyrène a été conquise en 13/634 et peu après cette date, sous le règne d''Umar b. al Khattāb, la province de Syrie (al-Shāam) a été créée et placée sous l'autorité de son conquérant Abū 'Ubayda b. al-Jarrāh, administrateur de la région. La province a alors été subdivisée en quatre districts, ou circonscriptions (militaires et administratives) appelées *ajnād* (singulier: *jund*) qui peuvent être considérées comme le résultat de l'affirmation de la nouvelle géographie du pouvoir politique, social et économique de l'époque. Al-Malik prendra en charge la division du Bilad al-Sham en unités régionales, dont le contrôle sera confié aux membres de sa famille, en particulier à ses fils.

On peut parler de quatre sous-régions: Damas (chef-lieu: Damas), Homs (chef-lieu: Homs), al-Urdunn (chef-lieu: Tabariyya) et al-Filastīn (chef-lieu Lydda, puis al-Ramla); la Palmyrène faisait donc partie du jund de Homs, à l'exception de sa pointe sud-ouest qui appartenait à celui de Damas (Fig.1). Vers 60/680, durant le règne de Yazīd b. Mu'āwiya, un cinquième *jund*, Qinnasrīn, a été créé au détriment de celui de Homs, en détachant sa partie septentrionale, autour d'Antioche, Alep et jusqu'à l'Euphrate en incluant Manbij et al-Rūsafa. Chaque *jund* disposait d'un trésor provincial qui pouvait être dépensé sur le territoire de pertinence, ce qui permettait, par conséquent, la présence économique de l'État, à l'aide des fonds provenant du régime fiscal dans chaque district établi par le califat. La répartition des richesses au sein de la hiérarchie des Omeyyades a ainsi fortement contribué à la réhabilitation massive des zones urbaines et rurales, notamment des infrastructures commerciales et de la production agricole (Walmsley 2000: 342-343).

Il y a donc un lien très évident entre la dynastie omeyyade et l'espace syrien qui devient alors un espace de pouvoir privilégié. Du point de vue des sources narratives islamiques, notre perception est très faible parce que les sites ne sont que très rarement traités dans les textes et, en effet, nous pouvons en attribuer la connaissance surtout à la recherche archéologique.

Dans l'espace syrien au IIème/VIIIème siècle, un vaste programme architectural est mis en place: des nouvelles villes voient la lumière, comme Qasr al-Hair al-Sharqi (Grabar 1978) et Madinat al-Far (Haase 2001: 11, 7-20; 2006: 54-60; 2008: 395-402), et les anciens sites urbains sont enrichis et réorganisés par la construction de nouveaux bâtiments comme à Palmyre et à al-Rusāfa. Par ailleurs, à ce même titre, dans le territoire extra-urbain la construction des *qusūr*, avec leur complexité structurale et fonctionnelle, modifie défini-

Studia Chaburensia 4 (2014), pp. 275–284.

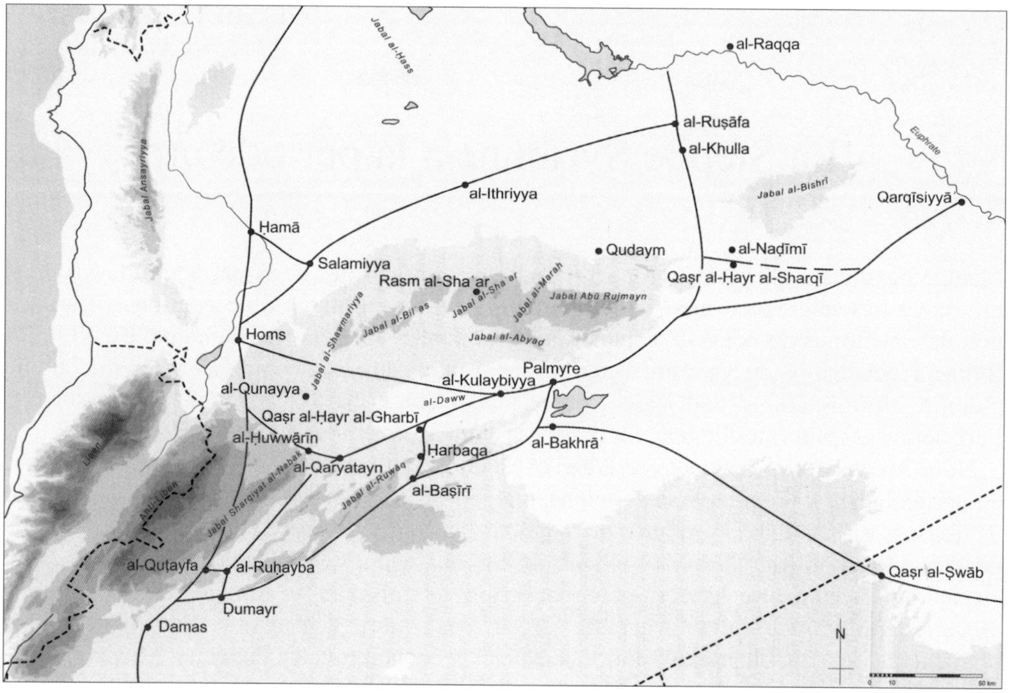

Fig. 1: La Palmirène Ommeyade.

tivement le paysage de la steppe syrienne. Tout cela a abouti à la création d'un nouveau paysage califal.

Cette explosion architecturale, pendant laquelle les califes parviennent à un «patronage architectural», comme l'a récemment expliqué A. Borrut, permet, par l'assignation des aires spécifiques au sein de l'espace syrien, la naissance d'une conception patrimoniale du pouvoir et, dans ce cadre, l'instauration d'une "souveraineté patrimoniale" (Borrut 2011: 391-411).

A ce titre, la région de la Palmyrène, faisant partie du *jund* de Homs, fut posée sous le contrôle d'Hisham.

On peut donc parler d'une pratique d'appropriation familiale de l'espace et de l'exercice de ce pouvoir (Décobert 2010: 229) et, en effet, Hisham (724-743) profondément ancré en Palmyrène déjà avant son accession au trône, poussa jusqu'au bout la logique du système en s'établissant prioritairement à al-Rusafa. Le septième calife marwanide contrôlait la partie la plus orientale de la Syrie où il a promu la création des infrastructures dans les grands centres urbains comme Raqqa et Rusafa; en revanche, les deux grands *qusur*, (situés dans la Palmyrène occidentale et orientale) et les structures hydrauliques témoignent des interventions extraordinaires dans le territoire rural. Nous savons, en effet, qu'Hisham avait reçu en fief la partie syrienne de l'Euphrate et qu'il avait décidé d'y vivre. Il avait ainsi continué, en qualité de calife, à concentrer ses activités sur le territoire (Borrut 2011: 442-443). Deux acteurs principaux jouaient donc un rôle très important dans la région, tant du point de vue de l'occupation, que de la création d'un véritable paysage califal omeyyade: le calife Hisham et son frère Maslama, qui contrôlait la région du Balis, au nord-est du *jund* de Qinnasrin.

Le monde Urbain

En ce qui concerne la steppe syrienne et la Palmyrène, on peut dire que la présence omeyyade dans la région est bien évidente. Le monde urbain connaît quant à lui la fondation de villes nouvelles et l'intervention islamique dans les centres existants, comme en témoignent, entre autre, le *suq* construit à Palmyre, entre la fin du VIIème et le début du VIIIème siècle (Walmsley 2000: 274), et une mosquée vraisemblablement congrégationnelle, vu que la ville à l'époque omeyyade jouait le rôle de centre urbain important et florissant. Tout cela est probablement commandé par Hisham et doit être connecté au développement du rôle commercial des villes. Ceci se vérifie également à Rusafah et à Raqqa qui ont connu une renaissance économique et culturelle exactement sous le gouvernement omeyyade qui, bien évidemment, avait créé les prémisses pour un développement urbain croissant. Les fouilles menées un peu partout dans la ville de Palmyre (As'ad et Stepniowski 1989: 205-223) témoignent d'une occupation continue entre l'époque protobyzantine et le début de l'époque abbasside.

D'un point de vue général, nous pouvons constater que toutes les villes omeyyades, qu'elles soient d'origine antique ou nouvellement fondées, témoignent de l'implication des califes marwanides dans de grands projets de développement et de réaménagement urbains. En outre, les mosquées représentaient l'affirmation d'une nouvelle religion et d'un nouveau pouvoir politique, tandis que les marchés étaient liés au développement économique sans lequel le nouveau pouvoir aurait eu peine à s'imposer.

Le cas Al-Bakhrā (Genequand 2012: 69-94) est également très intéressant. Situé à 21 km au sud de Palmyre, sur la route de la *caput viae* d'une section de la *Strata Diocleziana* dont seuls quelques vestiges en subsistent aujourd'hui, il s'agit du lieu où le calife al-Walid b. Yazīd a été assassiné. C'est en outre un des établissements aristocratiques omeyyades les plus anciens qui soit connu archéologiquement. Genequand décrit Al-Bakhra comme un établissement intermédiaire entre un *qasr* et une ville neuve avec, en fait, un aspect proto-urbain annonciateur des villes neuves planifiées plus formellement (comme notamment Qasr al-Hair al-Sharqi, la citadelle de 'Amman ou 'Anjar). A l'origine, il ne s'agit pas de sites conçus pour assumer le même rôle que les villes déjà existantes, mais ce sont des établissements qui ont aussi un très fort caractère aristocratique et destinés à des princes, à des gouverneurs ou à des hauts chefs militaires. Ils abritent leur entourage, leur administration et leurs troupes, sans pour autant manquer d'installation à caractère économique et commercial.

En effet, les villes nouvelles sont construites selon un modèle idéalisé et un format réduit afin de répondre aux exigences de leur fonction. Il s'agit de véritables centres urbains, parfois appellés "*madīna*" dans les textes, mais il ne revêtent pas l'aspect traditionnel des villes (Northedge 2000: 64). Au sein de cette catégorie des villes nouvelles, et en ce qui concerne la Palmyrène, il convient de considérer Hisn Maslama/ Madinat al-Far et Qasr al-Hair al-Sharqi.

La première, Madinat al-Fār, est mentionnée par Muqaddāsi, Yāqūt et d'autres géographes anciens. Le site est connu pour avoir été la résidence de Maslama b. 'Abd al-Malik et a en effet été appelé Isn Maslama, expression également employée pour la ville de Amman (Northedge 2000: 68)[1] par le poète al-Ahwas al-Ansārī (m. 86/705) (Al-Ahwas al-Ansārī, *Dīwān*, no. 91) qui parle de "*hisn 'Amman*" (le château ou fort de 'Amman).

1 Le Northedge a bien expliqué les origines des termes "*madīnā*" et "*isn*" dans son article de 1994 "*Archaeology and New Urban Settlement in Early Islamic Syria and Iraq*" et dans son étude "*Entre Amman et Baghdad*".

Fondée par le général militaire et prince omeyyade Maslama, gouverneur de la province entre 709 et 719, la ville est située dans la Jazīra syrienne, entre Hârran, au nord, et al-Raqqa au sud, et a abrité des généraux ainsi qu'une partie de leur armée. L'eau de la rivière affluente Balikh y était conduite par un système de canaux en l'irriguant, tout comme les territoires agricoles qui étaient à Maslama, situés aux alentours de la ville. Les sources littéraires décrivent en effet une vallée bien irriguée, riche en villages et fermes agricoles. En outre, la région a également été affectée par d'importants projets visant au développement de terres agricoles, en particulier dans la zone de l'Euphrate: notamment le Nahr Maslamah, commandé par Maslamah ibn Abd al-Malik, constitue un effort massif pour l'irrigation près de Néocésarée (Dibsi Faraj), destiné à diriger l'eau vers la rive sud de la rivière (Wilkinson 1975: 319-338). Le cas de Hisn Maslamah n'est pas un *unicum* car les propriétés foncières des élites arabes semblent avoir été très répandues dans le Bilad al-Sham, qui devinrent plus tard la propriété de la dynastie des Abbasides présentant une continuité de pratique significative par les membres de la famille califale, leur permettant d'exercer une sorte de féodalisme sur les territoires appartenant à l'empire.

La deuxième ville nouvelle, Qasr al-Hair al Sharqi, occupe une position ambiguë et peut appartenir aussi bien à la catégorie des établissements aristocratiques qu'à celle des villes neuves, bien que située dans un environnement typique de *qūsur*. L'épigraphe de fondation dans la mosquée congrégationnelle (Clermont-Ganneau 1900: 285-292; Combe, Sauvaget and Wiet 1931: n. 28; Grabar *et al.* 1978: 12, 191) devrait en effet, correspondre à l'ensemble du site et non à une seule de ses composantes. Il s'agit donc d'un établissement de type urbain pourvu, en outre, de maisons reflétant différent *status*, ce qui correspond bien à la réalité de l'organisation sociale. Les fonctions politiques sont identifiées grâce à la présence du palais califal et du *dar al-imāra* (dans la grande enceinte); les fonctions religieuses sont attestées par la présence d'une mosquée congrégationnelle; les fonctions économiques sont identifiées en raison de la présence d'une agriculture à grande échelle documentée dans les deux enclos irrigués et de celle des installations industrielles (notamment l'huilerie et le moulin hydraulique: Fig. 2; Fig.3); en outre, les fonctions résidentielles semblent évidentes en raison de la présence de maisons aristocratiques (dans la grande enceinte) et de l'habitat vernaculaire des établissement nord et est, qui mettent l'accent sur une hiérarchie dans les structures d'habitat, reflétant une situation similaire à celle des grandes villes. En revanche, il convient de remarquer l'absence de marché, composante fondamentale de tous les établissements islamiques à caractère urbain (Genequand 2012: 95-159).

Le monde Extra-urbain

En ce qui concerne la présence omeyyade sur le territoire, mais en s'éloignant du concept de «ville», on doit citer Qudaym, situé à 65 km au nord de Palmyre, qui présente toutes les caractéristiques d'un établissement que l'on pourrait qualifier d'aristocratique. On peut en effet noter la présence d'une grande enceinte en briques crues, d'une mosquée, d'un vaste réservoir, d'une adduction d'eau par *qanāt*, et de quelques maisons en très mauvais état de conservation, situées entre le château et la citerne (Genequand 2012: 174-180).

Un autre site très intéressant est celui d'al-Qunayya, qui ne se trouve pas à proprement parler dans la Palmyrène, mais à proximité de l'Anti-Liban. Néanmoins, nous le citons car il se rapporte vraisemblablement à la même série d'implantation d'établissements aristocratiques omeyyades. Il s'agit d'un complexe composé d'une résidence, d'un bâtiment annexe, d'une adduction d'eau, d'un réservoir et d'un enclos agricole, composantes déjà identifiées sur les autres sites (Genequand 2012: 180-184).

Fig. 2: Qasr al-Hair al-Sharqi, le moulin hydraulique et les constructions adjacentes (d'après Genequand 2002: fig. 364)

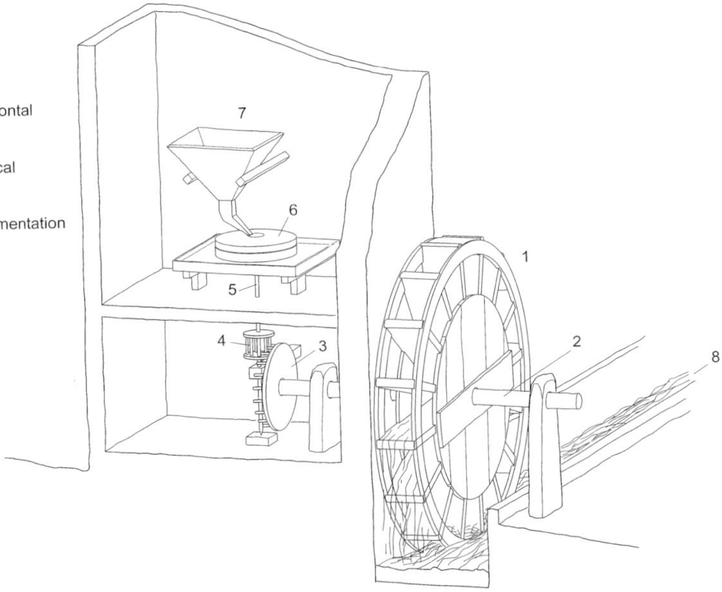

Fig. 3: Qasr al-Hair al-Sharqi, Schéma de fonctionnement d'un Moulin hydraulique à roué vertical (d'après Genequand 2002: fig. 363)

Le site de Rasm al-Sha'ar, à environ 55 km au nord-ouest de Palmyre, à proximité du sommet de la montagne, représente un autre cas explicatif de la présence omeyyade sur le territoire. Il s'agit d'un petit fortin qui, grâce à sa position dominante, devait revêtir un rôle de contrôle et de gestion des activités pastorales, et être une sorte de satellite d'un établissement omeyyade plus important (Palmyre, ou peut-être un autre établissement de la plaine d'al-Daww ou des piémonts nord des Palmyrénides) (Genequand 2012: 184-186).

Qasr al-Swab, situé aux confins méridionaux de la Palmyrène, constitue également un cas intéressant. Le site présente un important *wādī*, deux enceintes quadrangulaires, un grand nombre de bâtiments de plus petites dimensions et une petite mosquée à proximité de la porte de la deuxième enceinte qui, même si nous ne disposons d'aucun indice de datation, le site fait penser à un établissement omeyyade (Genequand 2012: 186-187). Des incertitudes à l'égard de la datation subsistent pour d'autres sites, mais nous jugeons néanmoins utile de les mentionner, dans la mesure où ils présentent des similarités structurelles et fonctionnelles qui nous font pencher vers l'identification omeyyade. Ils permettent donc de compléter l'étude de l'enracinement omeyyade sur la région; il s'agit des sites d'al-Basīri (Genequand 2012: 187-188), situé à 75 Km au sud-ouest de Palmyre, et d'al-Kulaybiyya (Genequand 2012: 188), situé à 30 km à l'ouest de Palmyre.

En outre, les prospections menées par une équipe norvégienne dans les montagnes au nord-ouest de Palmyre, ont révélé plusieurs petits sites occupés durant l'Antiquité tardive et/ou durant l'époque omeyyade. Les sites repérés dans le Jebel Abyad et dans le Jebel Chaar ont été étudiés par la mission norvégienne; les sites localisés dans le Jebel Abu Rigmen ont quant à eux été identifiés grâce à l'analyse d'images satellites. Il s'agit de petites implantations rurales, liées soit à l'exploitation des terres agricoles, soit à la proximité de la route menant à al-Ithriyya. Elles semblent nettement moins nombreuses que celles des époques romaine et protobyzantine. La céramique, enfin, montre la même tendance, c'est-à-dire une continuité dans la période omeyyade et sans changement important à l'égard de l'architecture et de la composition des sites (Meyer 2011: 5).

Les *Qusūr*

Dans le domaine extra-urbain, nous avons notamment les *qusūr*, dont l'édification dans les ajnad pourrait en fait être l'expression d'un "projet familial" de grande ampleur par ʻAbd al-Malik, qui délègue sa mobilité à ses fils, créant une sorte de souveraineté patrimoniale rendant possible une forme d'«itinérance partagée» dans le territoire (Borrut 2011: 405-411).

En ce qui concerne les qusūr, il faut relever l'extraordinaire multifonctionnalité documentée par l'archéologie. Il s'agissait de lieux importants et stratégiques de repos pour les pèlerins, de centre de production dédiés à l'exploitation des terres agricoles, mais, c'était aussi, dans certains cas, principalement des centres officiels du pouvoir dans lesquels l'exercice de la royauté politique se traduit concrètement par la construction de pièces sans doute déléguées à cela. Un autre point de vue envisageable pour observer le phénomène des *qusur* est celui du concept de translation urbaine dans les zones rurales, nettement détectable en particulier dans le cas où, comme à Qasr al-Hair al-Sharqi, la proximité d'un grand centre urbain, ainsi que la taille et la composition du *qasr*, permettent d'identifier l'ensemble architectural avec un lieu alternatif pour le calife et sa cour (Randazzo 2011: 452-456). Il s'agit de centres autonomes, non seulement du point de vue de la survie réelle des individus qui y vivaient, comme cela est attesté dans tous les cas, mais aussi du point de vue de l'administration, de l'exercice du pouvoir royal, de l'achèvement des services religieux, de l'expression du luxe et de la dimension des divertissements. A cet égard, il est significatif que les deux *qusūr,* commandés par Hisham dans la Palmyrène, apparaissent avec des mosquées congrégationnelles, des thermes

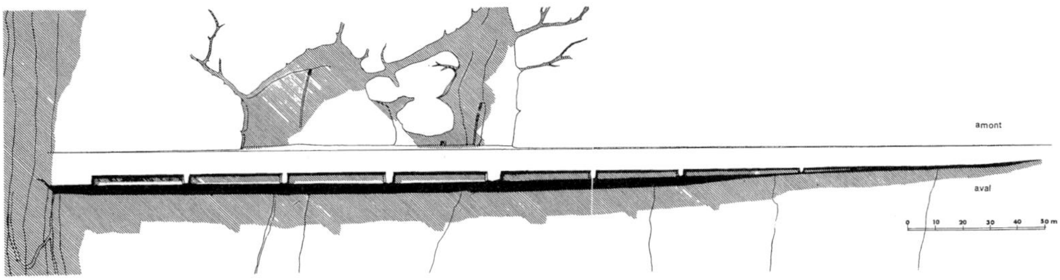

Fig. 4: Plan du Barrage de Harbaqa (Schlumberger 196)

et des terres cultivées et clôturées. Le rôle politique acquitté par le deux *qusūr* Qasr al-Hair ash-Sharqi et par Qasr al-Hair al-Gharbi dans la région est absolument évident compte tenu de leurs tailles, de leur décoration architecturale, et surtout des espaces intérieurs et extérieurs qui les constituent et qui font penser à une sorte de ville ou, du moins, à des centres fonctionnant en tant que tels.

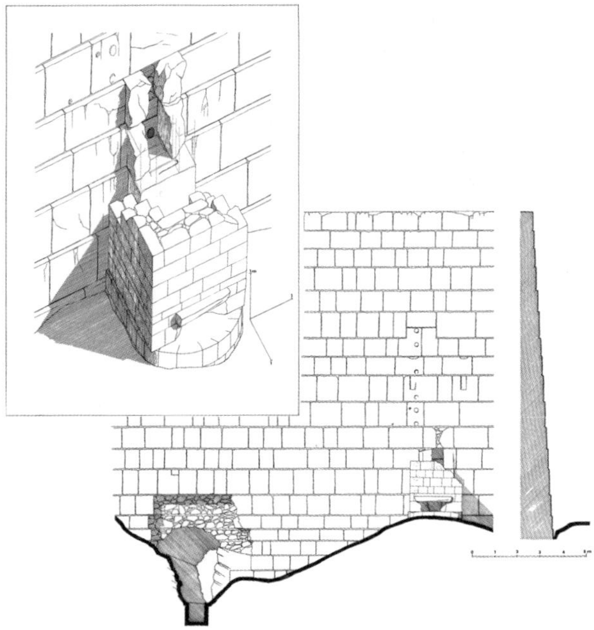

Fig. 5: Plan du Barrage de Harbaqa, detail de l'un des exutoires et la tourelle ajoutèe en avant (Schlumberger 196)

La localisation même des *qusūr* réalisés par Hisham indique qu'il était bien conscient à l'époque de leur position stratégique par rapport aux grandes routes commerciales et de communication. Ces qusūr revêtent une valeur économique et sociale naturellement énorme, dans la mesure où cela leur permettait de travailler comme points d'arrêt pour les caravanes transitant entre la zone centrale de l'Euphrate, la Jazira et la partie occidentale du Bilad al-Sham. Les connexions avec les principaux itinéraires et les grands centres révèlent la volonté de concevoir ces complexes comme faisant partie intégrante du programme d'implantation omeyyade. En fait, sur la route reliant l'Euphrate à Damas, on peut rejoindre les sites d'al-Sukna et d'Arak. Ce dernier a livré des vestiges archéologiques indiquant clairement une occupation omeyyade et abbaside (Walmsley 2000: 274-283). Grâce à cette même route, on pouvait ensuite rejoindre l'importante ville commerciale de Palmyre, où sont attestées d'importantes phases d'occupation omeyyades et, enfin, le site de Qasr al-Hair al Gharbi, probablement la limite ouest des territoires de Hisham, à partir de laquelle on pouvait continuer vers le sud.

L'extraordinaire exemple de ce dernier *qasr*, construit dans la steppe syrienne durant la première moitié du II/VIII siècle, a probablement été conçu comme un lieu de résidence secondaire du calife Hishām b. 'Abd al Malik, ce qui n'exclut néanmoins pas d'autres fonctions. L'un des rôles des palais omeyyades de la steppe et, donc, de Qasr al-Hair al Gharbi, était de servir de point de rencontre entre un pouvoir califal très mobile et les chefs des principales tribus du nord de l'Arabie qui comptaient parmi leurs plus forts alliés, les chefs de la tribu de Kalb, bien établie dans la région (Fück and Dixon 1998: 492-494). Le site a aussi joué un rôle agricole attesté par les diverses installations hydro-agricoles et industrielles, dont le poids économique réel est difficile à estimer. Le système hydraulique devait être impressionnant et avait deux rôles principaux: l'approvisionnement en eau du palais et des thermes voisines ainsi que l'irrigation d'un grand enclos agricole. La collecte des eaux de pluies hivernales se faisait à l'aide de deux barrages placés en travers de cours d'eau temporaires: le célèbre complexe de Harbaqa (Fig. 4; Fig. 5). Le transport de l'eau était assuré par un aqueduc de 18 km de long, doté de plusieurs ramifications (un moulin hydraulique est en effet placé sur une dérivation de l'aqueduc principal prévue à cet effet, entre le réservoir et l'enclos agricole). Situé à 16,5 km du palais de Qasr al-Hair al-Gharbi et à 18 km de l'enclos agricole, le barrage, connusous le nom de Harbaqa, constitue l'origine de l'ensemble du système hydraulique. Ce barrage est implanté en travers d'une gorge le long d'un cours d'eau temporaire, à proximité du débouché nord de la passe d'al-Bārida. Le complexe de Harbaqa se compose de deux barrages. Le premier, de 345 m de longueur et de 20,50 m de hauteur, pour une largeur à sa base de 18 m, est le plus grand barrage du Proche-Orient avant l'époque moderne. Le second barrage, beaucoup moins imposant, mais non dénué d'intérêt, est placé directement en amont de l'enclos agricole et relié à ce dernier par trois aqueducs. Il prend la forme d'un mur semi-circulaire et est peu élevé (moins de 1,50 m, pour une largeur de 2.75 m). L'eau rejoignait l'enclos agricole, qui termine le système hydraulique et mesure 46,41 hectares. Cet enclos est délimité par un mur composé d'un soubassement de pierres et d'une élévation de briques crues. Deux portes donnent accès à cet enclos. La principale est placée sur le petit côté oriental, à proximité d'un des angles, en direction du palais et du *khan*, et c'est juste à côté de cette porte que l'aqueduc arrivant du grand barrage de Harbaqa entre dans l'enclos. En revanche, les entrées des aqueducs provenant du petit barrage étaient placées à intervalles irréguliers, plus au sud de la porte principale de l'enclos, sur le même côté. L'intérieur de l'enclos était subdivisé en parcelles carrées par des canaux en terre, aujourd'hui disparus.

On peut donc remarquer, comment l'occupation systématique de la steppe syrienne met l'accent sur la relation dialectique entre le pouvoir officiel et le particularisme tribal, politiquement actif dans la région. Ceci a nécessité la construction de lieux pour accomplir des formes de dialogue de nature officielle pouvant représenter également des symboles concrets de la nouvelle forme de pouvoir dans une zone dans laquelle, évidemment, il était nécessaire de confirmer l'identité de la nouvelle classe dirigeante.

D'un point de vue général, on peut parler d'un processus réel d'implantation, mené systématiquement par la dynastie des Omeyyades, grâce à des programmes de construction spécifiques, d'une occupation et d'un enracinement sur le territoire par le prince qui, par sa gestion personnelle des terres, réalisait de véritables programmes d'implantation tant d'un point de vue urbain qu'extra-urbain.

Bibliographie

As'ad, K. and F.M. Stepniowski

1989 The Umayyad Suq in Palmyra, *Damaszener Mitteilungen* 4.

Bacharach, L. J.

1996 Marwanid Umayyad Building Activities: Speculations on Patronage, *Muqarnas,* vol. XIII.

Borrut, A.

2011 *Entre memoire et pouvoir: l'espace syrien sous les derniers Omeyyades et les premiers Abbasides*, Leiden-Boston: Brill.

Clermont-Ganneau, C.,

1900 Une inscription du calife Hicham (an 110 de l'Hégire), *Recueil d'archéologie orientale*, Vol.III, Paris: E. Leroux.

Combe E., Sauvaget J., Wiet, G.,

1931 *Répertoire chronologique d'épigraphie arabe*, Vol.I, Le Caire: IFAO.

Décobert, C.,

2010 Notule sur le patrimonialisme omeyyade. In A. Borrut and P. M. Cobb (eds.), *Umayyad Legacies. Medieval Memories from Syria to Spain*, Leiden-Boston: Brill.

Fück, J.W. and A.A. Dixon

1998 Kalb b. Wabara, *Encyclopedia of Islam*, Vol. IV. Brill.

Genequand, D.

2012 *Les établissements des élites omeyyades en Palmyrène et au Proche-Orient.* Beyrouth.

Grabar, O., R. Holod, J. Knutstad and W. Trusdale

1978 *A City in the Desert: Qasr al-Hayr East*, Harvard.

Haase, C. P.

2001 Une ville du débuts de l'Islam d'après les fouilles effectuèes à Madinat al-Far (Syrie du Nord), *Archéologie Islamique* 11: 7-20

2006 The Excavations at Mādināt al-Fār/Hisn Maslama on the Balikh Road. In H. Kennedy (ed.), *Muslim Military Architecture in Greater Syria. From the Coming of Islam to the ottoman Period*, Leiden-Boston: Brill, 54-60.

2008 Public and Domestic Architecture- the Case of Mādināt al-Fār/Hisn Maslama. In K. Bartl and A. al-Razzaq Moaz (eds.), *Residences, Castles, Settlements. Trasformation Process from Late Antiquity to Early Islam in Bilad al-Sham*, Rahden/Westf.: VML, 395-402.

Meyer, J. C.

2011 *Palmyrena. Palmyra and the surrounding territory joint Syrian-norwegian project, surface survey north of Palmyra jebel Merah, April and May 2011.* Report Historical Period.

Northedge, A.

2000 *Entre Amman et Samarra, l'Archéologie et les élites au début de l'Islam (VII-IX siècles)*, Synthèse des travaux soumis pour l'obtention de l'habilitation à diriger les recherches. Paris, Université de Paris I.

Randazzo, E.

2011 *I Qusūr Omayyadi: Circolazione Interna, Utilizzazione degli Spazi, Funzioni Sociali.* Thèse de doctorat de l'Université "Sapienza" de Rome (Scuola di Dottorato in Archeologia e Antichità Post-Classiche. XXIII Ciclo).

Schlumberger, D.
1986 *Qasr al-Heir al-Gharbi*. Paris: Geuthner (BAH 120).
Walmsley, A.
2000 Production, Exchange and Regional Trade in the Islamic East Mediterranean:Old Structures, New System?. In I. L. Hansen and C. Wichkam (ed.), *The long Eight Century*, Leiden, Boston, Köln, Brill: 265-343.
Wilkinson T.J.,
1975 The Physical Environmentof Dibsi Faraj; A preliminary Study. In R. P. Harper (ed.), *Excavation at Dibsi, Faraj, Northern Syria, 1972-1974: a Preliminary Note on the Site and its Monuments,* Dumbarton Oaks Papers 29.
Sources
al-Muqaddasi, Shams al-Din Muhammad b. Ahmad, *Ahsan al-taqāsīm fī ma'rifat al-aqālīm*: éd. M.J. De Goeje, Leiden, Brill, 1906 (BGA III).
Yāqūt b. 'Abdallah al-Hamawī, *Mu'jam al-buldān* : éd. F. Wüstenfeld, Leipzig, Brockaus, 6vols., 1866-1873.